Lecture Notes in Physics

For information about Vols. 1–131, please contact your bookseller or Springer-Verlag.

Vol. 132: Systems Far from Equilibrium. Proceedings, 1980. Edited by L. Garrido. XV, 403 pages. 1980.

Vol. 133: Narrow Gap Semiconductors Physics and Applications. Proceedings, 1979. Edited by W. Zawadzki. X, 572 pages. 1980.

Vol. 134: γγ Collisions. Proceedings, 1980. Edited by G. Cochard and P. Kessler. XIII, 400 pages. 1980.

Vol. 135: Group Theoretical Methods in Physics. Proceedings, 1980. Edited by K. B. Wolf. XXVI, 629 pages. 1980.

Vol. 136: The Role of Coherent Structures in Modelling Turbulence and Mixing. Proceedings 1980. Edited by J. Jimenez. XIII, 393 pages. 1981.

Vol. 137: From Collective States to Quarks in Nuclei. Edited by H. Arenhövel and A. M. Saruis. VII, 414 pages. 1981.

Vol. 138: The Many-Body Problem. Proceedings 1980. Edited by R. Guardiola and J. Ros. V, 374 pages. 1981.

Vol. 139: H. D. Doebner, Differential Geometric Methods in Mathematical Physics. Proceedings 1981. VII, 329 pages. 1981.

Vol. 140: P. Kramer, M. Saraceno, Geometry of the Time-Dependent Variational Principle in Quantum Mechanics. IV, 98 pages. 1981.

Vol. 141: Seventh International Conference on Numerical Methods in Fluid Dynamics. Proceedings. Edited by W. C. Reynolds and R. W. MacCormack. VIII, 485 pages. 1981.

Vol. 142: Recent Progress in Many-Body Theories. Proceedings. Edited by J. G. Zabolitzky, M. de Llano, M. Fortes and J. W. Clark. VIII, 479 pages. 1981.

Vol. 143: Present Status and Aims of Quantum Electrodynamics. Proceedings, 1980. Edited by G. Gräff, E. Klempt and G. Werth. VI, 302 pages. 1981.

Vol. 144: Topics in Nuclear Physics I. A Comprehensive Review of Recent Developments. Edited by T.T.S. Kuo and S.S.M. Wong. XX, 567 pages. 1981.

Vol. 145: Topics in Nuclear Physics II. A Comprehensive Review of Recent Developments. Proceedings 1980/81. Edited by T. T. S. Kuo and S. S. M. Wong. VIII, 571-1.082 pages. 1981.

Vol. 146: B. J. West, On the Simpler Aspects of Nonlinear Fluctuating. Deep Gravity Waves. VI, 341 pages. 1981.

Vol. 147: J. Messer, Temperature Dependent Thomas-Fermi Theory. IX, 131 pages. 1981.

Vol. 148: Advances in Fluid Mechanics. Proceedings, 1980. Edited by E. Krause. VII, 361 pages. 1981.

Vol. 149: Disordered Systems and Localization. Proceedings, 1981. Edited by C. Castellani, C. Castro, and L. Peliti. XII, 308 pages. 1981.

Vol. 150: N. Straumann, Allgemeine Relativitätstheorie und relativistische Astrophysik. VII, 418 Seiten. 1981.

Vol. 151: Integrable Quantum Field Theory. Proceedings, 1981. Edited by J. Hietarinta and C. Montonen. V, 251 pages. 1982.

Vol. 152: Physics of Narrow Gap Semiconductors. Proceedings, 1981. Edited by E. Gornik, H. Heinrich and L. Palmetshofer. XIII, 485 pages. 1982.

Vol. 153: Mathematical Problems in Theoretical Physics. Proceedings, 1981. Edited by R. Schrader, R. Seiler, and D.A. Uhlenbrock. XII, 429 pages. 1982.

Vol. 154: Macroscopic Properties of Disordered Media. Proceedings, 1981. Edited by R. Burridge, S. Childress, and G. Papanicolaou. VII, 307 pages. 1982.

Vol. 155: Quantum Optics. Proceedings, 1981. Edited by C.A. Engelbrecht. VIII, 329 pages. 1982.

Vol. 156: Resonances in Heavy Ion Reactions. Proceedings, 1981. Edited by K.A. Eberhard. XII, 448 pages. 1982.

Vol. 157: P. Niyogi, Integral Equation Method in Transonic Flow. XI, 189 pages. 1982.

Vol. 158: Dynamics of Nuclear Fission and Related Collective Phenomena. Proceedings, 1981. Edited by P. David, T. Mayer-Kuckuk, and A. van der Woude. X, 462 pages. 1982.

Vol. 159: E. Seiler, Gauge Theories as a Problem of Constructive Quantum Field Theory and Statistical Mechanics. V, 192 pages. 1982.

Vol. 160: Unified Theories of Elementary Particles. Critical Assessment and Prospects. Proceedings, 1981. Edited by P. Breitenlohner and H.P. Dürr. VI, 217 pages. 1982.

Vol. 161: Interacting Bosons in Nuclei. Proceedings, 1981. Edited by J.S. Dehesa, J.M.G. Gomez, and J. Ros. V, 209 pages. 1982.

Vol. 162: Relativistic Action at a Distance: Classical and Quantum Aspects. Proceedings, 1981. Edited by J. Llosa. X, 263 pages. 1982.

Vol. 163: J. S. Darrozes, C. Francois, Mécanique des Fluides Incompressibles. XIX, 459 pages. 1982.

Vol. 164: Stability of Thermodynamic Systems. Proceedings, 1981. Edited by J. Casas-Vázquez and G. Lebon. VII, 321 pages. 1982.

Vol. 165: N. Mukunda, H. van Dam, L.C. Biedenharn, Relativistic Models of Extended Hadrons Obeying a Mass-Spin Trajectory Constraint. Edited by A. Böhm and J.D. Dollard. VI, 163 pages. 1982.

Vol. 166: Computer Simulation of Solids. Edited by C.R.A. Catlow and W.C. Mackrodt. XII, 320 pages. 1982.

Vol. 167: G. Fieck, Symmetry of Polycentric Systems. VI, 137 pages, 1982.

Vol. 168: Heavy-Ion Collisions. Proceedings, 1982. Edited by G. Madurga and M. Lozano. VI, 429 pages. 1982.

Vol. 169: K. Sundermeyer, Constrained Dynamics. IV, 318 pages. 1982.

Vol. 170: Eighth International Conference on Numerical Methods in Fluid Dynamics. Proceedings, 1982. Edited by E. Krause. X, 569 pages. 1982.

Vol. 171: Time-Dependent Hartree-Fock and Beyond. Proceedings, 1982. Edited by K. Goeke and P.-G. Reinhard. VIII, 426 pages. 1982.

Vol. 172: Ionic Liquids, Molten Salts and Polyelectrolytes. Proceedings, 1982. Edited by K.-H. Bennemann, F. Brouers, and D. Quitmann. VII, 253 pages. 1982.

Lecture Notes in Physics

208

Supersymmetry and Supergravity Nonperturbative QCD

Proceedings of the Winter School
Held in Mahabaleshwar, India, January 5–19, 1984

Edited by Probir Roy and Virendra Singh

Springer-Verlag
Berlin Heidelberg GmbH 1984

Editors

Probir Roy
Virendra Singh
Tata Institute of Fundamental Research
National Centre of the Government of India
for Nuclear Science and Mathematics
Homi Bhabha Road, Bombay 400 005, India

ISBN 978-3-540-13390-2 ISBN 978-3-540-38942-2 (eBook)
DOI 10.1007/978-3-540-38942-2

Originally published by Springer-Verlag Berlin Heidelberg New York Tokyo in 1984

2153/3140-543210

PREFACE

Every few years the Tata Institute of Fundamental Research (Bombay) has been organizing a summer or winter school on some aspects of Theoretical Particle Physics. This tradition started in 1961 under the direction of the late Homi J. Bhabha. The latest in this series was held at Mahabaleshwar from the 5th to the 19th January (1984) and these proceedings contain written versions of most of the lectures delivered there.

The aim at this winter school was to focus on recent advances in the two areas of supersymmetry $\frac{1}{m}$ supergravity and nonperturbative QCD. The organizers meant to introduce research workers at various levels to developments in these current mainstreams. This is reflected in the present volume. Introductory and pedagogical lectures are followed by in-depth detailed treatments as well as technical and specialized discussions.

In the first of the two areas cited above, S. Joglekar introduces the basics of global supersymmetry. The following course by S. Ferrara covers local supersymmetry and its coupling to Yang-Mills theories including extension to the N=2 case. On the other hand, the lectures delivered by P. Nath stress the link with unification of the N=1 supergravity theory and the model-independent phenomenologically interesting features of the subject. The supplementary lectures of L. Hall, S. Raby, R.N. Mohapatra and R. Godbole are directed at specific topics of current interest.

In the area of nonperturbative QCD, the course by T. Eguchi is a pedagogical overview of the subject covering 1/N expansion techniques as well as lattice QCD. The lectures by S. Wadia stress the nonperturbative link between QCD and chiral symmetric models. The contributions from G. Bhanot and M. Atiyah focus on more specific topics, namely numerical simulations in QCD and the topological basis of anomalies. Finally, the last course by J. Maharana is in some sense a bridge between the two areas of our title in that it discusses the link between supersymmetry and stochastic quantization as well as issues related to nonperturbative dynamical breakdown of supersymmetry.

This school was organized by the Tata Institute of Fundamental Research. Financial aid was also received from the Department of Science and Technology of the Government of India, the Yamada Foundation of Japan and the U.S. National Science Foundation. We thank these organizations for their generosity. The authorities of The Club, Mahabaleshwar, deserve thanks for their cooperation. We also thank our lecturers and participants for their enthusiastic response. We are grateful to the other members of the organizing committee and the superintendent for their assistance. The careful typing of S.K. Bhonslay and the proof-reading by S. Mukhi deserve mention.

Probir Roy
Virendra Singh

TABLE OF CONTENTS

* Lecturer

INTRODUCTION TO GLOBAL SUPERSYMMETRY

Satish D. Joglekar
Department of Physics
Indian Institute of Technology
Kanpur 208016, India

Preface

The purpose of these lectures is to acquaint with supersymmetry those participants who are not already familiar with the basics of global supersymmetry. As such they are introductory in scope. We shall try to emphasize the structure of supersymmetry (and not, for instance, exactness of numerical coefficients in expressions). The review articles used in preparing these notes are listed at the end. No other references are given.

I: Supersymmetry algebra and Particle supermultiplets:

(A) Mathematical Preliminaries

(a) We use the metric $\eta_{\mu\nu}$ = diagonal $(1,-1,-1,-1)$

(b) As usual γ matrices satisfy: $\{\gamma_\mu, \gamma_\nu\} = 2\eta_{\mu\nu}$

$\sigma_{\mu\nu} \equiv \frac{i}{2}(\gamma_\mu\gamma_\nu - \gamma_\nu\gamma_\mu)$

$\gamma_5 \equiv \gamma_o\gamma_1\gamma_2\gamma_3$

(c) The charge conjugation matrix C is defined by

$$C^{-1}\gamma_\mu C = -\gamma_\mu^T$$

and satisfies $C^T = -C$.

We may further impose $C^{-1} = C = C^\dagger$.

[For example, in the notation of Bjorken and Drell for γ-matrices, $C = \gamma_o\gamma_2$.]

(d) Any 4x4 matrix can be expanded in a basis consisting of 16 linearly independent matrices of which 10 can be chosen to be symmetric and 6 can be chosen to be antisymmetric. Now, taking into account the above-stated properties of γ matrices and of C, it is easy to see that C, $\gamma_5 C$, $i\gamma_\mu\gamma_5 C$ are antisymmetric and provide the basis for 6 linearly independent antisymmetric matrices, while $\gamma_\mu C$, $\sigma_{\mu\nu}C$ provide the basis for 10 symmetric matrices.

(e) The charge conjugate of a spinor is defined by $\psi = C\bar{\psi}^T$. For a Majorana spinor, $\psi = \psi = C\bar{\psi}^T$.

(f) Chiral projections are defined by $\psi_\pm = \frac{1\pm i\gamma_5}{2}\psi$. For a Majorana spinor $\psi_\mp = C\bar{\psi}_\pm^T$.

(g) Anticommuting C numbers such as ϵ, θ have the property that they anticommute among themselves $\{\varepsilon,\theta\} = 0$, while they commute with commuting C-numbers x,y... i.e. $[\varepsilon,x] = 0$. In particular, $\varepsilon^2 = \theta^2 = 0$.

(h) Due to the property $\varepsilon^2 = 0$, a function of ε has a Taylor expansion that terminates.

$$f(x,\varepsilon) = f_o(x) + \varepsilon f_1(x).$$

(B) Wess-Zumino Model

Ten years ago Wess and Zumino invented a Lagrangian with a remarkable kind of new symmetry. The symmetry transformation transformed bosons into fermions and vice versa. These are now called supersymmetry (SUSY) transformations. Their Lagrangian was in terms of two scalar fields A, B, a Majorana field ψ and two auxiliary fields F and G. We exhibit the quadratic terms in the Lagrangian

$$\mathcal{L}[A,B, \;,F,G] = \frac{1}{2}\,\partial_\mu A\partial^\mu A + \frac{1}{2}\,\partial_\mu B\partial^\mu B + \frac{1}{2}\,i\bar{\psi}\not{\partial}\psi + \frac{1}{2}\,F^2 + \frac{1}{2}\,G^2.$$

This Lagrangian is invariant under the following transformations on fields characterized by a constant anticommuting Majorana spinor ε.

$$\delta A(x) = \frac{1}{\sqrt{2}}\,\bar{\varepsilon}\psi(x),$$

$$\delta B(x) = \frac{1}{\sqrt{2}}\,\bar{\varepsilon}\gamma_5\psi(x),$$

$$\delta\psi(x) = -\frac{i}{\sqrt{2}}\,(\not{\partial}A-\gamma_5\not{\partial}B) - i(F+\gamma_5 G)\varepsilon,$$

$$\delta F(x) = \frac{\bar{\varepsilon}}{\sqrt{2}}\,\not{\partial}\psi,$$

$$\delta G = \frac{\bar{\varepsilon}}{\sqrt{2}}\,\gamma_5\not{\partial}\psi.$$

This invariance leads naturally to a conserved current. This current is a vector spinor:

$$J^\mu_\alpha = \frac{1}{2}\,[\not{\partial}(A+\gamma_5 B)\gamma^\mu\psi]_\alpha.$$

Correspondingly, one has spinor charges in the theory:

$$Q_\alpha = \int J^o_\alpha(x)\,d^3\vec{x}.$$

(C) Super-Poincare Algebra

We would like to see, from first principles, what kind of algebra would be formed by the supersymmetry charges introduced in the last section. Let us characterize SUSY transformations as $e^{i\bar{\varepsilon}Q}$. Here ε is the same spinor that appeared in $\delta A = \frac{\bar{\varepsilon}}{\sqrt{2}}\psi$. Hence Dim $\varepsilon = -\frac{1}{2}$ (Dim $\equiv$ Dimension of). Now $\bar{\varepsilon}Q$ must be dimensionless, spinless, and commuting. Hence

(i) Dim $Q = \frac{1}{2}$,

(ii) Q must carry spin $\frac{1}{2}$,

(iii) Q is an anticommuting object.

Super-poincare Algebra is an extension of the usual Poincare algebra viz.

$$[P_\mu, P_\nu] = 0,$$

$$\frac{1}{i}[P_\mu, M_{\nu\lambda}] = \eta_{\mu\nu}P_\lambda - \eta_{\mu\lambda}P_\nu,$$

$$\frac{1}{i}[M_{\mu\nu}, M_{K\lambda}] = \eta_{\nu K}M_{\mu\lambda} - \eta_{\nu\lambda}M_{\mu K} + \eta_{\mu\lambda}M_{\nu K} - \eta_{\nu K}M_{\nu\lambda}.$$

To find the place of Q's in this algebra, we note firstly that Q must transform as a spinor under Lorentz transformations, viz.

$$e^{i\omega^{\mu\nu}M_{\mu\nu}}Q\, e^{-i\omega^{\mu\nu}M_{\mu\nu}} = e^{-\frac{i}{2}\sigma_{\mu\nu}\omega^{\mu\nu}}Q.$$

Taking an infinitesimal transformation, we have

$$[Q_\alpha, M_{\mu\nu}] = \frac{1}{2}(\sigma_{\mu\nu}Q)_\alpha = \frac{1}{2}(\sigma_{\mu\nu})_{\alpha\beta}Q_\beta,$$

Next we consider $[Q_\alpha, P_\mu]$. This commutator has dimension $\frac{3}{2}$. There are no operators of dimension $\frac{3}{2}$ under consideration (Dim $P = 1$ Dim $M_{\mu\nu}$ = Dim $\vec{r} \times \vec{p} = 0$). Hence we set

$$[Q_\alpha, P_\mu] = 0.$$

This simply says that the SUSY transformations and spacetime translations do not interfere with each other. (Later on we shall see that Q and P will become generators of translations in superspace. Thus generators of all supertranslations commute among themselves).

Now the commutator of two infinitesimal SUSY transformation $e^{i\bar{\varepsilon}_1 Q}$ and $e^{i\bar{\varepsilon}_2 Q}$ contains $[\bar{\varepsilon}_1 Q, \bar{\varepsilon}_2 Q]$ which naturally contains the anticommutator $\{Q_\alpha, Q_\beta\}$, as ε's are anticommuting. From dimensional arguments

$\{Q_\alpha, Q_\beta\}$ must have dimension 1. Moreover, $\{Q_\alpha, Q_\beta\}$ is an (α,β) element of a 4x4 matrix that is symmetric in α,β. Recalling that $\gamma_\mu C$ and $\sigma_{\mu\nu}C$ provide the basis for 4x4 symmetric matrices we can write

$$\{Q_\alpha, Q_\beta\} = A(\gamma^\mu C)_{\alpha\beta}P_\mu + B(\sigma^{\mu\nu}C)_{\alpha\beta}M_{\mu\nu},$$

where Dim B = 1, Dim A = 0. But now consider

$$[\{Q_\alpha, Q_\beta\}, P_\mu] = \{Q_\alpha, [Q_\beta, P_\mu]\} + \{[Q_\alpha, P_\mu], Q_\beta\} = 0.$$

Thus

$$B(\sigma^{\lambda K}C)_{\alpha\beta}[M_{\lambda K}, P_\mu] = 0$$

which implies that B = 0. We may further fix the (as yet arbitrary) normalization of Q by requiring

$$\{Q_\alpha, Q_\beta\} = -(\gamma^\mu C)_{\alpha\beta}P_\mu.$$

[The negative sign is however not arbitrary since will be related to the norm of states which must be positive.] Thus the commutator of two infinitesimal SUSY transformations is a translation:

$$[\bar{\varepsilon}_1 Q, \bar{\varepsilon}_2 Q] = \bar{\varepsilon}_1\gamma^\mu\varepsilon_2 P_\mu.$$

[Note however that this translation is not by an ordinary number as $(\bar{\varepsilon}_1\gamma^\mu\varepsilon_2)^5 = 0$].

We thus have the following super-Poincare algebra.

$$[P_\mu, P_\nu] = 0,$$

$$\frac{1}{i}[P_\mu, M_{\nu\lambda}] = \eta_{\mu\nu}P_\lambda - \eta_{\mu\lambda}P_\nu,$$

$$\frac{1}{i}[M_{\mu\nu}, M_{K\lambda}] = \eta_{\nu K}M_{\mu\lambda} - \eta_{\nu\lambda}M_{\mu K} + \eta_{\mu\lambda}M_{\nu K} - \eta_{\mu K}M_{\nu\lambda},$$

$$[Q_\alpha, P_\mu] = 0,$$

$$[Q_\alpha, M_{\mu\nu}] = \frac{1}{2}(\sigma_{\mu\nu})_{\alpha\beta}Q_\beta,$$

$$\{Q_\alpha, Q_\beta\} = -(\gamma^\mu C)_{\alpha\beta}P_\mu. \qquad (1.1)$$

The last relation can be written in a different form. On unitary representations $e^{i\bar{\varepsilon}Q}$ must be unitary or $\bar{\varepsilon}Q$ must be hermitian. As ε is a Majorana spinor, Q must also be so; i.e.

$$Q = C\bar{Q}^T, \ \bar{Q} = -Q^T C.$$

In terms of Q and $\bar{Q}$, the anticommutator can be written as

$$\{Q_\alpha, \bar{Q}_\beta\} = (\gamma^\mu)_{\alpha\beta} P_\mu .$$

(D) Two component notation

In the literature, both the 4-component and 2-component notations are used. In this section we shall express the super-Poincare algebra in two component notation. To make contact with that notation, it is convenient to go to a particular representation of γ-matrices called the Weyl representation. In this representation

$$\gamma^\mu = \begin{pmatrix} 0 & \sigma^\mu \\ \bar{\sigma}^\mu & 0 \end{pmatrix},$$

where

$$\sigma^\mu = (1, \vec{\sigma}); \ \bar{\sigma}^\mu = (1, -\vec{\sigma}).$$

Then

$$i\gamma_5 = i\gamma_o\gamma_1\gamma_2\gamma_3 = \begin{pmatrix} I & 0 \\ 0 & -I \end{pmatrix},$$

$$\gamma_+ = \begin{pmatrix} I & 0 \\ 0 & 0 \end{pmatrix}, \ \gamma_- = \begin{pmatrix} 0 & 0 \\ 0 & I \end{pmatrix}.$$

Thus $\gamma_\pm$ have simple forms in this representation. Moreover,

$$\sigma^{ij} = \begin{pmatrix} \sigma^k & 0 \\ 0 & \sigma^k \end{pmatrix} \varepsilon_{ijk},$$

$$\sigma^{oi} = -i \begin{pmatrix} \sigma^i & 0 \\ 0 & -\sigma^k \end{pmatrix}.$$

We can define chirality projections of Q as

$$Q_\pm = \gamma_\pm Q$$

They satisfy $\bar{Q}_\pm = -Q^T_\mp C$. Hence, instead of Q_+ and Q_-, we may choose Q_+ and $\bar{Q}_+$ as independent charges.

In terms of Q_+ and $\bar{Q}_+$, the algebra reads

$$\{Q_+, \bar{Q}_-\} = 0 = \{Q_-, \bar{Q}_+\} ,$$

$$\{Q_{+\alpha}, \bar{Q}_{+\beta}\} = (\gamma_+ \gamma^\mu)_{\alpha\beta} P_\mu .$$

Noting the structure of $\gamma_\pm$, we may parametrize Q_+ and $\bar{Q}_+$ as

$$Q_+ = \begin{pmatrix} Q_1 \\ Q_2 \\ 0 \\ 0 \end{pmatrix} , \quad \bar{Q}_+ = \begin{pmatrix} 0 \\ 0 \\ \bar{Q}_{\dot{1}} \\ \bar{Q}_{\dot{2}} \end{pmatrix} .$$

In terms of Q_α and $\bar{Q}_{\dot{\alpha}}$, the algebra reads

$$\{Q_\alpha, Q_\beta\} = 0 = \{\bar{Q}_{\dot{\alpha}}, \bar{Q}_{\dot{\beta}}\} = [Q_\alpha, P_\mu] = [\bar{Q}_{\dot{\alpha}}, P_\mu],$$

$$\{Q_\alpha, \bar{Q}_{\dot{\beta}}\} = (\sigma^\mu)_{\alpha\dot{\beta}} P_\mu .$$

(E) Particle Supermultiplets

The particle supermultiplets for supersymmetry can be constructed by Wigner's method of Induced Representations. We note that $[Q, P_\mu]=0$ implies that the spinor charges leave the 4 momentum of a state unchanged. Thus they leave the manifold of states of a given 4-momentum invariant.

Let $|\,>$ be any state in the manifold with 4-momentum p_μ. Then for such a state $P_\mu|\,> = p_\mu|\,>$ implies

$$\{Q_\alpha, \bar{Q}_{\dot{\beta}}\} = (\sigma^\mu)_{\alpha\dot{\beta}} p_\mu ,$$

i.e. the RHS becomes a pure number when restricted to this manifold. Let us now discuss two cases separately.

(i) Massive case: Here the particles in the supermultiplets are massive and we can always choose a Lorentz frame such that $p_\mu = (M, \vec{0})$. Then

$$\{Q_\alpha, \bar{Q}_{\dot{\beta}}\} = \delta_{\alpha\dot{\beta}} M,$$

Or, by rescaling Q and $\bar{Q}$, we have

$$\{Q_\alpha, \bar{Q}_{\dot{\beta}}\} = \delta_{\alpha\dot{\beta}}; \; \{Q_\alpha, Q_\beta\} = 0 = \{\bar{Q}_{\dot{\alpha}}, \bar{Q}_{\dot{\beta}}\}.$$

Thus Q's and $\bar{Q}$'s satisfy the algebra of creation and annihilation operators for a fermion field and hence can be used to build a 4-dimensional Fock space. We note, further, that there must exist a state $|\chi\rangle$ which satisfies $\bar{Q}_{\dot{\alpha}}|\chi\rangle = 0 \quad \alpha = 1,2$. This can be seen by considering any arbitrary state $|\psi\rangle$ and showing that such a state $|\chi\rangle$ can be constructed from it. [This is left as an exercise with the hint that the reader should consider the state $\bar{Q}_{\dot{1}}\bar{Q}_{\dot{2}}|\psi\rangle$ and that $\bar{Q}_{\dot{1}}^2=0=\bar{Q}_{\dot{2}}^2$.]

Let the state $|\chi\rangle$ have angular momentum quantum numbers J and J_3. Thus $|\chi\rangle \equiv |MJJ_3\rangle$. We can now construct three more new states $Q_1|MJJ_3\rangle$, $Q_2|MJJ_3\rangle$, $Q_1Q_2|MJJ_3\rangle$. [Here observe that $Q_1^2 = Q_2^2 = 0$]. To find the spin content of these states we note that Q_1, Q_2 carry spin $\frac{1}{2}$. Further information is obtained from

$$[Q_\alpha, M_{\mu\nu}] = \frac{1}{2}(\sigma_{\mu\nu}Q)_\alpha.$$

Hence, for $J_3 = M_{12}$,

$$[Q_\alpha, J_3] = \frac{1}{2}(\sigma_{12}Q)_\alpha.$$

Taking chiral projections, we have

$$[Q_{+\alpha}, J_3] = \frac{1}{2}(\sigma_{12}Q_+)_\alpha.$$

Noting that in the Weyl representation

$$\sigma_{12} = \begin{pmatrix} \sigma_3 & 0 \\ 0 & \sigma_3 \end{pmatrix},$$

we have - in the two-component notation -

$$[Q_\alpha, J_3] = (\sigma_3)_{\alpha\beta}Q_\beta.$$

and thus

$$[Q_1, J_3] = \frac{1}{2}Q_1 \;:\; Q_1 \text{ lowers } J_3 \text{ by } \frac{1}{2}.$$

$$[Q_2, J_3] = -\frac{1}{2}Q_2 \;:\; Q_2 \text{ raises } J_3 \text{ by } \frac{1}{2}.$$

It can further be verified that

$$[Q_1Q_2 - Q_2Q_1, J_i] = 0 \quad i = 1,2,3.$$

Therefore, $Q_1Q_2 = \frac{1}{2}(Q_1Q_2 - Q_2Q_1)$ does not alter the J and J_3 quantum numbers. We thus have

$$Q_1|MJJ_3> = a|M,J - \frac{1}{2}, J_3 - \frac{1}{2}> + b|M, J + \frac{1}{2}, J_3 - \frac{1}{2}>,$$

$$Q_2|M,J,J_3> = a'|M,J - \frac{1}{2}, J_3 + \frac{1}{2}> + b'|M,J + \frac{1}{2}, J_3 + \frac{1}{2}>,$$

$$Q_1Q_2|M,J,J_3> = C|M,J,J_3>.$$

Hence the supermultiplet contains in it spins of J, $J + \frac{1}{2}$, $J - \frac{1}{2}$ and J respectively.

(ii) Massless case: In this case, one can choose the 4-momentum to be $p_\mu = (p,0,0,p)$. We thus have the algebra,

$$\{Q_\alpha, \bar{Q}_{\dot{\beta}}\} = p(\sigma^0 + \sigma^3)_{\alpha\dot{\beta}} = 2p\begin{pmatrix}1 & 0\\ 0 & 0\end{pmatrix}_{\alpha\dot{\beta}}.$$

Thus, rescaling Q's and $\bar{Q}$'s, we have

$$\{Q_\alpha, Q_\beta\} = 0, \ \{\bar{Q}_{\dot{\alpha}}\ \bar{Q}_{\dot{\beta}}\} = 0$$

$$\{Q_1, \bar{Q}_{\dot{1}}\} = 1, \ \{Q_2, \bar{Q}_2\} = 0 \text{ and } \{Q_\alpha, \bar{Q}_{\dot{\beta}}\} = 0, \ \alpha \neq \beta.$$

For massless particles, states can be described in terms of helicity in a Lorentz-invariant way. Start with a state $|\lambda>$ assumed to satisfy

$$\bar{Q}_{\dot{\alpha}}|\lambda> = 0, \ \alpha = 1,2,$$

and construct four states $|\lambda>$, $Q_1|\lambda>$, $Q_2|\lambda>$ and $Q_1Q_2|\lambda>$. However, on account of $\{Q_2, \bar{Q}_{\dot{2}}\}$ being zero, the norms of the states $Q_2|\lambda>$ and $Q_1Q_2|\lambda>$ vanish. Thus the physical particle multiplet consists of $|\lambda>$ and $Q_1|\lambda>$ with helicities λ and $\lambda - \frac{1}{2}$ respectively.

If parity is included, the multiplet will consist of four states: $\pm\ |\lambda>$, $\pm\ |\lambda - \frac{1}{2}>$.

Examples:

$\lambda = 1$	states	$\pm\ \|1>$, Photon	$\pm\ \|\frac{1}{2}>$ photino
$\lambda = 2$	states	$\|\pm 2>$, Graviton	$\|\pm\frac{3}{2}>$ Gravitino

(F) Supersymmetry algebra with N > 1.

If the number of supersymmetry charges is greater than one, one can have a fusion of an internal symmetry group G and supersymmetry. A straightforward generalization of the SUSY algebra in this case is (in 2-component notation) the following

$$\{Q^L_\alpha, Q^M_\beta\} = 0 = \{\bar{Q}^L_{\dot\alpha}, \bar{Q}^M_{\dot\beta}\},$$

$$\{Q^L_\alpha, \bar{Q}^M_{\dot\beta}\} = \delta^{LM}(\sigma^\mu)_{\alpha\dot\beta}P^\mu,$$

$$[Q^L_\alpha, P_\mu] = 0 = [\bar{Q}^L_{\dot\alpha}, P_\mu],$$

$$[B_\ell, B_m] = if_{\ell mk}B_k$$

$$[Q^L_\alpha, B_\ell] = iS^{LM}_\ell Q^M_\alpha.$$

Here B_ℓ are generators of the internal symmetry group G and Q^L transform according to a certain representation of the group G, S^{LM}_α being hermitian matrices. This, however, is not most general. The maximal generalization of the SUSY algebra for N > 1 allows what are called central charges Z^{LM}:

$$\{Q^L_\alpha, Q^M_\beta\} = \varepsilon_{\alpha\beta}Z^{LM} .$$

But we shall choose not to deal with them.

(G) Particle supermultiplets for the case N > 1

We shall illustrate the construction of supermultiplets by considering massive states only. Here again one starts with a state $|\chi\rangle$ satisfying

$$\bar{Q}^L_{\dot\alpha}|\chi\rangle = 0 \qquad \dot\alpha = 1,2,\ L = 1,2,\ldots,N$$

and constructs a set of states

$$(Q^1_1)^{r^1_1}(Q^1_2)^{r^1_2} \ldots (Q^N_1)^{r^N_1}(Q^N_2)^{r^N_2}\ |\chi\rangle.$$

The r's can take two values r=0, r=1. As there are 2N operators there will be 2^{2N} such states.

II: Superspace, Superfields and Supermultiplets:

(A) Superspace, Superfields

The superspace provides us with an elegant formalism for supersymmetry in which the tensor calculus for supermultiplets becomes especially simple. It is a space with four ordinary dimensions x^o, x^1, x^2, x^3 and a number of anticommuting dimensions. We would like the ε appearing in the field transformations to be translations in superspace. Since ε is a Majorana spinor, it contains four independent (real) components. Hence the superspace consists of four ordinary dimensions and four anticommuting dimensions θ's, θ being a Majorana spinor. A superfield is a function of the superspace coordinates x and θ, i.e. $\Phi(x,\theta)$. Because the θ_α's are anticommuting and have 4 independent components, $\theta_\alpha\theta_\beta\theta_\gamma\theta_\delta\theta_\lambda \equiv 0$. Hence the Taylor expansion of Φ in powers of θ terminates:

$$\Phi(x,\theta) = \phi_o(x) + \phi_1^\alpha\theta_\alpha + \frac{1}{2!}\phi_2^{\alpha\beta}\theta_\alpha\theta_\beta + \frac{1}{3!}\phi_3^{\alpha\beta\gamma}\theta_\alpha\theta_\beta\theta_\gamma + \frac{1}{4!}\phi_4^{\alpha\beta\gamma\delta}\theta_\alpha\theta_\beta\theta_\gamma\theta_\delta .$$

Thus a single superfield contains a number of ordinary fields $\phi_o \dots \phi_4$.

In order to see the spin content of the fields in a single scalar superfield, it is convenient to expand it in terms of monomials of θ having definite Lorentz transformation properties.

Here we make a number of observations.

(i) Since θ is a Majorana spinor, θ and $\bar\theta$ are related to each other through $\theta = C\bar\theta^T$. Hence either θ_α or $\bar\theta_\alpha$ can be chosen to be independent linear monomials.

(ii) As to the bilinear monomials, we note that on account of the Majorana nature of θ and the fact that $\sigma_{\mu\nu}C$ and $\gamma_\mu C$ are symmetric matrices, $\bar\theta\sigma_{\mu\nu}\theta = 0 = \bar\theta\gamma_\mu\theta$.

(iii) Quartic monomials such as $\bar\theta\gamma_\mu\gamma_5\theta\ \bar\theta\gamma^\mu\gamma_5\theta$ or $(\bar\theta\gamma_5\theta)^2$ are not independent of $(\bar\theta\theta)^2$ as can be shown by using the Fierz identity. A similar remark applies to cubic monomials.

Thus we tabulate as follows:

No.of indices	No.of components	θ-monomials
0	1	1
1	4	θ or $\bar\theta$ (in all 4)
2	6	$\bar\theta\theta, \bar\theta\gamma_5\theta, \bar\theta i\gamma^\mu\gamma_5\theta$ (1+1+4=6)
3	4	$\bar\theta\theta\bar\theta$
4	1	$(\bar\theta\theta)^2$.

A general complex superfield $\Phi(x,\theta)$ will be called a general scalar superfield. A general scalar superfield can be, then, expanded in terms of θ-monomials having definite Lorentz transformation properties as follows:

$$\Phi(x,\theta) = A(x) + \bar\theta\psi(x) + \frac{1}{4}\bar\theta\theta F(x) + \frac{1}{4}\bar\theta\gamma_5\theta G(x) + \frac{1}{4}\bar\theta i\gamma^\nu\gamma_5\theta V_\nu(x)$$
$$+ \frac{1}{4}\bar\theta\theta\bar\theta\chi(x) + \frac{1}{32}(\bar\theta\theta)^2 D(x). \qquad (2.1)$$

Bosonic fields: A, F, G, V_ν, D.
Fermionic fields: ψ, χ.
Thus a general scalar field contains 16 complex fields among which fermionic and bosonic fields are equal in number.

(B) Two component notation

As before, the transition from 4-component to 2-component notation is made by going to the Weyl representation of γ-matrices in which the chiral projection operators look particularly simple. We let

$$\theta_+ = \begin{pmatrix}\theta_1\\ \theta_2\\ 0\\ 0\end{pmatrix}, \quad \bar\theta_+ = \theta_-^T C = \begin{pmatrix}0\\ 0\\ \bar\theta_{\dot 1}\\ \bar\theta_{\dot 2}\end{pmatrix}.$$

Thus the superspace consists of $(x,\theta,\bar\theta)$ where θ and $\bar\theta$ are 2-component spinors which have been chosen to be independent ones instead of the four component θ.

We introduce $\varepsilon_{\dot\alpha\dot\beta} = \varepsilon_{\alpha\beta} = \begin{pmatrix}0 & -1\\ 1 & 0\end{pmatrix}$, $\varepsilon^{\alpha\beta} = \begin{pmatrix}0 & 1\\ -1 & 0\end{pmatrix} = \varepsilon^{\dot\alpha\dot\beta}$,

$$\theta^\alpha = \varepsilon^{\alpha\beta}\theta_\beta \qquad \bar\theta^{\dot\beta} = \varepsilon^{\dot\beta\dot\alpha}\bar\theta_{\dot\alpha}$$

$$\theta\theta \equiv \theta^\alpha\theta_\alpha \qquad \bar\theta\bar\theta \equiv \bar\theta_{\dot\alpha}\bar\theta^{\dot\alpha}.$$

Then, in terms of these, one can show that $\bar\theta\theta$ and $\bar\theta\gamma_5\theta$ are linear combinations of $\theta\theta$ and $\bar\theta\bar\theta$ whereas $\bar\theta i\gamma^\mu\gamma_5\theta$ is proportional to $\theta\sigma^\mu\bar\theta$ and so on. The table for θ-monomials now reads as follows:

No.of indices	No.of components	θ-monomials
0	1	1
1	4	θ_α, $\bar\theta_{\dot\alpha}$
2	6	$\theta\theta$, $\bar\theta\bar\theta$, $\theta\sigma^\mu\bar\theta$
3	4	$\theta\theta\bar\theta$, $\bar\theta\bar\theta\theta$
4	1	$\theta\theta\bar\theta\bar\theta$.

The expansion of a scalar superfield $\Phi(x,\theta,\bar{\theta})$ in two component notation reads:

$$\Phi(x,\theta,\bar{\theta}) = A(x) + \theta\phi(x) + \bar{\theta}\bar{\phi}(x) + \theta\theta m(x) + \bar{\theta}\bar{\theta}n(x) + \theta\sigma^{\mu}\bar{\theta}v_{\mu}(x)$$

$$+ \theta\theta\bar{\theta}\bar{\lambda}(x) + \bar{\theta}\bar{\theta}\theta\lambda(x) + \theta\theta\bar{\theta}\bar{\theta}d(x).$$

The expansion coefficient fields are related to the corresponding 4-component quantities. Here ϕ and $\bar{\phi}$ are related to ψ, m and n to F and G, $\bar{\lambda}$ and λ to χ. v_{μ} and d are proportional to V_{μ} and D.

(C) <u>Transformations in superspace</u>

Consider an infinitesimal inhomogeneous Lorentz transformation

$$\delta x^{\mu} = b^{\mu} + \omega^{\mu}{}_{\nu}x^{\nu}.$$

As θ is a Majorana spinor, we must simultaneously associate a Lorentz transformation on the θ's in accordance with its spinorial character:

$$\delta\theta = - \frac{i}{4}\,\omega^{\mu\nu}\sigma_{\mu\nu}\theta.$$

Next, we consider the new transformations viz supertranslations. As mentioned earlier, SUSY charges induce translations in the superspace anticommuting coordinate θ:

$$\delta\theta = \varepsilon.$$

From the fact that the commutator of two spinorial transformations is an ordinary translation, it should not be too surprising if translations on θ are associated with ordinary translations on x^{μ}, viz.

$$\delta x^{\mu} = \frac{i}{2}\,\bar{\varepsilon}\gamma^{\mu}\theta.$$

Thus the most general infinitesimal transformations in superspace are:

$$\delta x^{\mu} = b^{\mu} + \omega^{\mu}{}_{\nu}x^{\nu} + \frac{i}{2}\,\bar{\varepsilon}\gamma^{\mu}\theta,$$

$$\delta\theta = \varepsilon - \frac{i}{4}\,\omega^{\mu\nu}\sigma_{\mu\nu}\theta.$$

It can be shown that the above transformations form a closed algebra. i.e. the commutation of two such infinitesimal transformations is another infinitesimal transformation of the same kind.

(D) Connection between supertranslations and supersymmetry transformations on fields

This connection is established as follows. Consider a supertranslation

$$\theta \to \theta + \varepsilon,$$

$$x^\mu \to x^\mu + \frac{i}{2}\bar{\varepsilon}\gamma^\mu\theta.$$

Under this transformation

$$\Phi(x,\theta) \to \Phi(x + \frac{i}{2}\bar{\varepsilon}\gamma^\mu\theta,\ \theta+\varepsilon) \equiv \Phi(x,\theta) + \delta\Phi(x,\theta;\varepsilon).$$

Field variations under supersymmetry transformations are contained in $\delta\Phi(x,\theta;\varepsilon)$ viz

$$\delta\Phi(x,\theta) = \delta A(x) + \bar{\theta}\delta\psi(x) + \frac{1}{4}\bar{\theta}\theta\delta F(x) + \ldots .$$

By carrying out the above operation explicitly, we obtain the following expressions for field transformations:

$$\delta A(x) = \bar{\varepsilon}\psi(x),$$

$$\delta\psi(x) = \frac{1}{2}(F+\gamma_5 G+i\gamma^\mu\gamma_5\gamma_\mu - i\not{\partial}A)\varepsilon,$$

$$\delta F(x) = \frac{1}{2}\bar{\varepsilon}\chi - \frac{1}{2}\bar{\varepsilon}i\not{\partial}\psi,$$

$$\delta G = \frac{1}{2}\bar{\varepsilon}\gamma_5\chi + \frac{1}{2}\bar{\varepsilon}i\not{\partial}\gamma_5\psi,$$

$$\delta V_\nu = \frac{1}{2}\bar{\varepsilon}i\gamma_\nu\gamma_5\chi + \frac{1}{2}\bar{\varepsilon}i\not{\partial}i\gamma_\mu\gamma_5\psi,$$

$$\delta\chi = \frac{1}{2}(D-i\not{\partial}F+\gamma_5 i\not{\partial}G-i\gamma^\nu\gamma_5 i\not{\partial}V_\nu)\varepsilon,$$

$$\delta\bar{D} = -\bar{\varepsilon}i\not{\partial}\chi.$$

These are general transformation laws for an arbitrary complex scalar superfield. Transformation laws for particular cases such as for chiral and vector fields can be obtained from these.

We now make a number of observations:

(i) The general transformation law is of the form:

$$\delta(\text{Field}) = \bar{\varepsilon}(\text{next fields}) + \bar{\varepsilon}\partial^{\mu}(\text{earlier fields}),$$

where "next fields" are fields accompanied by more θ's in the expansion of $\Phi(x,\theta)$ and "earlier fields" are those accompanied by fewer θ's.
(ii) δD is always a 4-gradient.
(iii) If V,χ,D are identically 4-gradients, then δF and δG are also 4-gradients.

(E) Superspace realizations of the algebra

The elements of the super-Poincare algebra have the following superspace realizations:

$$P_\mu = i\partial_\mu,$$

$$M_{\mu\nu} = \chi_\mu\partial_\nu - \chi_\nu\partial_\mu + \frac{i}{2}\bar{\theta}\sigma_{\mu\nu}\frac{\partial}{\partial\bar{\theta}},$$

$$Q = \frac{\partial}{\partial\bar{\theta}} + \frac{i}{2}(\not{\partial}\theta).$$

(F) Chiral and Vector superfields

Dimensionwise, it is clear that all physical fields in Φ cannot correspond to physical fields. Noting that the dimension of θ and the dimension of ε both equal $-\frac{1}{2}$, we obtain the following dimensions for fields from the expansion of $\Phi(x,\theta)$.
(i) If Dim $\Phi = 1$, then

$$\text{Dim}(A) = 1,\ \text{Dim}\ \psi = \frac{3}{2},\ \text{Dim}(F,G,V_\mu) = 2,\ \text{Dim}\ \chi = \frac{5}{2},\ \text{Dim}\ D = 3.$$

Thus only A and ψ have canonical dimensions and correspond to physical particles. Other fields must be either auxiliary fields or related to A and ψ.
(ii) If Dim $\Phi = 0$,

$$\text{Dim}\ A = 0,\ \text{Dim}\ \psi = \frac{3}{2},\ \text{Dim}\ V_\mu = 1,\ \text{Dim}\ \chi = \frac{3}{2},\ \text{Dim}\ D = 2.$$

Thus V_μ and χ have canonical dimensions. Other fields must be either related to these or auxiliary or, as it turns out, gauge degrees of freedom.

Thus it is clear that the general scalar superfield is reducible.

We know the existence of a massless supermultiplet with states of helicities $|\lambda>$ and $<\lambda + \frac{1}{2}>$. For $\lambda=0$, this corresponds to a scalar field and a fermion of definite chirality. Hence there must be a superfield that contains a fermion of definite chirality. Consider the expansion for $\Phi(x,\theta)$ given earlier in Eq.(2.1). Suppose we require $\psi \equiv \psi_+$ i.e. $\psi_- \equiv 0$. This, by itself, is not sufficient because supersymmetry transformations on ψ could induce a ψ_- term. Hence we must also require $\delta\psi_- = 0$. This puts the following restrictions:

$$F_- = F+iG = 0, \; V_\mu = i\partial_\mu A.$$

Again these equations must be valid even after a supersymmetry transformation is performed, i.e. we must have $\delta F_- = 0$ $\delta(V_\mu - i\partial_\mu A) = 0$ and so on. These lead to further restrictions

$$\chi_+ = 0, \; \chi_- = -i\not\partial\psi_+ \, , \; D = -\partial^2 A.$$

We can now write the resultant superfield as:

$$\Phi_+(x,\theta) = A_+(x) + \bar{\theta}_-\psi_+(x) + \frac{1}{2}\bar{\theta}_-\theta_+F_+(x) + \frac{1}{4}\bar{\theta}i\gamma^\nu\gamma_5\theta i\partial^\nu A_+ - \frac{i}{4}\bar{\theta}\theta\bar{\theta}_+\not\partial\psi_+ - \frac{1}{32}(\bar{\theta}\theta)^2\partial^2 A_+. \tag{2.2}$$

But this is just the Taylor expansion of

$$\Phi_+(x,\theta) = A_+(z) + \bar{\theta}_-\psi_+(z) + \frac{1}{2}\bar{\theta}_-\theta_+F_+(z) \equiv \hat{\phi}(z,\theta_+),$$

where

$$z^\mu = x^\mu - \frac{1}{4}\bar{\theta}\gamma^\mu\gamma_5\theta.$$

Similarly, one can construct a superfield that contains ψ_- only. This is of the form:

$$\Phi_-(x,\theta) = \hat{\phi}(z^*,\theta_-) = A_-(z^*) + \bar{\theta}_+\psi_-(z^*) + \frac{1}{2}\bar{\theta}_+\theta_-F_-(z^*),$$

where

$$z^* = x^\mu + \frac{1}{4}\bar{\theta}\gamma^\mu\gamma_5\theta.$$

As Φ_+, Φ_- (when expressed in terms of z, z^*) depends only upon θ_+, θ_-, i.e. the chiral projections of θ, they are referred to as chiral superfields. (actually chiral and antichiral superfields.). We shall call them chiral scalar fields of positive and negative kind respectively. The transformation properties of chiral scalar fields under supersymmetry transformations are as under:

$$\delta A_\pm = \bar{\varepsilon}_\mp \psi_\pm ,$$

$$\delta \psi_\pm = F_\pm \varepsilon_\pm - i\partial\!\!\!/ A_\pm \varepsilon_\mp ,$$

$$\delta F_\pm = - \bar{\varepsilon}_\pm i\partial\!\!\!/ \psi_\pm = \text{a 4-gradient.}$$

(G) Properties of chiral fields

(i) The complex conjugate of a positive chiral field is a negative chiral field:

$$\Phi_+^*(x,\theta) = \hat{\phi}(z,\theta_+)^* = \hat{\phi}'(z^*,\theta_-) = \Phi_-'(x,\theta).$$

Note here that θ_- and θ_+^* are related by $\theta_- = C\gamma_0^T\theta_+^*$.

(ii) The product of chiral fields of the same kind is a chiral field of the same kind:

$$\Phi_+^{(1)}(x,\theta)\Phi_+^{(2)}(x,\theta) = \hat{\phi}^{(1)}(z,\theta_+)\hat{\phi}^{(2)}(z,\theta_+) = \hat{\phi}^{(3)}(z,\theta_+) = \Phi_+^{(3)}(x,\theta).$$

(iii) The product of a positive chiral field and a negative chiral field is not a chiral field:

$$\Phi_+(x,\theta)\Phi_-(x,\theta) = \hat{\phi}(z,\theta_+)\hat{\phi}'(z^*,\theta_-) = \hat{\phi}''(z,z^*,\theta_+,\theta_-).$$

(iv) Chiral fields satisfy a differential condition, namely

$$\left.\frac{\partial}{\partial\theta_-}\right|_z \Phi_+(x,\theta) = 0 = \left.\frac{\partial}{\partial\theta_-}\right|_z \hat{\phi}(z,\theta_+).$$

Also

$$\left.\frac{\partial}{\partial\theta_+}\right|_z \Phi_-(x,\theta) = 0 = \left.\frac{\partial}{\partial\theta_+}\right|_z \hat{\phi}'(z^*,\theta_-).$$

(H) Covariant derivatives

The above differential conditions can be transformed into equations containing $\left.\frac{\partial}{\partial\theta}\right|_x$ and derivatives with respect to x. These conditions then read:

$$\gamma_- D \equiv D_- \Phi_+(x,\theta) = 0,$$

$$\gamma_+ D \equiv D_+ \Phi_-(x,\theta) = 0,$$

where

$$D \equiv \frac{\partial}{\partial \bar{\theta}} - \frac{i}{2}(\not\partial \theta).$$

D behaves as a spinor with respect to Lorentz transformations and it anticommutes with supersymmetry generators:

$$\{D_\alpha, Q_\beta\} = 0.$$

Because of this, D is called the covariant derivative. $D_\mp \Phi_\pm = 0$ are covariant conditions because if Φ_+ satisfies $D_- \Phi_+ = 0$, the supersymmetry-transformed superfield $\Phi_+ + \delta\Phi_+$ also satisfies $D_-(\Phi_+ + \delta\Phi_+) = 0$. This is since D commutes through Q. Thus a general scalar superfield is reduced by imposing covariant constraints on it. D_α have a supersymmetry algebra of their own:

$$\{D_\alpha, D_\beta\} = -(\gamma^\mu C)_{\alpha\beta}(i\partial_\mu) = -i(\not\partial C)_{\alpha\beta}.$$

In 2-component notation, D takes the following form:

$$D_+ \rightarrow D_\alpha = \frac{\partial}{\partial \theta^\alpha} + i\,\sigma^\mu_{\alpha\dot{\alpha}}\,\bar{\theta}^{\dot{\alpha}}\,\partial_\mu,$$

$$D_- \rightarrow \bar{D}_{\dot{\alpha}} = -\frac{\partial}{\partial \bar{\theta}^{\dot{\alpha}}} - i\theta^\alpha\,\sigma^\mu_{\alpha\dot{\alpha}}\,\partial_\mu.$$

Chiral fields satisfy

$$\bar{D}_{\dot{\alpha}}\,\Phi_+(x,\theta,\bar{\theta}) = 0,$$

$$D_\alpha\,\Phi_-(x,\theta,\bar{\theta}) = 0.$$

The algebra of D's reads

$$\{D_\alpha, \bar{D}_{\dot{\beta}}\} = (\sigma^\mu)_{\alpha\dot{\beta}}(i\partial_\mu).$$

(I) Vector superfields

A vector superfield is a general scalar superfield satisfying the reality condition:

$$\Phi(x,\theta) = \Phi^*(x,\theta).$$

This condition requires that A, V_μ, D are real (hermitian) fields and that ψ and χ are Majorana spinors. This is the smallest multiplet containing a vector field V_μ and hence is called a vector superfield.

Note that, if we let Dim $\Phi = 0$, we find

$$\text{Dim } A = 0,\ \text{Dim } \psi = \frac{1}{2}\ ,\ \text{Dim } F = \text{Dim } G = 1,$$

$$\text{Dim } V_\mu = 1,\ \text{Dim } \chi = \frac{3}{2}\ ,\ \text{Dim } D = 2.$$

Thus, this is a multiplet containing a vector field V_μ and a spinor field χ. When used as a gauge multiplet, A, ψ, F, G turn out to be gauge degrees of freedom while D turns out to be an auxiliary field.

(J) Example of a spinor multiplet

Superfields need not be scalars. One can have a set of superfields transforming as a multiplet under homogeneous Lorentz transformations:

$$\Phi(x,\theta) \rightarrow \Phi'(x',\theta') = D(\Lambda)\Phi(x,\theta),$$

where $D(\Lambda)$ is $D(j_1,j_2)$ or a direct sum of such representations.

We now give an example of a spinor multiplet that is chiral with respect to both the spinor index and internal structure, viz:

$$(\gamma_-)_{\alpha\beta}\psi_{++\beta} = 0,\ D_{-\alpha}\psi_{++\beta} = 0.$$

Such a superfield appears when we consider field strengths of gauge theory. As the superfield is chiral, it can be expanded as

$$\psi_{++} = \lambda_+(z) + (D(z) + \frac{1}{2}\,\sigma_{\mu\nu}F^{\mu\nu}(z))\theta_+ + \frac{1}{2}\,\bar{\theta}_-\theta_+V_+(z),$$

where λ_+ and V_+ are positive chirality spinors, D is a scalar and $F_{\mu\nu}$ a self-dual antisymmetric tensor. The transformation law is as follows:

$$\delta\lambda_+ = (D + \frac{1}{2}\sigma^{\mu\nu}F_{\mu\nu})\varepsilon_+,$$

$$\delta D = -\frac{1}{2}\bar{\varepsilon}_-V_+ - \frac{1}{2}\bar{\varepsilon}_+ i\partial\!\!\!/\lambda_+,$$

$$\delta F_{\mu\nu} = -\frac{1}{2}\bar{\varepsilon}_-\sigma_{\mu\nu}V_+ - \frac{1}{2}\bar{\varepsilon}_+ i\partial\!\!\!/\sigma_{\mu\nu}\lambda_+,$$

$$\delta V_+ = (D + \frac{1}{2}\sigma^{\mu\nu}F_{\mu\nu}) i\partial\!\!\!/\varepsilon_-.$$

III. Construction of supersymmetric Lagrangians

Having shown the sort of field multiplets that are supported by the supersymmetry algebra, we turn to the construction of supersymmetric actions.

(A) Nonexistence of supersinglet Lagrangian densities

First we note that there are no supersymmetric invariant Lagrange densities. This is a direct consequence of the anticommutation relation of Eq.(1.1). To see this, suppose there were such a Lagrangian density (x,θ). Invariance under supersymmetry transformations would imply the relation

$$e^{i\bar{\varepsilon}Q}\mathcal{L}(x,\theta)e^{-i\bar{\varepsilon}Q} = \mathcal{L}(x,\theta).$$

Considering infinitesimal transformations, we would then have

$$[\bar{\varepsilon}Q, \mathcal{L}(x,\theta)] = 0$$

and hence

$$[[\bar{\varepsilon}_1 Q, \bar{\varepsilon}_2 Q], \mathcal{L}(x,\theta)] = + [[\bar{\varepsilon}_1 Q, \mathcal{L}(x,\theta)], \bar{\varepsilon}_2 Q] - [[\bar{\varepsilon}_2 Q, \mathcal{L}(x,\theta)], \bar{\varepsilon}_1 Q] = 0.$$

However, using

$[\bar{\varepsilon}_1 Q, \bar{\varepsilon}_2 Q] = \bar{\varepsilon}_1\gamma^\mu\varepsilon_2 P_\mu$, we find

$$[[\bar{\varepsilon}_1 Q, \bar{\varepsilon}_2 Q], \mathcal{L}(x,\theta)] = [\bar{\varepsilon}_1\gamma^\mu\varepsilon_2 P_\mu, \mathcal{L}(x,\theta)] = \bar{\varepsilon}_1\gamma^\mu\varepsilon_2 i\partial_\mu\mathcal{L}(x,\theta) \neq 0$$

Thus we reach a contradiction. So there can be no supersymmetry-invariant Lagrangian densities.

We must then search for Lagrangian densities that change by a 4-divergence under supersymmetry transformations. We know that the F component of a chiral scalar superfield and the D component of a general scalar superfield are just such objects. Further $\mathcal{L}$ must be hermitian. Hence we can guess a general structure:

$$\mathcal{L} = (\text{chiral superfield})_F + \text{h.c.}$$
$$+ (\text{vector superfield})_D. \qquad (3.1)$$

It is possible to argue that this exhausts all possible (linear) Lagrange densities.

(B) <u>Construction of Lagrangian density for a single chiral superfield $\Phi_+(x,\theta)$.</u>

Now $\mathcal{L}$ must be of the form

$$\mathcal{L} = (\tilde{\Phi}_+)_F + \text{h.c.} + (\tilde{\Phi})_D,$$

where $\tilde{\Phi}_+$ and $\tilde{\Phi}$ are certain composite chiral and vector superfields constructed out of $\Phi_+(x,\theta)$. We note that, for renormalizability, one must have

$$\text{Dim } D = \text{Dim } \tilde{\Phi} + 2 < 4,$$

$$\text{Dim } F_+ = \text{Dim } \tilde{\Phi}_+ + 1 < 4.$$

Thus $\tilde{\Phi}$ and $\tilde{\Phi}_+$ are local composite vector and chiral fields of dimensions upto 2 and 3 respectively. Now, since Dim Φ_+ = Dim $A_+ = 1$, we have the general forms:

$$\tilde{\Phi} = a\Phi_+^*\Phi_+,$$

$$\tilde{\Phi}_+ = s\Phi_+ + M\Phi_+^2 + g\Phi_+^3.$$

Thus

$$\mathcal{L} = a(\Phi_+^*\Phi_+)_D + (s\Phi_+ + M\Phi_+^2 + g\Phi_+^3)_F + \text{h.c.},$$

where the constant a can be determined from the requirement of correct normalization for the kinetic energy terms for the scalar and the fermian fields. Since the D-component is the coefficient of $(\bar{\theta}\theta)^2$ term, it is easy to see from the structure of Φ_+ in Eq.(2.2) that the kinetic energy term $a(\Phi_+^*\Phi_+)_D$ yields

$$\mathcal{L}_K = -\frac{1}{4} A_+^* \partial^2 A_+ + \text{c.c.} + \frac{1}{2} |\partial A_+|^2 + \bar{\psi}_+ i\not{\partial}\psi_+ + |F_+|^2$$

$$= |\partial A_+|^2 + \bar{\psi}_+ i\not{\partial}\psi_+ + |F_+|^2 + \text{gradient terms.}$$

On the other hand, the superpotential term yields

$$\mathcal{L}_{pot} = s(F_+ + \text{h.c.}) + M[(2A_+F_+ + \psi_+^T C^{-1}\psi_+) + \text{h.c.}]$$

$$+ g[(F_+A_+^2 + 3\psi_+^T C^{-1}\psi_+A_+) + \text{h.c.}].$$

This Lagrangian has a form invariance under $\Phi_+ \to \Phi_+ + C$ where C is a space-time independent constant. Under this transformation the form of the Lagrangian remains unchanged except to induce the following transformations on the parameters:

$$s \to s + MC + gC^2,$$

$$M \to M + 2gC,$$

$$g \to g.$$

Hence the Lagrangian is essentially a function of two invariants: g and $s - M^2/4g$. Because of this, one can set $s = 0$ or $M = 0$ without loss of generality.

(C) <u>Construction of Lagrangian given a number of chiral scalar fields</u>

Suppose we are given n positive chiral scalar fields $\Phi_{+1}, \ldots, \Phi_{+n}$ and m negative ones $\Phi_{-,n+1}, \ldots, \Phi_{-,n+m}$. That is equivalent to being given a set of positive chiral scalar fields. $\Phi_{+1}, \ldots \Phi_{+n}, \Phi_{+,n+1} \ldots, \Phi_{+,n+m}$ where $\Phi_{+,n+i} = \Phi_{-,n+i}^*$. The most general Lagrangian is obtained as before:

$$\mathcal{L} = a\left(\sum_{i=1}^{n+m} \Phi_{+i}^*\Phi_{+i}\right)_D + [s_i\Phi_{+i} + M_{ij}\Phi_{+i}\Phi_{+j} + g_{ijk}\Phi_{+i}\Phi_{+j}\Phi_{+k}]_F + \text{h.c.},$$

where the coefficients are consistent with any internal symmetry that may be present.

(D) Construction of gauge field Lagrangian: Abelian case:

To construct gauge interactions we must have a superfield containing a vector field which we already know, viz. the vector superfield $v(x,\theta)$. Recall that $v(x,\theta) = v^*(x,\theta)$. Thus it has the following expansion

$$v(x,\theta) = A_1 + \bar{\theta}\psi_1 + \frac{1}{4}\bar{\theta}\theta F_1 + \frac{1}{4}\bar{\theta}\gamma_5\theta G_1 + \frac{1}{4}\bar{\theta}i\gamma^\mu\gamma_5\theta V_\mu + \frac{1}{4}\bar{\theta}\theta\bar{\theta}\chi_1 + \frac{1}{32}(\bar{\theta}\theta)^2 D_1 .$$

The reality condition requires that A_1, F_1, G_1, V_μ, D, are real fields while ψ_1 and χ_1 are Majorana spinors.

Next, we must seek a supersymmetric generalization of gauge transformations on the vector superfield $v(x,\theta)$. It must have the following properties:

(i) It should preserve the vector nature of the superfield.

(ii) It should yield the correct gauge transformation for V_μ, viz. $V_\mu \to V_\mu + \partial_\mu\Lambda$.

These conditions restrict the transformation to have the following form:

$$v \to v - \frac{i}{2}(\Lambda_+ - \Lambda_+^*),$$

where the factor of $\frac{i}{2}$ is inserted for convenience. Λ_+ is a chiral superfield and as such has the following form:

$$\Lambda_+ = A_+ + \bar{\theta}_-\psi_+ + \frac{1}{2}\bar{\theta}_-\theta_+F_+ + \frac{1}{4}\bar{\theta}i\gamma^\mu\gamma_5\theta i\partial_\mu A_+ - \frac{i}{4}\bar{\theta}\theta\bar{\theta}_+\not{\partial}\psi_+ - \frac{1}{32}(\bar{\theta}\theta)^2\partial^2 A_+ .$$

Hence Λ_+^* is given by

$$\Lambda_+^* = A_+^* + \bar{\theta}_+(\psi_+)^C + \frac{1}{2}\bar{\theta}_+\theta_-F_+^* - \frac{1}{4}\bar{\theta}i\gamma^\mu\gamma_5\theta i\partial_\mu A_+^* - \frac{i}{4}\bar{\theta}\theta\bar{\theta}_-\not{\partial}(\psi_+)^C - \frac{1}{32}(\bar{\theta}\theta)^2\partial^2 A_+^* .$$

Thus, under this transformation, we find that

$$V_\mu \to V_\mu + \frac{1}{2}(\partial_\mu A_+ + \partial_\mu A_+^*) = V_\mu + \partial_\mu(\mathrm{Re}A_+)$$

which is a correct gauge transformation for V_μ. For the other components of $v(x,\theta)$ we find:

$$A_1 \to A_1 - \frac{i}{2}(A_+ - A_+^*) = A_1 + \mathrm{Im}A_+,$$

$$\psi_1 \to \psi_1 - \frac{i(\psi_+ - \psi_+^C)}{2} = \psi_1 + \text{a Majorana spinor},$$

$$F_1 - iG_1 \quad F_1 - iG_1 - \frac{i}{2} F_+,$$

$$F_1 + iG_1 \quad F_1 + iG_1 + \frac{i}{2} F_+^*,$$

$$\chi_1 \to \chi_1 - \frac{i\partial\!\!\!/}{2}(-i\psi_+ + i\psi_+^C) = \chi_1 - \frac{1\partial\!\!\!/}{2}(\psi_+ - \psi_+^C),$$

$$D_1 \to D_1 - \frac{i}{2}\partial^2(A_+ - A_+^*) = D_1 + \text{a 4-divergence}.$$

We note the following properties of the above gauge transformations:
(i) For the transformations of the components A_1, ψ_1, F_1, G_1, we find that they can all be made equal to zero by a suitable choice of the gauge transformation Λ_+ i.e. by properly choosing $\mathrm{Im}A_+$, ψ_+, F_+. Such gauges are called Wess-Zumino gauges. In these gauges v has the structure:

$$v = \frac{1}{4}\bar{\theta}i\gamma^\mu\gamma_5 V_\mu + \frac{1}{4}\bar{\theta}\theta\bar{\theta}\chi_1 + \frac{1}{32}(\bar{\theta}\theta)^2 D_1.$$

(ii) However, the above gauge fixing only fixes $\mathrm{Im}A_+$ leaving undetermined $\mathrm{Re}A_+$ which appears in the transformation law for V_μ. Thus within Wess-Zumino gauges, there is still the freedom of making ordinary gauge transformation $V_\mu \to V_\mu + \partial_\mu\Lambda$.
(iii) D_1 changes by a 4-divergence under a gauge transformation. Thus $\int D_1 d^4x$ is a gauge-invariant quantity and thus can be lead to a gauge-invariant term in the Lagrangian.

Having discovered the supersymmetric analog of gauge transformations, we must construct a supersymmetric analog of the field strengths f(v). It must satisfy the following properties:

(i) It must be linear in the vector superfield $v(x,\theta)$.
(ii) It must satisfy: $f(v - \frac{i}{2}(\Lambda_+ - \Lambda_+^*)) = f(v)$.
(iii) It must contain $|\partial_\mu V_\nu - \partial_\nu V_\mu|$ as a component.

We exhibit the construction in steps. Consider $v - \frac{i}{2}(\Lambda_+ - \Lambda_+^*)$. We must invent a differential operator that will remove the last two terms. We know that the operation of $D_{+\alpha}$ will remove Λ_+^* because $D_{+\alpha}\Lambda_+^* = 0$ (Λ_+^* is a negative chiral scalar field.)
Thus

$$D_{+\alpha}(v - \frac{i}{2}\Lambda_+ + \frac{i}{2}\Lambda_+^*) = D_{+\alpha}(V - \frac{i}{2}\Lambda_+).$$

Next, act with $D_{-\beta}$ and, using $D_{-\beta}\Lambda_+ = 0$, we have

$$D_{-\beta}D_{+\alpha}(v - \frac{i}{2}\Lambda_+ + \frac{i}{2}\Lambda_+^*) = D_{-\beta}D_{+\alpha}v - \frac{i}{2}\left\{D_{-\beta}, D_{+\alpha}\right\}\Lambda_+$$

$$= D_{-\beta}D_{+\alpha}v - \frac{i}{2}(\gamma_+\gamma^\mu C)_{\alpha\beta}i\partial_\mu\Lambda_+.$$

A further operation of $D_{-\gamma}$ removes the last term because $D_{-\gamma}\partial_\mu\Lambda_+ = \partial_\mu D_{-\gamma}\Lambda_+ = 0$. Thus

$$D_{-\gamma}D_{-\beta}D_{+\alpha}(v - \frac{i}{2}\Lambda_+ + \frac{i}{2}\Lambda_+^*) = D_{-\gamma}D_{-\beta}D_{+\alpha}v.$$

Now there is essentially just one operator of the form $D_{-\gamma}D_{-\beta}$ as D_- contains only two independent components. We may choose it to be the Lorentz invariant form $\overline{D}_+D_- = -D_{-\gamma}D_{-\beta}C_{\gamma\beta}$. We have thus the supersymmetric field strength

$$\Psi_{\alpha++} = -\frac{i}{\sqrt{2}}\overline{D}_+D_-D_{+\alpha}v,$$

which is invariant under $v \to v - \frac{i}{2}\Lambda_+ + \frac{i}{2}\Lambda_+^*$. Further, as $(D_-)^3 = 0$, it satisfies

$$D_-\Psi_{\alpha++} = 0$$

and also

$$(\gamma_-)_{\alpha\beta}\Psi_{\beta++} = 0$$

Thus $\Psi_{\alpha++}$ is a chiral spinor field of positive chirality with respect to both internal structure as well as the external index α. It is of the kind already exhibited in Sec.II J.

One could have alternatively started by operating D_- and could have constructed

$$\Psi_{\alpha--} = -\frac{i}{\sqrt{2}}\overline{D}_-D_+D_{-\alpha}v.$$

$\Psi_{\alpha--}$ is a chiral scalar field of negative chirality with respect to both internal structure and the external index. It can be shown that Ψ_{--} and Ψ_{++} are charge conjugates, i.e.

$$\Psi_{--} = C\Psi_{++}^{T}.$$

In Wess-Zumino gauges $\Psi_{\alpha++}$ has the structure

$$\Psi_{++}(x,\theta) = \frac{\gamma_+\chi_1(z)}{\sqrt{2}} + \frac{i}{2\sqrt{2}}\,[D_1(z) + \sigma^{\mu\nu}(\partial_\mu V_\nu(z) - \partial_\nu V_\mu(z)]\theta_+ + \frac{1}{2\sqrt{2}}\,\bar\theta_-\theta_+[-i\gamma_+(\not\partial - \not V)\chi]$$

and thus clearly involves $F_{\mu\nu}$. By applying the general formula (3.1), the gauge invariant Lagrangian for abelian gauge fields can be constructed:

$$L = a(\bar\Psi_{--}\Psi_{++})_F + \text{h.c.} + \xi D_1(x).$$

Here we have utilized the fact that $\int D_1(x)d^4x$ is gauge invariant.

(E) Construction of abelian supersymmetric gauge theories with matter fields

Suppose we are given n chiral fields Φ_{+i}, $i = 1,2,\ldots n$ which belong to representations of a global U(1) symmetry. They transform under this U(1) transformation as

$$\Phi_{+\ell} \to \Phi'_{+\ell} = e^{+it_\ell\eta}\Phi_{+\ell},$$

where t_ℓ are respective U(1) charges. The Lagrangian, which is invariant under these global U(1) transformations as well as supersymmetric, is of the form:

$$L = a\left(\sum_\ell \Phi^\dagger_{+\ell}\Phi_{+\ell}\right)_D + (M_{ij}\Phi_{+i}\Phi_{+j} + g_{ijk}\Phi_{+i}\Phi_{+j}\Phi_{+k})_F + \text{h.c.}$$

Here M_{ij} and g_{ijk} respect global U(1)-invariance. Thus $M_{ij} = 0$ unless $t_i+t_j = 0$ and $g_{ijk} = 0$ unless $t_i+t_j+t_k = 0$.

We would like to make this Lagrangian into a local Lagrangian by introducing a local gauge field. First, we must construct local transformations. This is easily done by making η a function of x, but that is not enough. $e^{+it_\ell\eta(x)}\Phi_{+\ell}$ is not a chiral field. Besides, if η depends on x, a supersymmetry transformation will make it dependent on θ as well. Thus, generally, η must be a function of both x and θ; in addition, it must be a chiral field of positive chirality if it is to preserve the chiral nature of $\Phi_{+\ell}$. Thus the correct local U(1) transformations are

$$\Phi_{+\ell} \rightarrow \Phi'_{+\ell} = e^{+it_\ell \Lambda_+(x,\theta)} \Phi_{+\ell}(x,\theta).$$

We now come to the question of constructing a gauge-invariant as well as supersymmetric action. As in the case of ordinary gauge theories, the potential term is already invariant under local U(1) transformations on account of the conditions on M_{ij} and g_{ijk} stated earlier. Further, the kinetic terms are not invariant. This is since

$$\Phi^\dagger_{+\ell}\Phi_{+\ell} \rightarrow \Phi^{\dagger\prime}_{+\ell}\Phi'_{+\ell} = \Phi^\dagger_{+\ell}\Phi_{+\ell} e^{-it_\ell(\Lambda^*_+ - \Lambda_+)}.$$

Now recall the gauge transformation on $v(x,\theta)$:

$$v \rightarrow v' = v - \frac{i}{2}(\Lambda_+ - \Lambda^*_+).$$

Thus the kinetic energy term can be made gauge-invariant by being modified to

$$a\left(\sum_\ell \Phi^\dagger_{+\ell}\, e^{2t_\ell v}\, \Phi_{+\ell}\right)_D$$

Thus the gauge-invariant supersymmetric Lagrangian for system of gauge and chiral fields is

$$\begin{aligned} L = {} & a(\overline{\psi}_{--}\psi_{++})_F + \text{h.c.} + \xi D_1 \\ & + a'\left(\sum_\ell \Phi^\dagger_{+\ell} e^{2t_\ell v} \Phi_{+\ell}\right)_D + (M_{ij}\Phi_{+i}\Phi_{+j} + g_{ijk}\Phi_{+i}\Phi_{+j}\Phi_{+k})_F + \text{h.c.} \end{aligned}$$

(F) Non abelian supersymmetric gauge theories

Internal symmetries and supersymmetry can be combined in a straightforward way. Suppose an internal symmetry group G is given. Superfields are then combined into sets that support representations of G, e.g.

$$\Phi_+(x,\theta) = \begin{pmatrix} \Phi_{+1}(x,\theta) \\ \Phi_{+2}(x,\theta) \\ \vdots \\ \Phi_{+n}(x,\theta) \end{pmatrix}$$

Under global transformations the column vector $\Phi_\pm$ transforms as

$$\Phi_\pm \rightarrow \Phi'_\pm = e^{i\Lambda^k B^k_\pm} \Phi_\pm,$$

where $B_{\pm k}$ are respective representations of the generators of the group G. We now wish to construct local gauge transformations. As in the abelian case, the parameters Λ^k should be changed to chiral superfields themselves. Thus a local gauge transformation on Φ_+ reads:

$$\Phi_+ \to \exp(i\Lambda_+^k(x,\theta)B_+^k)\Phi_+ \ .$$

This implies the following transformation law for Φ_+^*

$$\Phi_+^* \to \exp(-i\Lambda_+^*(x,\theta)B_+^k)\Phi_+ .$$

(Here B_+^k are hermitian). This suggests that we define local gauge transformations for Φ_- as

$$\Phi_- \to \exp(-i\Lambda_+^*(x,\theta)B_-^k)\Phi_- \ ,$$

where B_-^k are representations of generators of G corresponding to the column vector Φ_-.

Recall now that the most general Lagrangian for n positive chiral fields is

$$L = a\Big(\sum_\ell \Phi_{+\ell}\Phi_{+\ell}\Big)_D + (M_{ij}\Phi_{+j}\Phi_{+k} + g_{ijk}\Phi_{+i}\Phi_{+j}\Phi_{+k})_F + \text{h.c.},$$

where M_{ij} and g_{ijk} are restricted by global invariance under G. This restriction, as in the abelian case, makes the potential term invariant under local transformations also. The kinetic terms, however, are not locally gauge invariant. As in the abelian case, we see that they can be made invariant under local gauge transformations via the introduction of gauge superfields $v^k(x,\theta)$ which transform as an adjoint representation of the global group G. The kinetic term is modified as before, namely

$$\Phi_+^\dagger\Phi_+ \to \Phi_+^\dagger\, e^{2v^kB_+^k}\,\Phi_+ \equiv \Phi_+^\dagger\, e^{2v}\Phi_+,$$

provided v changes under gauge transformations as:

$$e^{2v} \to e^{i\Lambda_+^{k*}B^k}\, e^{2v}\, e^{-i\Lambda_+^kB^k} .$$

The latter can be expressed in short, with $\Lambda_+ = \Lambda_+^kB^k$, as

$$e^{2v} \to e^{2v'} = e^{i\Lambda_+^*}\, e^{2v}\, e^{-i\Lambda_+} .$$

This implies (since $e^{-2v}e^{2v} = 1$) that

$$e^{-2v} \rightarrow e^{-2v'} = e^{i\Lambda_+}e^{-2v}e^{-i\Lambda_+^*} .$$

Having secured the gauge transformation in the nonabelian case, we must now obtain supersymmetric analog of field strengths. As in ordinary nonabelian gauge theories, the supersymmetric field strengths will be not gauge-invariant but gauge-covariant. They can be obtained by a process similar to the one we used for the abelian case. We shall only give the final result:

$$\Psi_{\alpha++} = -\frac{i}{2\sqrt{2}}\bar{D}_+D_-[e^{-2v}D_{+\alpha}e^{2v}].$$

It transforms covariantly under a gauge transformation, viz.

$$\Psi_{++} \rightarrow \Psi'_{++} = e^{i\Lambda_+}\ \Psi_{++}\ e^{-i\Lambda_+} .$$

In a similar manner we could construct

$$\Psi_{\alpha--} = -\frac{i}{2\sqrt{2}}\bar{D}_-D_+[e^{-2v}D_{-\alpha}e^{2v}].$$

This transforms as

$$\Psi_{--} \rightarrow \Psi' = e^{i\Lambda_+^\dagger}\ \Psi_{--}\ e^{-i\Lambda_+^\dagger} .$$

It can be shown that

$$\Psi_{--} = C\bar{\Psi}^T_{++}$$

[In the above expressions $v = v^kB^k$, $\Lambda_+ = \Lambda_+^kB^k$ where B^k transform as the adjoint representation of the gauge group]. Thus the Lagrangian density for the gauge fields can be written as

$$L_g = a''\mathrm{tr}(\bar{\Psi}_{--}\Psi_{++})_F + \mathrm{h.c.}$$

In addition, if the group contains invariant abelian subgroups, generated by $\sum_k \xi^k_{(r)}B^k$ $(r = 1,2,\ldots,p)$, then one can add the D-term to the Lagrangian,

$$L_\xi = \sum_{r=1}^{p}(\sum_k \xi^k_{(r)}v^k)_D.$$

In the nonabelian case the gauge transformations on v are more complicated. They can be expressed as an infinite series in Λ_+:

$$v \to v + \frac{i}{2}(\Lambda_+^* - \Lambda_+) + \ldots$$

The presence of the term $\frac{i}{2}(\Lambda_+^* - \Lambda_+)$, as in the abelian case, makes it reasonable to infer that Wess-Zumino gauges exist in the nonabelian case also. In the Wess Zumino gauges v has the form

$$v = \frac{1}{4}\bar{\theta}i\gamma^\mu\gamma_5\theta V_\mu + \frac{1}{4}\bar{\theta}\theta\bar{\theta}\chi_1 + \frac{1}{32}(\bar{\theta}\theta)^2 D_1 .$$

Thus it is evident that here also $v^3=0$ and $\mathcal{L}_g$ reduces to a usual polynomial type of Lagrangian which is manifestly renormalizable. In Wess-Zumino gauges, Ψ_{++} is given by

$$\Psi_{++} = \frac{\gamma_+\chi_1(z)}{\sqrt{2}} + \frac{i}{2\sqrt{2}}[D_1(z) + \sigma^{\mu\nu}V_{\mu\nu}(z)]\theta_+ + \frac{1}{2\sqrt{2}}\bar{\theta}_-\theta_+[-i\gamma_+\nabla\!\!\!/\chi_1(z)],$$

where

$$V_{\mu\nu} = \partial_\mu V_\nu - \partial_\nu V_\mu - i[V_\mu, V_\nu],$$

$$\nabla_\mu\chi_1 = \partial_\mu\chi_1 - i[V_\mu, \chi_1].$$

Further, the Lagrangian L_g is given by

$$g^2\mathcal{L}_g = -\frac{1}{4}V^k_{\mu\nu}V^{k\mu\nu} + \frac{1}{2}\bar{\chi}^k_{1-}i\nabla\!\!\!/\chi^k_{1-} + \frac{1}{2}D^k_1 D^k_1,$$

where

$$\chi_{1-} = \gamma_-\chi_1 .$$

This is just the Yang-Mills theory with χ_{1-} in the adjoint representation. Thus such a gauge theory is automatically supersymmetric.

Finally let us write down the total Lagrangian involving a number of chiral fields represented by column vectors Φ_+, Φ_-:

$$\mathcal{L} = \mathcal{L}_g + \mathcal{L}_\xi + \left(\Phi_+^\dagger e^{2v^k B^k_+}\Phi_+ + \Phi_-^\dagger e^{2v^k B^k_-}\Phi_-\right)_D$$
$$+ (s_i\Phi_i + M_{ij}\Phi_{+i}\Phi_{+j} + g_{ijk}\Phi_{+i}\Phi_{+j}\Phi_{+k})_F$$
$$+ \text{similar terms for } \Phi_-$$
$$+ \text{h.c.}$$

References

1. Supersymmetry and superfields: A. Salam and J. Strathdee, Fortschritte der Physik 26, 57-142 (1978).
2. Supersymmetry: P. Fayet and S. Ferrara, Phys. Rep. 32C, 249-334 (1977).
3. Supersymmetry and supergravity: J. Wess and J. Bagger (1981) Princeton University Press.

N=1 AND N=2 SUPERGRAVITIES COUPLED TO YANG-MILLS THEORIES: GENERAL COUPLINGS AND SPONTANEOUS SYMMETRY BREAKING

S.Ferrara
CERN, Geneva, Switzerland

1. INTRODUCTION

The aim of these lectures is to describe, at an introductory level, supersymmetric gauge theories encompassing the fundamental particle interactions coupled to N=1 supergravity [1]. A preliminary study of spontaneous supersymmetry breaking in N=2 supergravity is also reported.

In the past year these theories have received considerable attention [2] because they give us the possibility of describing realistic models for low-energy particle physics, generalizing the Standard Model in a supersymmetric framework. Investigation of spontaneous supersymmetric breaking in Lagrangian field theory in recent times, and its application to models for elementary particle interactions, have revealed that the supersymmetry-breaking scale M_{SB}^2, defined as the goldstino decay constant, can be possibly as large as [3]

$$M_{SB}^2 \simeq M_W M_P \tag{1.1}$$

i.e. intermediate between the Planck scale $M_P = 0(10^{19}$ GeV) and the weak scale $M_W = 0(100$ GeV).

Under these circumstances supergravity corrections to global supersymmetry can no longer be neglected since the gravitino-mass growth implies, via the super-Higgs effect, the (tree-level) relation [4,5]

$$m_{3/2} = \sqrt{\frac{8\pi}{3}}\, \frac{M_{SB}^2}{M_P} = 0(M_W) \tag{1.2}$$

and the intramultiplet splitting of masses, due to supergravity corrections, is typically [6,7]

$$\Delta m^2 = 0(m_{3/2}^2). \tag{1.3}$$

Equation (1.3) reflects the fact that the Goldberger-Treiman relation for supersymmetry,

$$\Delta m_i^2 = g_i M_{SB}^2 \tag{1.4}$$

can be fulfilled with a large M_{SB}^2 provided the goldstino coupling is very small,

$$g_i = 0(M_W/M_P) \qquad (1.5)$$

as naturally happens with a gravitational sector which causes the spontaneous supersymmetry breaking [5].

We refer the reader to recent reviews [2] for phenomenological motivations to consider an effective supersymmetry-breaking scale $\Delta m^2 = 0(M_W^2)$. This is mainly related to the so-called hierarchy problem of gauge unified theories (GUTs) of electroweak and strong interactions [7].

From a more fundamental point of view, supersymmetry nowadays seems the only existing symmetry principle which may keep scalar (Higgs) masses naturally small and therefore avoid unnatural fine-tuning for the Higgs scalar v.e.v.'s which are required in order to keep the relation $M_W << M$ (M being the GUT scale or the Planck scale) stable under radiative corrections.

The present lectures are organized as follows: Sections 2,3 and 4 are used to derive the final form of locally supersymmetric N=1 supergravity-matter coupled systems from general invariance principles and tensor calculus for supersymmetric field theories.

In Section 5 the super-Higgs effect is described, and in Section 6 the low-energy limit of locally supersymmetric Lagrangians, with M_P (Planck mass) $\rightarrow \infty$ and $m_{3/2}$ (gravitino mass) kept fixed, is derived. In Section 7, new examples of the super-Higgs sector of the theory are derived. One uses the idea of "flat" potentials; the other idea is a locally supersymmetric generalization of the Fayet-Iliopoulos mechanism.

Section 8 reports simple tree-level examples of the generalization of the Standard Model for the electroweak theory to local supersymmetry. In Section 9 N=2 extended supergravity theories are considered and the relation between the N=1 and N=2 scalar potential derived. The connection with possible truncation of higher N extended supergravity is shortly discussed. In Section 10 the Higgs and superHiggs effect for N=2 pure Yang-Mills theories "minimally coupled" to N=2 supergravity is studied. All gauge internal symmetry breaking patterns for unitary groups which also break [9] local supersymmetry are found.

Finally in Section 11 the N=2 SuperHiggs effect without cosmological constant is discussed and N=2 flat potentials are introduced.

2. TENSOR CALCULUS

The action and transformation laws of locally supersymmetric systems are constructed by the use of the N=1 tensor calculus of supergravity [8]. A major simplification occurs by employing the so-called conformal tensor calculus which uses, for the gravitational sector, the Weyl multiplet, i.e. the multiplet which only contains the gauge fields of conformal supergravity [9].

The N=1 Weyl multiplet contains the following set of fields:

$$(e_{m\mu},\ \psi_{\mu a},\ A_\mu,\ b_\mu) \tag{2.1}$$

which are respectively the gauge fields for general coordinate transformations (vierbein), for Q supersymmetry (gravitino), for chiral U(1) (R-symmetry) and dilatations.

The additional gauge fields $\omega_{\mu mn}$, $\Phi_{\mu a}$, $f_{m\mu}$ for local Lorentz rotations, special supersymmetry, and special conformal transformations M_{mn}, S_a, K_m, respectively, can be solved in terms of the gauge fields in Eq.(2.1) by means of the conventional constraints[F1)]

$$\begin{aligned} &R^m_{\mu\nu}(P) = 0 \\ &\gamma^\mu R_{\mu\nu}(Q) = 0 \\ &\hat{R}_{\mu\nu mn}(M)e^{\nu n} + \frac{i}{3}\hat{R}_{\mu m}(A) = 0 \\ &(\tilde{T}_{mn} = \frac{1}{2}\varepsilon_{mnpq}T^{pq}) \end{aligned} \tag{2.2}$$

$\hat{R}_{\mu\nu mn}$ denotes the fully covariant Lorentz connection (Riemann tensor). The curvatures are covariant, but since the constraints in Eqs.(2.2) are not Q-invariant, the variations $\delta_Q\omega_{\mu nn}$, $\delta_Q\Phi_{\mu a}$, $\delta_Q f_{m\mu}$ are not simply given by the group rule. In the conformal tensor calculus we use two basic multiplets other than the one of Eq.(2.1). The scalar (or chiral) multiplet $S = (z, x_L, h)$ with Weyl weight w (w is the Weyl weight of the first component z of S) transforming as

$$\begin{aligned} \delta Z &= \bar{\varepsilon}_L\chi_L - \frac{1}{3}iW\alpha Z \\ \delta\chi_L &= \frac{1}{2}\hat{\not{\partial}}Z\varepsilon_R + \frac{1}{2}h\varepsilon_L + WZ\eta_L + i\alpha(\frac{1}{2} - \frac{1}{3}W)\chi_L \\ \delta h &= \bar{\varepsilon}_R\hat{\not{\partial}}\chi_L + 2\bar{\eta}_L\chi_L(1-W) + i\alpha(1 - \frac{1}{3}W)h \end{aligned} \tag{2.3}$$

where ε is a space-time dependent anticommuting Weyl (or Majorana) spinor corresponding to Q transformations, and α is the space-time dependent parameter of the U(1) chiral group of the superconformal algebra, normalized such that for the U(1) gauge field A_μ

$$\delta_\alpha A_\mu = \partial_\mu \alpha \tag{2.4}$$

In Eq.(2.3) η is a space-time dependent anticommuting Weyl (or Majorana) spinor corresponding to special S_a supersymmetry transformations.

In Eq.(2.3) $\hat{D}_\mu$ denotes the superconformal covariant derivative.

The second multiplet is a real vector multiplet V: V = (C, S_L, $\not{H}$, B_m, λ_R, D). Its transformation is

$$\begin{aligned}
\delta G &= \frac{1}{2} i\bar{\varepsilon}_L S_L - \frac{1}{2} i\bar{\varepsilon}_R S_R \\
\delta S_L &= \frac{1}{2} i\not{H}\varepsilon_L - \frac{1}{2} \not{B}\varepsilon_R - \frac{1}{2} i\hat{\not{D}}C\varepsilon_R - iwC\eta_L + \frac{1}{2} i\alpha S_L \\
\delta\not{H} &= -i\bar{\varepsilon}_R\hat{\not{D}}S_L - i\bar{\varepsilon}_R\lambda_R + i(-2+w)\bar{\eta}_L S_L + i\alpha\not{H} \\
\delta B_m &= (-\frac{1}{2}\bar{\varepsilon}_L\hat{D}_m S_L - \frac{1}{2}\bar{\varepsilon}_L\gamma_m\lambda_R + \frac{1}{2}(1+w)\bar{\eta}_L\gamma_m S_R) + h.c. \\
\delta\lambda_R &= \frac{1}{2}(\sigma.\hat{F} - iD)\varepsilon_R + \frac{1}{2} w(i\not{H}\eta_R - \not{B}\eta_L + i\hat{\not{D}}C\eta_L) + \frac{1}{2} i\alpha\lambda_R - w\not{\Lambda}_K S_L \\
\delta D &= (\frac{1}{2} i\bar{\varepsilon}_L\hat{\not{D}}\lambda_R + iw\bar{\eta}_L(\lambda_L + \frac{1}{2}\hat{\not{D}}S_R) - w\Lambda_{Km}\hat{D}_m C) + h.c. \\
\hat{F}_{mn} &= 2\hat{D}_{[m}B_{n]} + i\varepsilon_{mnpq}\hat{D}_p\hat{D}_q C, \quad [mn] = \frac{1}{2}(mn - nm)
\end{aligned} \tag{2.5}$$

where Λ_{Km} is the space-time dependent parameter for special conformal transformations.

The scalar multiplet S, for Weyl weight w = 3, has its last component h with Weyl weight w = 4 and it can be extended to a superconformal density,

$$e^{-1}L_{S(w=3)} = h + \bar{\psi}_R\cdot\gamma\chi_L + \bar{\psi}_{\mu R}\sigma_{\mu\nu}\psi_{\nu R}Z \tag{2.6}$$

Analogously we can embed a vector multiplet V with w = 2 into a scalar multiplet S_V with w = 3, whose components are

$$(-\not{H}^*, -i\lambda_L - i\hat{\not{D}}S_R, D + \hat{\Box}C + i\hat{D}_m B^m) \tag{2.7}$$

Then we can apply Eq.(2.6) to the multiplet S_V to obtain a density formula for the vector multiplet with w = 2:

$$e^{-1}L_{V(w=2)} = D + \hat{\Box}C + (\frac{1}{2} i\bar{\psi}.\gamma(\lambda_R+\hat{\not{D}}S_L) - \frac{1}{2}\bar{\psi}_{\mu L}\sigma_{\mu\nu}\psi_{\nu L}\mathcal{H} + h.c.) \tag{2.8}$$

One can also get a vector embedding of a scalar multiplet S with $w = 0$ into a vector multiplet V_S with $w = 0$ whose components are

$$V_S = [i(Z-Z^*),\ 2\chi_L,\ -2ih,\ -\hat{D}_m(Z+Z^*),\ 0,\ 0] \tag{2.9}$$

The multiplication rules of scalar and vector multiplets are given respectively by

$$S_1S_2 = (Z_1Z_2,\ Z_1\chi_{2L}+Z_2\chi_{1L},\ Z_1h_2+Z_2h_1-2\bar{\chi}_{1L}\chi_{2L}) \tag{2.10}$$

$$V_1V_2 = (C_1C_2, C_1S_{2L}+C_2S_{1L},\ C_1\mathcal{H}_2 + C_2\mathcal{H}_1 - \bar{S}_{1L}S_{2L},$$

$$C_1B_{m2} + \frac{1}{2} i\bar{S}_{1L}\gamma_mS_{2R} + (1\leftrightarrow 2),\ C_1\lambda_{2R} + \frac{1}{2}(-\hat{\not{D}}C_1S_{2L}$$

$$+ \mathcal{H}_1S_{2R} - i\not{B}_1S_{2L}) + (1\leftrightarrow 2),\ C_1D_2-\bar{S}_{1L}\lambda_{2L}-\bar{S}_{1R}\lambda_{2R}$$

$$+ \frac{1}{2}\mathcal{H}_1\mathcal{H}_2^* - \frac{1}{2}B_1.B_2 - \frac{1}{2}\hat{D}_mC_1\hat{D}^mC_2 - \frac{1}{2}\bar{S}_{1L}\hat{\not{D}}S_{2R} - \frac{1}{2}\bar{S}_{1R}\hat{\not{D}}S_{2L}+(1\leftrightarrow 2);$$

$$\frac{1}{2}(S_1\bar{S}_2+\bar{S}_1S_2) = (\frac{1}{2}Z_1Z_2^*,\ -iZ_2^*\chi_{1L},\ -Z_2^*h_1,\ \frac{1}{2}i(Z_2^*\hat{D}_mZ_1-$$

$$-Z_1\hat{D}_mZ_2^*) - i\bar{\chi}_{2R}\gamma_m\chi_{1L},\ -ih_1\chi_{2R} + i\hat{\not{D}}Z_2^*\chi_{1L},$$

$$h_1h_2^* - \hat{D}_mZ_1\hat{D}^mZ_2^* - \bar{\chi}_{1L}\overleftrightarrow{\not{D}}\chi_{1R}) + (1\leftrightarrow 2) \tag{2.11}$$

the resulting multiplets having Weyl weight $w_1 + w_2$.

The rules we have given so far are sufficient for constructing all superconformal invariant actions. However, if we want to restrict our tensor calculus to Poincaré supersymmetry only, we have to introduce Poincaré rules which will enable us to construct Poincaré supersymmetric local densities for arbitrary Weyl weight of the fields involved. This is most easily achieved by introducing a new compensating chiral multiplet S_o with Weyl weight $w = 1$, and then to use it to fix a special superconformal gauge. This choice of compensating multiplet corresponds to the so-called old minimal formulation of N=1 supergravity originally proposed in Refs.7. The connection of this formulation with the new minimal formulation of Sohnius and West [10], which uses a different compensating multiplet to fix the superconformal gauge, has been given in general elsewhere [11].

The chiral compensating multiplet has components

$$S_o = (a,\ \xi,\ \frac{1}{3}\ u) \qquad (2.12)$$

where u = S - iP is the complex auxiliary field of Poincaré supergravity defined by Cremmer et al. [5].

To get Poincaré supergravity from the superconformal tensor calculus we fix a superconformal gauge by fixing the gauges of dilatation, U(1) chiral rotation, special S-supersymmetry transformations, and conformal boosts by the conditions

$$a = 1,\ \xi = 0,\ b_\mu = 0 \qquad (2.13)$$

These conditions are preserved under a newly defined Poincaré Q local supersymmetry with the parameter

$$\delta^P(\varepsilon) = \delta_Q(\varepsilon) + \delta_S(\eta_L = -\frac{1}{6}\ u\varepsilon_L - \frac{1}{6}\ i\not{A}\varepsilon_R)$$
$$+ \delta_K(\Lambda_{K\mu} = \frac{1}{4}\ \bar{\varepsilon}\phi_\mu - \frac{1}{4}\ \bar{\eta}\psi_\mu) \qquad (2.14)$$

Equation (2.14) provides the rule for computing a Poincaré local supersymmetry transformation in terms of the local superconformal rules.

To compute the density formula for a chiral multiplet S of Weyl weight w = 0 we first construct the new chiral multiplet SS_o^3 which Weyl weight w = 3. Since in the Poincaré gauge defined by Eq.(2.13), using Eq.(2.10), we have

$$S_o^n = (1,\ 0,\ \frac{1}{3}\ nu) \qquad (2.15)$$

We obtain

$$SS_o^n = (Z,\ \chi_L,\ h + \frac{1}{3}\ nuZ) \qquad (2.16)$$

Then the Poincaré density formula for S is given by

$$[SS_o^3]_{F,SC} = [S]_{F,\text{Poincaré}} =$$

$$L_{S(w=0)} = eh + euZ + e\bar{\psi}_R\cdot\gamma\chi_L + e\bar{\psi}_{\mu R}\sigma_{\mu\nu}\psi_{\nu R}Z \qquad (2.17)$$

For the vector multiplet V, we take a Weyl weight w = 0 and then introduce the compensating vector multiplet $S_o\bar{S}_o$ with w = 2 with components

$$S_o\bar{S}_o = (1,\ 0,\ -\frac{2}{3}u,\ -\frac{2}{3}A_m,\ 0,\ \frac{2}{9}(uu^*-A_mA^m)) \tag{2.18}$$

We then get a Poincare density formula for V by using a superconformal density formula for $VS_o\bar{S}_o$ with w = 2 by using the multiplication rule given by Eq.(2.11) in the Poincaré gauge [Eq.(2.13)]:

$$[VS_o\bar{S}_o]_{D,SC} = [V]_{D,\text{Poincaré}} =$$

$$e^{-1}\mathcal{L}_{V(w=0)} = D - \frac{1}{2}i\bar{\psi}\cdot\gamma\gamma_5\lambda - \frac{1}{3}(u^*\mathcal{H}+u\mathcal{H}^*) + \frac{2}{3}B_mA^m$$
$$-\frac{1}{3}i\bar{S}\gamma_5\gamma R^P + \frac{1}{4}i\varepsilon^{mnrs}\bar{\psi}_m\gamma_n\psi_r(B_s - \frac{1}{2}\bar{\psi}_sS) - \frac{2}{3}Ce^{-1}\mathcal{L}_{SG} \tag{2.19}$$

where

$$R_\mu^P = e^{-1}\varepsilon_{\mu\nu}{}^{\rho\sigma}\gamma_5\gamma^\nu[\partial_\rho + \frac{1}{2}\omega_{\rho mn}(e_1\psi)\sigma^{mn} + \frac{1}{2}i\gamma_5A_\rho$$
$$+\frac{1}{6}\gamma_\rho(\omega-i\not{A}\gamma_5)]\psi_\sigma \tag{2.20}$$

$$\mathcal{L}_{SG} = -\frac{1}{2}eR(e,\omega(e,\psi))$$
$$-\frac{1}{2}\varepsilon^{\mu\nu\rho\sigma}\bar{\psi}_\mu\gamma_5\gamma_\nu D_\rho(\omega(e,\psi))\psi_\sigma - \frac{1}{3}euu^* + \frac{1}{3}eA_mA^m \tag{2.20}$$

We remark that the Poincaré density formula as well as the Poincaré local supersymmetry transformations are w-independent. If S has weight w, then $S' = S'S_o^{-w}$ has w = 0. Equation (2.15) defines the components of S' with $h' = h - \frac{1}{3}wuz$. This defines the w-independent transformation rules of Ref.6. Analogously for a vector multiplet V with Weyl weight w, we define the new multiplet $V' = V(S_o\bar{S}_o)^{-w/2}$ with w = 0.

The components of V' are easily computed to be

$$V' - (C,\ S_L,\ \mathcal{H}',\ B_m',\ \lambda_R',\ D') \tag{2.21}$$

with

$$\mathcal{H}' = \mathcal{H} + \frac{1}{3}wCu,\quad B_m' = B_m + \frac{1}{3}wCA_m$$
$$\lambda_R' = \lambda_R + \frac{1}{6}w(uS_R - i\not{A}S_L)$$
$$D' = D + \frac{1}{6}w(\mathcal{H}u^* + \mathcal{H}^*u - 2B_mA^m + \frac{1}{2}i\bar{S}\gamma_5\gamma\cdot R^P)$$
$$+\frac{1}{18}w^2C(uu^* - A_mA^m) \tag{2.22}$$

The density formulae for S' and V' as given by Eq.(2.17) (with h = h') and Eq.(2.19) (with $\not{H} = \not{H}'$, $B_m = B'_m$, $\lambda_R = \lambda'_R$, D = D') are local supersymmetric densities with respect to the following Poincaré local supersymmetry transformations:

$$\delta^P z = \bar{\varepsilon}_L \chi_L, \quad \delta^P \chi_L = \frac{1}{2} \hat{\not{D}}^P z \varepsilon_R + \frac{1}{2} h' \varepsilon_L$$

$$\delta^P h' = \bar{\varepsilon}_R \hat{\not{D}}^P \chi_L - \frac{1}{3} \bar{\varepsilon}_L \chi_L u - \frac{1}{6} i \bar{\varepsilon}_R \not{A} \chi_L \tag{2.23}$$

$$\delta^P C = \frac{1}{2} i \bar{\varepsilon}_L S_L + \text{h.c.}, \quad \delta^P S_L = \frac{1}{2} i \not{H}' \varepsilon_L - \frac{1}{2} \not{B}' \varepsilon_R - \frac{1}{2} i \hat{\not{D}}^P C \varepsilon_R$$

$$\delta^P \not{H}' = -i \bar{\varepsilon}_R \hat{\not{D}}^P S_L - i \bar{\varepsilon}_R \lambda'_R + \frac{1}{3} i \bar{S}_L \varepsilon_L u + \frac{1}{6} \bar{S}_L \not{A} \varepsilon_R$$

$$\delta^P B'_m = - \frac{1}{2} \bar{\varepsilon}_L (\hat{D}^P_m - \frac{1}{2} i A_m) S_L - \frac{1}{2} \bar{\varepsilon}_L \gamma_m \lambda'_R + \frac{1}{12} \bar{S}_R \gamma_m (u \varepsilon_L + i \not{A} \varepsilon_R) + \text{h.c.} \tag{2.24}$$

$$\delta^P \lambda'_R = \frac{1}{2} (\sigma . \hat{F}^P - i D') \varepsilon_R$$

$$\delta^P D' = \frac{1}{2} i \bar{\varepsilon}_L (\hat{\not{D}}^P - \frac{1}{2} i \not{A}) \lambda'_R + \text{h.c.}$$

with

$$\hat{F}^P_{mn} = 2 D_{[m} B'_{n]} + \bar{\psi}_{[m} \hat{D}^P_{n]} S + \bar{\psi}_{[m} \gamma_{n]} \lambda' + \bar{S} e^{\mu}_{[m} D_{n]} \psi_{\mu}$$

where the symbol [ ] denotes antisymmetrization.

We now extend our rules of tensor calculus by introducing a Yang-Mills group [6]. To maintain both supersymmetry and gauge invariance, the space-time dependent parameters of Yang-Mills transformations must be assigned to an entire chiral multiplet.

The Lie algebra-valued transformation parameters are

$$\Lambda = \tilde{g}^{\alpha} \Lambda^{\alpha} T^{\alpha j}_{i} \tag{2.25}$$

where $\tilde{g}^{\alpha}$ are gauge coupling constants, T^{α} are representation matrices for the generators of the group G, and Λ^{α} are scalar multiplets with Weyl weight w = 0. For a simple group, $\tilde{g}^{\alpha} = \tilde{g}$. We will often use the simpler notation $\tilde{g}$ for the set of gauge coupling constants $\tilde{g}^{\alpha}$ in the case of semisimple or not semisimple gauge groups G.

The components of the parameter multiplet Λ^{α} are denoted by

$$\Lambda^{\alpha} = (y^{\alpha},\ \rho^{\alpha},\ v^{\alpha}) \tag{2.26}$$

For the Yang-Mills connection (gauge potential) we introduce a Lie algebra-valued vector multiplet

$$V = \tilde{g}V^{\alpha}T_{i}^{\alpha j} \tag{2.27}$$

which, under a finite Yang-Mills variation, transforms as

$$e^{2V} \rightarrow e^{-i\bar{\Lambda}}\ e^{2V}\ e^{i\Lambda} \tag{2.28}$$

In the infinitesimal, Eq.(2.28) reads

$$\delta V = \frac{i}{2}\ i(\Lambda-\bar{\Lambda}) + \frac{1}{2}\ i[V,\Lambda+\bar{\Lambda}] + \frac{1}{6}\ i[V,[V,\Lambda-\bar{\Lambda}]] + O(V^3) \tag{2.29}$$

If we are interested in discussing a supersymmetric Yang-Mills theory, we can take gauge choices for all Λ transformations except the real scalar Re y which actually corresponds to the standard Yang-Mills parameter in ordinary space-time.

The Wess-Zumino gauge is defined by the following gauge choice:

$$C^{\alpha} = S^{\alpha} = \not{H}^{\alpha} = 0 \tag{2.30}$$

for the components of V^{α}.

The constraints in Eq.(2.30) are not invariant under supersymmetry and gauge transformations, since under these combined transformations we get

$$\begin{aligned} \delta C^{\alpha} &= \frac{1}{2}\ i(y-y^{*}) \\ \delta S_L^{\alpha} &= \rho_L^{\alpha} - \frac{1}{2}\ \not{B}\varepsilon_R \\ \delta\not{H}^{\alpha} &= -iv^{\alpha} - \frac{1}{2}\ i\bar{\varepsilon}_R\gamma^{\mu}\not{B}^{\alpha}\psi_{\mu R} - i\bar{\varepsilon}_R\lambda_R^{\alpha} \end{aligned} \tag{2.31}$$

Therefore the Wess-Zumino gauge is only invariant for $y = y^{*}$ and under a redefined supersymmetry transformation

$$\delta(\varepsilon) = \delta_Q(\varepsilon) + \delta(\rho_L^{\alpha} = \frac{1}{2}\ \not{B}^{\alpha}\varepsilon_R) + \delta(v^{\alpha} = -\ \frac{1}{2}\ \bar{\varepsilon}_R\gamma^{\mu}\not{B}^{\alpha}\psi_{\mu R} - \bar{\varepsilon}_R\lambda_R^{\alpha}) \tag{2.32}$$

By applying this rule to the components of V^{α} in the Wess-Zumino gauge

$$V^{\alpha}_{WZ} = (0,0,0,\lambda^{\alpha}_{R},B^{\alpha}_{\mu},D^{\alpha}) \qquad (2.33)$$

we get

$$\delta B_{\mu} = -\frac{1}{2}\bar{\varepsilon}_{L}\gamma_{\mu}\lambda_{R} - \frac{1}{2}\bar{\varepsilon}_{R}\gamma_{\mu}\lambda_{L} - D_{\mu}y$$

$$\delta\lambda_{R} = \frac{1}{2}(\sigma.\hat{F}-iD)\varepsilon_{R} + i[\lambda_{R},y] \qquad (2.34)$$

$$\delta D = \frac{1}{2}i\bar{\varepsilon}_{L}\hat{\not{D}}\lambda_{R} - \frac{1}{2}i\bar{\varepsilon}_{R}\hat{\not{D}}\lambda_{L} + i[D,y]$$

where $\hat{F}_{\mu\nu}$ and $\hat{D}_{\mu}$ are now covariant derivatives with respect to local supersymmetry and Yang-Mills transformations.

For chiral multiplets S^{i} transforming according to some (generally complex) representation of G, we have the finite transformation

$$S \rightarrow e^{-i\Lambda}S \qquad (2.35)$$

or in the infinitesimal

$$\delta S = -i\Lambda S \qquad (2.36)$$

Equation (2.36) when written in components reads:

$$\delta S' = -i(yz,\ y\chi_{L}+\rho_{L}z,\ yh+vz-\bar{\rho}_{L}\chi_{L}) \qquad (2.37)$$

Therefore in the transformation law given by Eq.(2.3) all derivatives are now replaced by Yang-Mills covariant derivatives, and in δh there is an extra term due to the last term in Eq.(2.32):

$$\delta' h = i\bar{\varepsilon}_{R}\lambda_{R}z = i\tilde{g}\bar{\varepsilon}_{R}\lambda^{\alpha}_{R}T^{\alpha j}_{i}z_{j} \qquad (2.38)$$

We end this section by giving the vector multiplication of two scalar multiplets which can actually be converted in a chiral multiplication of a chiral multiplet S_1 with the "kinetic multiplet" of S_2, $T(S_2)$, which is the curved generalization of the flat space superfield $\bar{D}\bar{D}\bar{S}$.

The components of the multiplet T(S) are

$$T(S_{i}) = h^{*i} + \frac{1}{3}u^{*}z^{*i},\ \tilde{\psi}^{i}_{L},\ \tilde{\mathcal{H}}^{i}) \qquad (2.39)$$

where

$$\tilde{\psi}^i_L = \hat{\not{D}}^P \chi^i_R + \frac{1}{6} i\not{A}\chi^i_R + \frac{1}{6}\gamma.R^P_R z^{*i} - i\tilde{g}z^{*j}T^{\alpha i}_j \lambda^\alpha_L \tag{2.40}$$

$$\tilde{\mathcal{H}}^i = \hat{\Box}^c z^{*i} - \frac{2}{3}u(h^{*i} + \frac{1}{3}u^* z^{*i}) + \tilde{g}z^{*j}T^{\alpha i}_j D^\alpha - 2i\tilde{g}\bar{\chi}^j_R T^{\alpha i}_j \lambda^\alpha_R \tag{2.41}$$

and $\hat{\Box}^c$ denotes the conformal d'Alembertian with Yang-Mills covariant derivatives.

Using the chiral density formula for the new chiral multiplet

$$C^i_j S_{1i} T(S_{2j}) + C^{i*}_j S_{2j} T(S_{1i}) \tag{2.42}$$

we get a new local density corresponding to the vector multiplication of chiral multiplets instead of their chiral multiplication as given by Eq.(2.10). The final result for the local density associated with the chiral multiplet Eq.(2.42) is

$$\begin{aligned}
&e^{-1}L(C^i_j S_{1i} T(S_{ij}) + C^{i*}_j S_{2j} T(S_{1i})) = \\
&C^i_j \Big\{ \frac{1}{6}(R + e^{-1}\varepsilon^{\mu\nu\rho\sigma}\bar{\psi}_\mu\gamma_5\gamma_\nu D_\rho\psi_\sigma) z_{1i} z^{*j}_2 - \hat{D}_\mu z_{1i}\hat{D}^\mu z^{*j}_2 \\
&- \bar{\chi}_{1Li}\not{D}\chi^j_{2R} - \bar{\chi}^j_{2R}\not{D}\chi_{1Li} + (h_{1i} + \frac{1}{3}uz_{1i})(h^{*j}_2 + \frac{1}{3}u^* z^{*j}_2) \\
&- \frac{1}{9}A^2_m z_{1i} z^{*j}_2 + \frac{1}{3} iA^m(-\bar{\chi}_{1Li}\gamma_m\chi^j_{2R} + z^{*j}_2\hat{D}_m z_{1i} - z_{1i}\hat{D}_m z^{*j}_2) \\
&+ \frac{1}{8} e^{-1}\varepsilon^{\mu\nu\rho\sigma}\bar{\psi}_\mu\gamma_\nu\psi_\rho(z_{1i}\hat{D}_\sigma z^{*j}_2 - z^{*j}_2\hat{D}_\sigma z_{1i}) + \\
&+ \frac{1}{3}\bar{\chi}_{1Li}\sigma^{\nu\mu}D_\mu\psi_\nu z^{*j}_2 + \frac{1}{3}\bar{\chi}^j_{2R}\sigma^{\nu\mu}D_\mu\psi_\nu z_{1i} + \\
&+ \bar{\chi}_{1Li}\sigma^{\nu\mu}\psi_\nu\hat{D}_\mu z^{*j}_2 + \bar{\chi}^j_{2R}\sigma^{\nu\mu}\psi_\nu\hat{D}_\mu z_{1i} - D_\mu\bar{\chi}_{1Li}\sigma^{\nu\mu}\psi_\nu z^{*j}_2 \\
&- D_\mu\bar{\chi}^j_{2R}\sigma^{\nu\mu}\psi_\nu z_{1i} + 2i\tilde{g}T^{\alpha k}_i\bar{\lambda}^\alpha_L\chi_{1Lk}z^{*j}_2 - 2i\tilde{g}\bar{\chi}^k_{2R}T^{\alpha j}_k\lambda^\alpha_R z_{1i} + \\
&+ \tilde{g}D^\alpha T^{\alpha k}_i z_{1k} z^{*j}_2 - \frac{1}{2} i\tilde{g}\bar{\psi}.\gamma\gamma_5\lambda^\alpha T^{\alpha k}_i z_{1k} z^{*j}_2 \Big\} + \text{h.c.}
\end{aligned} \tag{2.43}$$

where $C^i_j z_{1i} z^{*j}_2$ is invariant under G, i.e.

$$C^i_j T^{\alpha k}_i z_{1k} z^{*j}_2 - C^i_j T^{\alpha j}_k z_{1j} z^{*k}_2 = 0 \tag{2.44}$$

The other (antisymmetric) combination of the chiral multiplets,

$$C^i_j S_{1i} T(S_{2j}) - C^{i*}_j S_{2j} T(S_{1i}) \tag{2.45}$$

gives rise to a density formula which is a total derivative.

Equation (2.43) produces the kinetic terms of the chiral fields and their non-polynomial generalizations.

3. MANIFESTLY INVARIANT ACTIONS AND TRANSFORMATION LAWS

In this section we give the matter-coupled supergravity Lagrangian in its superspace form and its component form together with the transformation laws of the component fields under local Poincare supersymmetry transformations [6].

Let us start with the most general supersymmetric superspace action for a set of n chiral multiplets. This is given in flat superspace by

$$\int d^4x d^4\Theta \Phi(S,\bar{S}) + \int d^4x \mathrm{Re}\int d^2\Theta g(S) \tag{3.1}$$

where $\Phi(S,\bar{S})$ is an arbitrary real function of the chiral multiplets S_i and their complex conjugate $\bar{S}^i$ and $g(S)$ is a chiral function constructed out of the chiral multiplets S^i. The first component of $g(S)$, $g(z_i)$ is often called the superpotential. The first term in Eq.(3.1) gives rise to a D-type density, whilst the second term gives rise to an F-type density. They are, respectively, the non-polynomial generalizations of the scalar kinetic term $S_i\bar{S}^i$ of global supersymmetry and the self-interaction term (including the mass term) of the chiral multiplets S^i. If the chiral multiplets S_i transform according to some representation of a compact Lie group G, the action is assumed to be G-invariant. This requires the following group properties on the scalar functions $\Phi(z,z^*)$ and $g(z)$:

$$z^{*j}T^{\alpha i}_{j}\Phi_i = \Phi^j T^{\alpha i}_{j} z_i$$

$$g^j T^{\alpha i}_{j} z_i = 0 \tag{3.2}$$

$$\left\{\Phi^i = \frac{\partial\Phi}{\partial z_i}\,,\ \Phi_i = \frac{\partial\Phi}{\partial z^{*i}}\,,\ g^i = \frac{\partial g}{\partial z_i}\right\}$$

The extension of the action [Eq.(3.1)] to local supersymmetry and to a local Yang-Mills group G needs the following modification to Eq.(3.1) [12]:

$$\int d^4x d^4\Theta E[\Phi(S,\bar{S}e^{2V}) + \mathrm{Re}(\tfrac{1}{R}\, g(S))] \tag{3.3}$$

where V is the gauge vector multiplet of G, E is the superspace determinant, and R is the chiral scalar curvature superfield [12]. The component form of Eq.(3.3) can be computed with the rules of the tensor calculus developed in the previous section.

The remaining part of the Yang-Mills-supergravity coupling corresponds to the supergravity extension of the flat Yang-Mills action [12]:

$$\int d^4x \ \mathrm{Re}\int d^2\theta W_a^\alpha \varepsilon^{ab} W_b^\alpha \tag{3.4}$$

where W_a^α is the field-strength chiral multiplet.

We can extend Eq.(3.4), including non-polynomial interactions with the chiral multiplets S_i, as follows [6]:

$$\int d^4x \ \mathrm{Re}\int d^2\theta [f_{\alpha\beta}(S) W_a^\alpha \varepsilon^{ab} W_b^\beta] \tag{3.5}$$

where $f_{\alpha\beta}(S)$ is a chiral superfield transforming as the symmetric product of the adjoint representation of G.

The curved space generalization of Eq.(3.5) is

$$\int d^4x d^4\theta E \ \mathrm{Re}(\tfrac{1}{R} f_{\alpha\beta}(S) W_a^\alpha \varepsilon^{ab} W_b^\beta) \tag{3.6}$$

and we can compute Eq.(3.6) using the local density formula for chiral multiplets obtained in Section 2. In principle, under the requirement that the field-Lagrangian contains only first derivatives in fermion fields and second derivatives in the scalar fields, we could also have new invariants with higher power of the field strength multiplet W_a^α such as $f_{\alpha_1 \ldots \alpha_n}(S) W^{\alpha_1} \ldots W^{\alpha_n}$. We will not consider these terms here. All other possible modifications of Eqs.(3.3) and (3.6) would necessarily contain higher derivatives of boson and fermion fields. These terms cannot appear in the tree-level Lagrangian if we require normal propagation of the physical fields, but they may well appear in the effective action as a result of radiative corrections, in exact analogy to the high curvature terms which appear in the quantum loop expansion of the Einstein theory.

In global supersymmetry, if the gauge group G contains an Abelian factor U(1), there is an extra possible invariant - the so-called Fayet-Iliopoulos term [13]

$$\int d^4x d^4\theta V \tag{3.7}$$

The naive extension of Eq.(3.7) to curved space,

$$\int d^4x d^4\theta E V \tag{3.8}$$

is not gauge invariant, since under a U(1) transformation E → E and $V \to V + (\Lambda - \bar{\Lambda})$. Then we have to modify both Eq.(3.8) and the U(1) gauge transformation in order to preserve a U(1) gauge invariance [14].

The Fayet-Iliopoulos term can be introduced in curved superspace as the result of the gauging of an R-symmetry [15], which is defined as follows

$$
\begin{aligned}
&V_R \to V_R + i/g_R(\Lambda - \bar{\Lambda}) \\
&S_o \to e^{i\Lambda} S_o, \; \bar{S}_o \to e^{-i\bar{\Lambda}} \bar{S}_o \\
&S_i \to e^{-in_i\Lambda} S_i, \; \bar{S}^i \to e^{in_i\bar{\Lambda}} \bar{S}^i
\end{aligned} \tag{3.9}
$$

and V_R is the gauge-multiplet for this U(1) gauge group. The extension of Eqs.(3.3) and (3.5) to local R-symmetric interactions requires the following condition on the functions $\Phi(S_i\bar{S})$, $g(S)$, and $F_{\alpha\beta}(S)$ [11]:

$$
\begin{aligned}
&\Phi(S_i, \bar{S}^i) = \Phi(e^{-in_iq} S_i, \; e^{in_iq} \bar{S}^i) \\
&g(S_i) = e^{3iq}(e^{-in_iq} S_i) \\
&f_{\alpha\beta}(S_i) = f_{\alpha\beta}(e^{-in_iq} S_i)
\end{aligned} \tag{3.10}
$$

where q is a real constant parameter.

Equation (3.10) implies that Φ and $f_{\alpha\beta}$ are R-symmetric, but the superpotential g(z) must transform with a given phase under R-symmetry. This is due to the non-invariance of the chiral measure E/R in Eq.(3.3) or, equivalently, to the transformation property of the compensating multiplet S_o in the density formula given by Eq.(2.17).

Under the restriction (3.10) the most general Lagrangian density invariant under the Yang-Mills group G x $U_R(1)$ is [11]

$$
\begin{aligned}
&\int d^4x d^4\theta E[e^{-g_R V_R} \Phi(S_i, \bar{S}^i e^{n_i g_R V_R} e^{2V})] \\
&\qquad + \mathrm{Re}[\tfrac{1}{2}\, g(S_i) + f_{\alpha\beta}(S_i) W^\alpha W^\beta + f_R(S_i) W_R^2]
\end{aligned} \tag{3.11}
$$

or equivalently, using the density formulae of Section 2:

$$
\begin{aligned}
L = &-\tfrac{1}{2}\, [S_o e^{-g_R V_R} \bar{S}_o \Phi(S_i, \bar{S}^i e^{n_i g_R V_R} e^{2V})]_D \\
&+ [g(S_i) S_o^3]_F - [f_{\alpha\beta}(S_i) W^\alpha W^\beta]_F - [f_R(S_i) W_R^2]_F
\end{aligned} \tag{3.12}
$$

where W_R is the field-strength multiplet for V_R, $g_R = k^2\xi$, where ξ is a dimensional parameter (dim ξ = 2), and g_R is the R-gauge coupling constant. We will now derive the component expression of Eqs. (3.3) and (3.6), and comment at the end on their generalization to the case of a gauged R-symmetry [Eq. (3.12)].

In order to compute the following density formulae,

$$[S_o\bar{S}_o\Phi(S,\ \bar{S}e^{2V})]_D \tag{3.13}$$

$$[g(S)S_o^3]_F \tag{3.14}$$

$$[f_{\alpha\beta}W^\alpha W^\beta]_F = [fW^2]_F \tag{3.15}$$

we have to work out the components of the vector multiplet $\Phi(S,\bar{S}e^{2V})$ and of the chiral multiplets $g(S)$ and fW^2, respectively.

This can be done using the multiplication rules (2.10), (2.11), and (2.42), and the Wess-Zumino gauge choice for the vector multiplet V^α. In this gauge $V^\alpha V^\beta V^\gamma = 0$, and we get

$$\begin{aligned}\Phi(S,\bar{S}e^2\ ) &= \Phi(S,\bar{S}) + 2\tilde{g}\Phi_i\bar{S}^jT^{\alpha i}_jV^\alpha \\ &+ 2\tilde{g}^2(\Phi_{ij}\bar{S}^kT^{\alpha i}_k\bar{S}^\ell T^{\beta j}_\ell + \Phi_i\bar{S}^kT^{\alpha j}_kT^{\beta i}_j)V^\alpha V^\beta\end{aligned} \tag{3.16}$$

The components $C(\Phi)$, $S_L(\Phi)$, $\mathcal{H}(\Phi)$, $B_m(\Phi)$, $\lambda_R(\Phi)$, and $D(\Phi)$ of the local vector multiplet given in Eq. (3.16) are given by [6]

$$\begin{aligned}
C(\Phi) &= \Phi(z,z^*),\ S_L(\Phi) = -2i\Phi^i\chi_{Li} \\
\mathcal{H}(\Phi) &= -2\Phi^ih_i + 2\Phi^{ij}\bar{\chi}_{Li}\chi_{Lj},\ B_m(\Phi) = i\Phi^i\hat{D}_mz_i \\
&\quad - i\Phi_i\hat{D}_mz^{*i} - 2i\Phi_i{}^j\bar{\chi}^i_R\gamma_m\chi_{Lj} \\
\lambda_R(\Phi) &= -2i\Phi^j_ih_j\chi^i_R + 2i\Phi^{ij}_k\chi^k_R\bar{\chi}_{Li}\chi_{Lj} \\
&\quad + 2i\Phi_i{}^j\hat{\not{D}}z^{*i}\chi_{Lj} + 2\tilde{g}\lambda^\alpha_R\tilde{D}^\alpha \\
D(\Phi) &= 2\Phi^i_jh_ih^{*j} - 2\Phi^{ij}_k\bar{\chi}_{Li}\chi_{Lj}h^{*k} - 2\Phi^k_{ij}\bar{\chi}^i_R\chi^j_Rh_k \\
&\quad + 2\Phi^{ij}_{k\ell}\bar{\chi}_{Li}\chi_{Lj}\bar{\chi}^k_R\chi^\ell_R - 2\Phi^j_i\hat{D}_\mu z^{*i}\hat{D}^\mu z_j \\
&\quad - 2\Phi^j_i\bar{\chi}_{Lj}\hat{\overleftrightarrow{\not{D}}}\chi^i_R - 2(\Phi^i_{jk}\hat{D}_\mu z^{*k} - \Phi^{ik}_j\hat{D}_\mu z_k)\bar{\chi}_{Li}\gamma^\mu\chi^j_R
\end{aligned} \tag{3.17}$$

$$+ 2\tilde{g}D^\alpha\tilde{D}^\alpha + 4i\tilde{g}\bar{\lambda}^\alpha_L\tilde{D}^{\alpha i}\chi_{Li} - 4i\tilde{g}\bar{\lambda}^\alpha_R D^\alpha_i\chi^i_R$$

$$(\tilde{D}^\alpha = \Phi_i T^{\alpha i}_j z^{*j} = \Phi^i T^{\alpha j}_i z_j,\ \tilde{D}^{\alpha i} = \frac{\partial\tilde{D}^\alpha}{\partial z}\ \tilde{D}^\alpha_i = \frac{\partial\tilde{D}^\alpha}{\partial z^{*i}}) \tag{3.17}$$

(the derivatives $\hat{D}_\mu$ are supersymmetric and Yang-Mills covariant).

Analogously the components of the two chiral multiplets $g(S)$ and fW^2 are given respectively by [6]

$$g(S) = [g(z),\ \chi_L(g) = g^i\chi_{Li},\ h(g) = g^i h_i - g^{ij}\bar{\chi}_{Li}\chi_{Lj}] \tag{3.18}$$

$$fW^2 = [z(fW^2),\ \chi_L(fW^2),\ h(fW^2)] \tag{3.19}$$

with

$$z(fW^2) = -\frac{1}{2} f_{\alpha\beta}\bar{\lambda}^\alpha_L\lambda^\beta_L$$

$$\chi_L(fW^2) = \frac{1}{2} f_{\alpha\beta}(\sigma.\hat{F}^{-\alpha} - iD^\alpha)\lambda^\beta_L - \frac{1}{2} f^i_{\alpha\beta}\chi_{Li}\bar{\lambda}^\alpha_L\lambda^\beta_L$$

$$h(fW^2) = f_{\alpha\beta}(-\bar{\lambda}^\alpha_L\hat{\not{D}}\lambda^\beta_R - \frac{1}{2}\hat{F}^{-\alpha}_{\mu\nu}\hat{F}^{-\beta}_{\mu\nu} + \frac{1}{2} D^\alpha D^\beta)$$

$$+ f^i_{\alpha\beta}\bar{\chi}_{Li}(-\sigma.\hat{F}^{-\alpha} + iD^\alpha)\lambda^\beta_L$$

$$- \frac{1}{2} f^i_{\alpha\beta}h_i\bar{\lambda}^\alpha_L\lambda^\beta_L + \frac{1}{2} f^{ij}_{\alpha\beta}\bar{\chi}_{Li}\chi_{Lj}\bar{\lambda}^\alpha_L\lambda^\beta_L \tag{3.20}$$

Using the density formulae (2.17) and (2.19), we finally get

$$e^{-1}L(\Phi) = -\frac{1}{6}\Phi e^{-1}L_{SG} + \Phi^i_j(-\frac{1}{2}D_\mu z_i D^\mu z^{*j} - \bar{\chi}_{Li}\not{D}\chi^j_R$$

$$+ \frac{1}{2}h_i h^{*j}) - \Phi^{ij}_k\bar{\chi}_{Li}\chi_{Lj}h^{*k} + \Phi^{ij}_k\bar{\chi}_{Li}\not{D}z_j\chi^k_R +$$

$$+ \frac{1}{2}\Phi^{ij}_k\bar{\chi}_{Li}\chi_{Lj}\bar{\chi}^k_R\chi^\ell_R + \frac{1}{3}u^*(\Phi^i h_i - \Phi^{ij}\chi_{Li}\chi_{Lj})$$

$$+ \frac{1}{3}iA^\mu(\frac{1}{2}\Phi^i_j\bar{\chi}^j_R\gamma_\mu\chi_{Li} + \Phi^i(D_\mu z_i - \bar{\psi}_{\mu L}\chi_{Li}))$$

$$+ \Phi^i_j\bar{\psi}_{\mu L}\not{D}z^{*j}\gamma_\mu\chi_{Li} - \frac{4}{3}\Phi^i\bar{\chi}_{Li}\sigma^{\mu\nu}D_\mu\psi_{\nu L}$$

$$- \frac{1}{8}e^{-1}\varepsilon^{\mu\nu\rho\sigma}\bar{\psi}_\mu\gamma_\nu\psi_\rho(\Phi^i D_\sigma z_i + \frac{1}{2}\Phi^j_i\bar{\chi}^i_R\gamma_\sigma\chi_{Lj})$$

$$- \frac{1}{2}\Phi^i_j\bar{\psi}_{\mu R}\chi^j_R\bar{\psi}^\mu_L\chi_{Li} + \frac{1}{6}\Phi^i\bar{\chi}_{Li}(\bar{\psi}_{\mu L}\bar{\psi}.\gamma\psi^\mu$$

$$+ \frac{1}{2}\sigma^{\mu\nu}\psi^{\rho}_{L}\bar{\psi}_{\nu}\gamma_{\rho}\psi_{\mu} + \sigma^{\mu\nu}\psi_{\nu L}\bar{\psi}_{\mu}\gamma.\psi)$$

$$+ \frac{1}{2}\tilde{g}\Phi^{i}T^{\alpha j}_{i}z_{j}(D^{\alpha}+i\bar{\psi}_{L}.\gamma\lambda^{\alpha}_{R}) - 2i\tilde{g}\Phi^{j}_{i}T^{\alpha k}_{j}z_{k}\bar{\lambda}^{\alpha}_{R}\chi^{i}_{R} + \text{h.c.} \tag{3.21}$$

$$e^{-1}L(g) = -\frac{1}{2}g^{ij}\bar{\chi}_{Li}\chi_{Lj} + \frac{1}{2}g^{i}h_{i} + \frac{1}{2}gu + \frac{1}{2}\bar{\psi}_{R}.\gamma\chi_{Li}$$

$$+ \frac{1}{2}g\bar{\psi}_{\mu R}\sigma^{\mu\nu}\psi_{\nu R} + \text{h.c.} \tag{3.22}$$

$$e^{-1}L(fw^{2}) = \frac{1}{2}f_{\alpha\beta}(-\frac{1}{4}F^{\alpha}_{\mu\nu}F^{\beta}_{\mu\nu} - \frac{1}{2}\bar{\lambda}^{\alpha}\tilde{\not{D}}\lambda^{\beta} + \frac{1}{2}D^{\alpha}D^{\beta}$$

$$+ \frac{1}{4}F^{\alpha}_{\mu\nu}\bar{\psi}_{\rho}\sigma^{\mu\nu}\gamma^{\rho}\lambda^{\beta} + \frac{1}{4}F^{\alpha}_{\mu\nu}\tilde{F}^{\beta}_{\mu\nu} - \frac{1}{2}D_{\mu}(\bar{\lambda}^{\alpha}_{L}\gamma^{\mu}\lambda^{\beta}_{R}))$$

$$+ \frac{1}{2}f^{i}_{\alpha\beta}(\bar{\chi}_{Li}(-\sigma.\hat{F}^{-\alpha}+iD^{\alpha})\lambda^{\beta}_{L} - \frac{1}{2}h_{i}\bar{\lambda}^{\alpha}_{L}\lambda^{\beta}_{L}$$

$$- \frac{1}{2}\bar{\psi}_{R}.\gamma\chi_{Li}\bar{\lambda}^{\alpha}_{L}\lambda^{\beta}_{L}) + \frac{1}{4}f^{ij}_{\alpha\beta}\bar{\chi}_{Li}\chi_{Lj}\bar{\lambda}^{\alpha}_{L}\lambda^{\beta}_{L} + \text{h.c.} \tag{3.23}$$

Here $\tilde{D}_{\mu}\lambda^{\alpha}$ is a supersymmetric and Yang-Mills covariant derivative but without the $\psi_{\mu}D^{\alpha}$ term. The other derivatives are Yang-Mills and general coordinate covariants but have no ψ_{μ} torsion.

The over-all Lagrangian corresponding to the superspace expressions of Eqs.(3.3) and (3.6) is therefore given by [6]

$$L = L(\Phi) + L(g) + L(fw^{2}) \tag{3.24}$$

The three terms in Eq.(3.24) are separately invariant under local supersymmetry transformations.

These transformations have the following form, for the several fields involved in Eq.(3.24):

$$\delta B^{\alpha}_{\mu} = -\frac{1}{2}\bar{\varepsilon}_{L}\gamma_{\mu}\lambda^{\alpha}_{R} - \frac{1}{2}\bar{\varepsilon}_{R}\gamma_{\mu}\lambda^{\alpha}_{L}$$

$$\delta\lambda^{\alpha}_{R} = \frac{1}{2}\sigma^{\mu\nu}\hat{F}^{\alpha}_{\mu\nu}\varepsilon_{R} - \frac{1}{2}iD^{\alpha}\varepsilon_{R} \tag{3.25}$$

$$\delta D^{\alpha} = \frac{1}{2}i\varepsilon_{L}(\hat{\not{D}}^{P} - \frac{1}{2}i\not{A})\lambda_{R} - \frac{1}{2}i\bar{\varepsilon}_{R}(\hat{\not{D}}^{P} + \frac{1}{2}i\not{A})\lambda^{\alpha}_{L}$$

$$\delta z_{i} = \bar{\varepsilon}_{L}\chi_{Li}$$

$$\delta\chi_{Li} = \frac{1}{2}\hat{\not{D}}z_{i}\varepsilon_{R} + \frac{1}{2}h_{i}\varepsilon_{L}$$

$$\delta h_{i} = \bar{\varepsilon}_{R}(\hat{\not{D}}^{P} - \frac{1}{2}i\not{A})\chi_{Li} - \frac{1}{3}\bar{\chi}_{Li}(u\varepsilon_{L}+i\not{A}\varepsilon_{R}+i\tilde{g}\bar{\varepsilon}_{R}\lambda^{\alpha}_{R}T^{\alpha j}_{i}z_{j}$$

$$\delta e^m_\mu = \frac{1}{2}\,\bar{\varepsilon}_L\gamma^m\psi_{\mu R} + \frac{1}{2}\,\bar{\varepsilon}_R\gamma^m\psi_{\mu L}$$

$$\delta\psi_{\mu L} = (\partial_\mu + \frac{1}{2}\,\omega_{\mu mn}(e,\psi)\,\sigma^{mn} + \frac{1}{2}\,iA_\mu)\,\varepsilon_L$$

$$+ \frac{1}{6}\,\gamma_\mu(u^*\varepsilon_R - i\not{A}\varepsilon_L)$$

$$\delta A_\mu = \frac{3}{4}\,i\bar{\varepsilon}_L(R^P_\mu - \frac{1}{3}\,\gamma_\mu\gamma.R^P)_L + \text{h.c.}$$

$$\delta u = \frac{1}{2}\,\bar{\varepsilon}_R\gamma.R^P_L \tag{3.25}$$

The over-all Lagrangian given by Eq.(3.24) contains auxiliary field components of the multiplets S^i, V^α, and the supergravity multiplet

$$(e_{\mu m},\ \psi_{\mu a},\ u,\ A_m) \tag{3.26}$$

In addition, an appropriate Weyl rescaling of the vierbein field, as well as redefinitions of the fermion fields, have to be performed in the first term of Eq.(3.24), in its standard form [5]. These redefinitions, as well as the elimination of the auxiliary fields h_i, D^α, u, and A_m, will be done in the next section.

These manipulations will allow us to recast the final Lagrangian and transformation rules in a simple form in terms of the particle fields

$$(B^\alpha_m,\ \lambda^\alpha_R),\ (z_i,\ \chi_{Li}),\ (e_{\mu m},\ \psi_{\mu a}) \tag{3.27}$$

We end this section by showing, using simple superspace arguments, that the action given by Eq.(3.24) depends only on a particular combination of the function $\Phi(S,\bar{S})$ and g(S) [5].

To prove this, we make a super-Weyl rescaling on the super-vierbein so that its superdeterminant transforms as

$$E \rightarrow E'\exp(\Sigma + \bar{\Sigma}) \tag{3.28}$$

where Σ is a chiral parameter superfield. Under the same rescaling the second term in Eq.(3.3) transforms as

$$\int d^4x d^4\theta\ \mathrm{Re}\left(\frac{E'}{R'}\ (\exp 3\Sigma)g(S)\right) \tag{3.29}$$

while the Yang-Mills part [Eq.(3.6)] keeps the same form:

$$\int d^4x d^4\theta \ \mathrm{Re}\left\{\frac{E'}{R^\tau} f_{\alpha\beta}(S) W^\alpha W^\beta\right\} \tag{3.30}$$

If we now choose Σ such that

$$\exp 3\Sigma g(S) = 1, \text{ i.e. } \exp\Sigma = g(S)^{-1/3} \tag{3.31}$$

we finally obtain that Eq.(3.3) becomes

$$\int d^4x d^4\theta E' \exp \frac{1}{3} \mathcal{G}(S, \bar{S}e^{2V}) + \int d^4x d^4\theta \mathrm{Re}(\frac{E'}{R^\tau}) \tag{3.32}$$

where the function $\mathcal{G}$ is given by

$$\mathcal{G}(z,z^*) = 3\log\Phi(z,z^*) - \log|g(z)|^2 \tag{3.33}$$

so that

$$\exp \frac{1}{3}\mathcal{G}(z,z^*) = \frac{\Phi(z,z^*)}{|g(z)|^{2/3}} \tag{3.34}$$

The use of the function $\mathcal{G}(S,\bar{S})$ makes the introduction of the Fayet-Iliopoulos term in formula (3.11) trivial [11,16]. In fact, in this case we simply get

$$e^{\frac{1}{3}\mathcal{G}(S,\bar{S}e^{n_i g_R V_R} e^{2V})} = \frac{\Phi(S_i,\bar{S}e^{n_i g_R V_R} e^{2V})}{\left\{g^*(\bar{S}e^{n_i g_R V_R} e^{2V}) g(S)\right\}^{1/3}} \tag{3.35}$$

This means that we simply have to replace in the $\mathcal{G}$ superfield $\bar{S}e^{2V}$ with $\bar{S}e^{n_i g_R V_R} e^{2V}$, i.e. we have to covariantize with respect to the full gauge group $G \times U_R(1)$.

We note that this simple rule does not hold in the case of a vanishing superpotential, $g(S_i) = 0$, since in this case the rescaling given by formula (3.31) can no longer be performed [11]. At the end of the next section we will give simple substitution rules for the Lagrangian in terms of physical fields which will enable us to go from the general form of $\mathcal{G}(S,\bar{S})$ to the limiting case corresponding to $g(S) = 0$.

4. FINAL FORM OF THE LAGRANGIAN, TRANSFORMATION LAWS, AND GAUGED R-SYMMETRY

The over-all action of the coupled matter-Yang-Mills supergravity multiplets is given in Eq.(3.24).

In order to obtain the final form of the Lagrangian, we have to eliminate, by means of their field equations, the auxiliary fields h_i, D^α, u, and A_m of the multiplets involved.

It is important to notice that, after their elimination, only the total Lagrangian (3.24), but not the three separate pieces given by Eqs.(3.21) to (3.23) will be invariant under local supersymmetry transformations of the physical fields.

The part of the Lagrangian (3.24) which contains the auxiliary fields is

$$\begin{aligned} e^{-1}L_{AUX} &= \frac{1}{18}\,\Phi(uu^*-A_mA^m) + \frac{1}{2}\,\Phi^i_j h_i h^{*j} \\ &- \Phi^{ij}_k h^{*k}\bar{\chi}_{Li}\chi_{Lj} + \frac{1}{2}\,g^i h_i - \frac{1}{4}\,f^i_{\alpha\beta}h_i\bar{\lambda}^\alpha_L\lambda^\beta_L + \frac{1}{3}\,(\Phi^i h_i \\ &- \Phi^{ij}\bar{\chi}_{Li}\chi_{Lj} + \frac{3}{2}\,g^*) + \frac{1}{3}\,iA^\mu(\frac{1}{2}\,\Phi^i_j\bar{\chi}^j_R\gamma_\mu\chi_{Li} \\ &+ \Phi^i(D_\mu z_i-\bar{\psi}_{\mu L}\chi_{Li}) + \frac{3}{4}\,\bar{\lambda}^\alpha_L\gamma_\mu\lambda^\beta_R f_{\alpha\beta}) \\ &+ \frac{1}{2}\,\tilde{g}\Phi^i T^{\alpha j}_i z_j D^\alpha + \frac{1}{4}\,f_{\alpha\beta}D^\alpha D^\beta + \frac{1}{2}\,f^i_{\alpha\beta}\bar{\chi}_{Li}\lambda^\alpha_L D^\alpha + \text{h.c.} \end{aligned} \tag{4.1}$$

If we define the new expressions

$$J = 3\log - \frac{\Phi}{3} \tag{4.2}$$

and

$$\tilde{u} = u + J^i h_i \tag{4.3}$$

then Eq.(4.1) reads

$$\begin{aligned} e^{-1}L_{AUX} &= \frac{1}{18}\,\Phi(\tilde{u}\tilde{u}^*-A_mA^m) + \frac{1}{6}\,\Phi J^i_j h_i h^{*j} \\ &+ \frac{1}{2}\,g[\tilde{u}-h_i(J^i - \frac{g^i}{g})] - \frac{1}{g}\,u^*(J^{ij} + \frac{1}{3}\,J^iJ^j)\Phi\bar{\chi}_{Li}\chi_{Lj} \\ &- \frac{1}{3}\,\Phi h^{*k}\bar{\chi}_{Li}\chi_{Lj}(J^{ij}_k + \frac{2}{3}\,J^i_kJ^j) - \frac{1}{4}\,f^i_{\alpha\beta}h_i\bar{\lambda}^\alpha_L\lambda^\beta_L \\ &+ \frac{1}{3}\,iA^\mu(\frac{1}{2}\,\Phi^i_j\bar{\chi}^j_R\gamma_\mu\chi_{Li}+\Phi^i(D_\mu z_i-\bar{\psi}_{\mu L}\chi_{Li}) + \frac{3}{4}\,\bar{\lambda}^\alpha_L\gamma_\mu\lambda^\beta_R f_{\alpha\beta}) \\ &+ \frac{1}{2}\,\tilde{g}\Phi^i T^{\alpha j}_i z_j D^\alpha + \frac{1}{4}\,f_{\alpha\beta}D^\alpha D^\beta + \frac{i}{2}\,f^i_{\alpha\beta}\bar{\chi}_{Li}\lambda^\alpha_L D^\beta + \text{h.c.} \end{aligned} \tag{4.4}$$

The field equations for the auxiliary fields are

$$\tilde{u} = \frac{2}{\Phi}\,[-\,\frac{1}{2}\,g^* + \frac{1}{g}\,(J^{ij} + \frac{1}{3}\,J^iJ^j)\,\Phi\bar{\chi}_{Li}\chi_{Lj}]$$

$$\frac{1}{3}\,\Phi h_i J^i_k = -\,\frac{1}{2}\,g^*\,(\frac{g^*_k}{g^*} - J_k) + \frac{1}{3}\,\Phi(J^{ij}_k + \frac{2}{3}\,J^i_k J^j)\,\bar{\chi}_{Li}\chi_{Lj}$$

$$+ \frac{1}{4}\,f^*_{\alpha\beta k}\bar{\lambda}^\alpha_R\lambda^\beta_R$$

$$\frac{2}{3}\,\Phi A_\mu = \frac{1}{2}\,i\Phi^{i}_{j}\bar{\chi}^{j}_R\gamma_\mu\chi_{Li} + i\Phi^i(D_\mu z_i - \bar{\psi}_{\mu L}\chi_{Li})$$

$$+ \frac{3}{4}\,if_{\alpha\beta}\bar{\lambda}^\alpha_L\gamma_\mu\lambda^\beta_R + \text{h.c.}$$

$$-\,\mathrm{Re}f_{\alpha\beta}D^\beta = \tilde{g}\Phi^i T^{\alpha j}_i z_j + \frac{1}{2}\,if^i_{\alpha\beta}\bar{\chi}_{Li}\lambda^\beta_L - \frac{1}{2}\,if^*_{\alpha\beta i}\bar{\chi}^i_R\lambda^\beta_R \tag{4.5}$$

The insertion of Eq.(4.5) into Eq.(4.4) finally gives

$$e^{-1}L = e^{-1}L_{AUX} + e^{-1}\hat{L} \tag{4.6}$$

where

$$e^{-1}L_{AUX} = -\,\frac{g}{\Phi}\,\left|\frac{1}{2}\,g^* - \frac{1}{g}\,(J^{ij} + \frac{1}{3}\,J^iJ^j)\,\Phi\bar{\chi}_{Li}\chi_{Lj}\right|^2 - \frac{3}{\Phi}\,(J^{-1})^k_\ell$$

$$\times\,[\frac{1}{2}\,g^*\,(\frac{g^*_k}{g^*} - J_k) - \frac{1}{3}\,\Phi(J^{ij}_k + \frac{2}{3}\,J^i_kJ^j)\,\bar{\chi}_{Li}\chi_{Lj} - \frac{1}{4}\,f^*_{\alpha\beta k}\bar{\lambda}^\alpha_R\lambda^\beta_R]$$

$$\times\,[\frac{1}{2}\,g(\frac{g^\ell}{g} - J^\ell) - \frac{1}{3}\,\Phi(J^\ell_{mn} + \frac{2}{3}\,J^\ell_m J_n)\,\bar{\chi}^m_R\chi^n_R - \frac{1}{4}\,f^\ell_{\gamma\delta}\bar{\lambda}^\gamma_L\lambda^\delta_L]$$

$$-\,\frac{1}{4\Phi}\,[\Phi^{i}_{j}\bar{\chi}^{j}_R\gamma_\mu\chi_{Li} + \Phi^i(D_\mu z_i - \bar{\psi}_{\mu L}\chi_{Li}) - \Phi_i(D_\mu z^{*i} - \bar{\psi}_{\mu R}\chi^i_R)$$

$$+ \frac{3}{2}\,\mathrm{Re}f_{\alpha\beta}\bar{\lambda}^\alpha_L\gamma_\mu\lambda^\beta_R]^2 - \frac{1}{2}\,(\mathrm{Re}f)^{-1}_{\alpha\beta}(\tilde{g}\Phi^i T^{\alpha j}_i z_j + \frac{i}{2}\,f^i_{\alpha\gamma}\bar{\chi}_{Li}\lambda^\gamma_L$$

$$-\,\frac{1}{2}\,if^*_{\alpha\gamma i}\bar{\chi}^i_R\lambda^\gamma_R) \times (\tilde{g}\Phi^k T^{\beta\ell}_k z_\ell + \frac{1}{2}\,if^k_{\beta\delta}\bar{\chi}_{Lk}\lambda^\delta_L - \frac{1}{2}\,if^*_{\beta\delta k}\bar{\chi}^k_R\lambda^\delta_R)$$

$$e^{-1}\hat{L} = \frac{1}{12}\,\Phi[R(\omega(e)) + \bar{\psi}_\mu\gamma_5\gamma_\nu D_\rho\psi_\sigma\varepsilon^{\mu\nu\rho\sigma} + \frac{1}{16}\,(\bar{\psi}_\mu\gamma_\lambda\psi_\rho$$

$$+ 2\bar{\psi}_\lambda\gamma_\mu\psi_\rho)\bar{\psi}^\mu\gamma^\lambda\psi^\rho - \frac{1}{4}\,(\bar{\psi}_\mu\gamma.\psi)^2 + e^{-1}\partial_\mu(e\bar{\psi}.\gamma\psi^\mu)]$$

$$+ \Phi^i_j(-\,\frac{1}{2}\,D_\mu z_i D^\mu z^{*j} - \bar{\chi}_{Li}\not{D}\chi^j_R) + \Phi^{ij}_k\bar{\chi}_{Li}\not{D}z_j\chi^k_R$$

$$+ \Phi^i_j\bar{\psi}_{\mu L}\not{D}z^{*j}\gamma^\mu\chi_{Li} - \frac{4}{3}\,\Phi^i\bar{\chi}_{Li}\sigma^{\mu\nu}D_\mu\psi_{\nu L}$$

$$-\,\frac{1}{8}\,e^{-1}\varepsilon^{\mu\nu\rho\sigma}\bar{\psi}_\mu\gamma_\nu\psi_\rho(\Phi^i D_\sigma z_i + \frac{1}{2}\,\Phi^j_i\bar{\chi}^i_R\gamma_\sigma\chi_{Lj})$$

$$- \frac{1}{2}\, \Phi^{i}_{j}\bar{\psi}_{\mu L}\chi_{Li}\bar{\psi}^{\mu}_{R}\chi^{j}_{R} + \frac{1}{2}\, \Phi^{ij}_{k\ell}\bar{\chi}_{Li}\chi_{Lj}\bar{\chi}^{k}_{R}\chi^{\ell}_{R}$$

$$+ \frac{1}{6}\, \Phi^{i}\chi_{Li}(\psi_{\mu L}\bar{\psi}\cdot\gamma\psi^{\mu} + \frac{1}{2}\, \sigma^{\mu\nu}\psi^{\ell}_{L}\bar{\psi}_{\nu}\gamma_{\rho}\psi_{\mu} + \sigma^{\mu\nu}\psi_{\nu L}\bar{\psi}_{\mu}\gamma\cdot\psi)$$

$$+ \frac{1}{2}\, i\tilde{g}\Phi^{i}T^{\alpha j}_{i}z_{j}\bar{\psi}_{L}\gamma\lambda^{\alpha}_{k} - 2i\tilde{g}\Phi^{i}_{k}T^{\alpha j}_{i}z_{j}\bar{\lambda}^{\alpha}_{R}\chi^{k}_{R}$$

$$- \frac{1}{2}\, g^{ij}\bar{\chi}_{Li}\chi_{Lj} + \frac{1}{2}\, g^{i}\bar{\psi}_{R}\cdot\gamma\chi_{Li} + \frac{1}{2}\, g\bar{\psi}_{\mu R}\sigma^{\mu\nu}\psi_{\nu R}$$

$$+ \frac{1}{2}\, f_{\alpha\beta}(- \frac{1}{4}\, F^{\alpha}_{\mu\nu}F^{\beta}_{\mu\nu} - \frac{1}{2}\, \bar{\lambda}^{\alpha}\not{D}\lambda^{\beta} + \frac{1}{2}\, \bar{\lambda}^{\alpha}\gamma^{\mu}\sigma^{\rho\sigma}\psi_{\mu}F^{\beta}_{\rho\sigma}$$

$$+ \frac{1}{4}\, \bar{\lambda}^{\alpha}\gamma^{\mu}\sigma^{\rho\sigma}\psi_{\mu}\bar{\psi}_{\rho}\gamma_{\sigma}\lambda^{\alpha} - \frac{1}{16}\, \bar{\lambda}^{\alpha}\gamma^{\mu}\sigma^{mn}\lambda^{\beta}(2\bar{\psi}_{\mu}\gamma_{m}\psi_{n}$$

$$+ \bar{\psi}_{m}\gamma_{\mu}\psi_{n}) + \frac{1}{4}\, F^{\alpha}_{\mu\nu}\tilde{F}^{\beta}_{\mu\nu} - \frac{1}{2}\, D_{\mu}(\bar{\lambda}^{\alpha}_{L}\gamma^{\mu}\lambda^{\beta}_{R}))$$

$$+ \frac{1}{2}\, f^{i}_{\alpha\beta}(-\bar{\chi}_{Li}\sigma_{\mu\nu}\lambda^{\beta}_{L}(F^{-\alpha}_{\mu\nu} - \frac{1}{2}\, \bar{\psi}_{\rho L}\sigma_{\mu\nu}\gamma^{\rho}\lambda^{\alpha}_{R} + \frac{1}{2}\, \bar{\psi}_{R}\cdot\gamma\sigma_{\mu\nu}\lambda^{\alpha}_{L})$$

$$- \frac{1}{2}\, \bar{\psi}_{R}\cdot\gamma\chi_{Li}\bar{\lambda}^{\alpha}_{L}\lambda^{\beta}_{L}) + \frac{1}{4}\, f^{ij}_{\alpha\beta}\bar{\chi}_{Li}\chi_{Lj}\bar{\lambda}^{\alpha}_{L}\lambda^{\beta}_{L} + \text{h.c.} \qquad (4.7)$$

In order to recast the Einstein term and the Rarita-Schwinger action in canonical form, as well as to keep the gaugino and chiral fermion terms in a quasi-canonical form, the following Weyl rescaling on the vierbein and the fermion fields must be performed:

$$e_{m\mu} \to e_{m\mu}e^{\sigma}, \quad e \to e^{4\sigma}e$$

$$\lambda \to e^{-3\sigma/2}\lambda$$

$$\chi \to e^{-\sigma/2}\chi$$

$$\psi_{\mu} \to e^{\sigma/2}\psi_{\mu} \qquad (4.8)$$

with

$$e^{2\sigma} = - \frac{3}{\Phi}\, , \quad \sigma = - \frac{1}{6}\, J \qquad (4.9)$$

Under the Weyl rescaling of the vierbein we get the standard changes for the Einstein-curvature term and of the Lorentz connection:

$$\frac{1}{6}\, e\Phi R \to - \frac{1}{2}\, eR - \frac{3}{4}\, e(\partial_{\mu}\log\Phi)^{2} + \text{4-div.}$$

$$\omega_{\mu mn} \rightarrow \omega_{\mu mn} - 2e^{\nu}_{[m}e_{n]\mu}\partial_\nu\sigma \tag{4.10}$$

After the substitution (4.8), the total Lagrangian becomes the sum of a pure bosonic part and a fermionic part:

$$e^{-1}L = e^{-1}L_B + e^{-1}L_F \tag{4.11}$$

where

$$\begin{aligned} e^{-1}L_B = &- \frac{1}{2} R + J^i_j D_\mu z_i D_\mu z^{*j} - \frac{1}{4} \mathrm{Ref}_{\alpha\beta}F^\alpha_{\mu\nu}F^\beta_{\mu\nu} + \frac{1}{4} i\mathrm{Imf}_{\alpha\beta}F^\alpha_{\mu\nu}\tilde{F}^\beta_{\mu\nu} \\ &+ \frac{1}{4} |g|^2 e^{-J}(3+(J^{-1})^k_\ell(\frac{g^*_k}{g^*} - J_k)(\frac{g^\ell}{g} - J^\ell)) \\ &- \frac{1}{2} \tilde{g}^2 \mathrm{Ref}^{-1}_{\alpha\beta}(J^i T^{\alpha j}_i z_j)(J^k T^{\beta\ell}_k z_\ell) \end{aligned} \tag{4.12}$$

and

$$\begin{aligned} e^{-1}L_F = &\frac{1}{2} e^{3\sigma}(J^{ij} + \frac{1}{3} J^i J^j) g\bar{\chi}_{Li}\chi_{Lj} \\ &+ \frac{1}{2} g(\frac{g^\ell}{g} - J^\ell) e^{3\sigma} J^{-1k}_\ell (J^{ij}_k + \frac{2}{3} J^i_k J^j)\bar{\chi}_{Li}\chi_{Lj} \\ &+ \frac{1}{2} g(\frac{g^\ell}{g} - J^\ell) \frac{3}{4\Phi} e^\sigma J^{-1k}_\ell f^*_{\alpha\beta k}\bar{\lambda}^\alpha_R\lambda^\beta_R - \frac{1}{18} \Phi e^{2\sigma}(J^{ij} \\ &+ \frac{1}{3} J^i J^j)\bar{\chi}_{Li}\chi_{Lj}(J_{k\ell} + \frac{1}{3} J_k J_\ell)\bar{\chi}^k_R\chi^\ell_R - \frac{1}{6} \Phi e^{2\sigma} J^{-1k}_\ell (J^{ij}_k \\ &+ \frac{2}{3} J^i_k J^j)\bar{\chi}_{Li}\chi_{Lj}(J^\ell_{mn} + \frac{2}{3} J^\ell_m J_n)\bar{\chi}^m_R\chi^n_R \\ &- \frac{1}{4} f^\ell_{\alpha\beta}\bar{\lambda}^\alpha_L\lambda^\beta_L J^{-1k}_\ell (J^{ij}_k + \frac{2}{3} J^i_k J^j)\bar{\chi}_{Li}\chi_{Lj} \\ &- \frac{3}{32}\frac{1}{\Phi} e^{-2\sigma} J^{-1k}_\ell f^\ell_{\alpha\beta}\bar{\lambda}^\alpha_L\lambda^\beta_L f^*_{\gamma\delta k}\bar{\lambda}^\gamma_R\lambda^\delta_R \\ &- \frac{1}{2\Phi} e^{2\sigma}\phi^i_j\bar{\chi}^j_R\gamma^\mu\chi_{Li}\phi^k D_\mu z_k - \frac{1}{2\Phi} e^{2\sigma}\phi^i D_\mu z_i(\phi_J\bar{\psi}_{\mu R}\chi^j_R \\ &- \phi^j\bar{\psi}_{\mu L}\chi_{Lj}) - \frac{3}{4\Phi} \mathrm{Ref}_{\alpha\beta}\phi^i D_\mu z_i\bar{\lambda}^\alpha_L\gamma^\mu\lambda^\beta_R \\ &- \frac{1}{8\Phi}(\phi^i_j\bar{\chi}^j_R\gamma_m\chi_{Li})^2 + \frac{1}{2\Phi}\phi^i_j\bar{\chi}^j_R\gamma^\mu\chi_{Li}\phi^k\bar{\psi}_{\mu L}\chi_{Lk}e^{2\sigma} \\ &- \frac{3}{8\Phi}\phi^i_j\bar{\chi}^j_R\gamma^m\chi_{Li}\mathrm{Ref}_{\alpha\beta}\bar{\lambda}^\alpha_L\gamma_m\lambda^\beta_R + \frac{3}{4\Phi}\phi^i\bar{\psi}_{\mu L}\chi_{Li}\mathrm{Ref}_{\alpha\beta}\bar{\lambda}^\alpha_L\gamma^\mu\lambda^\beta_R \end{aligned}$$

$$+ \frac{1}{4\Phi} e^{2\sigma}\phi^j\bar{\psi}_{\mu L}\chi_{Lj}(\phi_i\bar{\psi}_{\mu R}\chi^i_R - \phi^i\bar{\psi}_{\mu L}\chi_{Li})$$

$$- \frac{9}{32\Phi} \mathrm{Re}f_{\alpha\beta}\mathrm{Re}f_{\gamma\delta}\bar{\lambda}^\alpha_L\gamma^m\lambda^\beta_R\bar{\lambda}^\gamma_L\gamma_m\lambda^\delta_R e^{-2\sigma}$$

$$- \frac{1}{2} i\mathrm{Re}f^{-1}_{\alpha\beta}\tilde{g}\phi^i T^{\alpha j}_i z_j f^k_{\beta\gamma}\bar{\chi}_{Lk}\lambda^\gamma_L e^{2\sigma}$$

$$+ \frac{1}{8} \mathrm{Re}f^{-1}_{\alpha\beta} f^i_{\alpha\gamma}\chi_{Li}\lambda^\gamma_L(f^j_{\beta\delta}\bar{\chi}_{Lj}\lambda^\delta_L - f^*_{\beta\delta j}\bar{\chi}^j_R\lambda^\delta_R)$$

$$- \frac{1}{4} \bar{\psi}_\mu\gamma_5\gamma_\nu D_\rho\psi_\sigma\sigma^{\mu\nu\rho\delta}e^{-1} - \frac{1}{64} (\bar{\psi}_\mu\gamma_\lambda\psi_\rho + 2\bar{\psi}_\lambda\gamma_\mu\psi_\rho)\bar{\psi}^\mu\gamma^\lambda\psi^\rho$$

$$+ \frac{1}{16} (\bar{\psi}_\mu\cdot\gamma\cdot\psi)^2 + 3 \frac{\phi^i_j}{\Phi} \bar{\chi}_{Li}\not{D}\chi^j_R - 3 \frac{\phi^{ij}_k}{\Phi} \bar{\chi}_{Li}\not{D}z_j\chi^k_R$$

$$- 3 \frac{\phi^i_j}{\Phi} \bar{\psi}_{\mu L}\not{D}z^{*j}\gamma_\mu\chi_{Li} + \frac{4\phi^i}{\Phi} \bar{\chi}_{Li}\sigma^{\mu\nu}D_\mu\psi_\nu$$

$$- \frac{3\phi^i}{\Phi} \bar{\chi}_{Li}\gamma^\nu\gamma^\mu\psi_{L\nu}D_\mu\sigma + \frac{3}{8} e^{-1}\varepsilon^{\mu\nu\rho\sigma}\bar{\psi}_\mu\gamma_\nu\psi_\rho(\frac{\phi^i}{\Phi} D_\sigma z_i$$

$$+ \frac{1}{2} \frac{\phi^j_i}{\Phi} \bar{\chi}^i_R\gamma_\sigma\chi_{Lj}) + \frac{3}{2} \frac{\phi^i_j}{\Phi} \bar{\psi}_{\mu L}\chi_{Li}\bar{\psi}_{\mu R}\chi^j_R$$

$$- \frac{3}{2} \frac{\phi^{ij}_k}{\Phi} \bar{\chi}_{Li}\chi_{Lj}\bar{\chi}^k_R\chi^\ell_R - \frac{1}{2} \phi^i\bar{\chi}_{Li}(\bar{\chi}_{\mu L}\bar{\psi}\cdot\gamma\psi^\mu$$

$$+ \frac{1}{2} \sigma_{\mu\nu}\psi_{\rho L}\bar{\psi}^\nu\gamma^\rho\psi^\mu + \sigma^{\mu\nu}\psi_{\nu L}\bar{\psi}_\mu\gamma\cdot\psi) - \frac{3}{2} i\tilde{g} \frac{\phi^i}{\Phi} T^{\alpha j}_i z_j\bar{\psi}_L\cdot\gamma\lambda^\alpha_R$$

$$+ 6i\tilde{g} \frac{\phi^j_i}{\Phi} T^{\alpha k}_j z_k\bar{\lambda}^\alpha_R\chi^i_R - \frac{1}{2} g^{ij}e^{3\sigma}\bar{\chi}_{Li}\chi_{Lj} + \frac{1}{2} g^i e^{3\sigma}\bar{\psi}_R\cdot\gamma\chi_{Li}$$

$$+ \frac{1}{2} ge^{3\sigma}\bar{\psi}_{\mu R}\sigma^{\mu\nu}\psi_{\nu R} + \frac{1}{2} \mathrm{Re}f_{\alpha\beta}(- \frac{1}{2} \bar{\lambda}^\alpha\not{D}\lambda^\beta + \frac{1}{2} \bar{\lambda}^\alpha\gamma^\mu\sigma^{\rho\sigma}\psi_\mu$$

$$\times (F^\beta_{\rho\sigma} + \frac{1}{2} \bar{\psi}_\rho\gamma_\sigma\lambda^\beta) - \frac{1}{32} \bar{\lambda}^\alpha\gamma_5\gamma_\nu\lambda^\beta e^{-1}\varepsilon^{\mu\nu\rho\sigma}\bar{\psi}_\mu\gamma_\rho\psi_\sigma)$$

$$- \frac{1}{4} i \,\mathrm{Im}f_{\alpha\beta}D^\mu(\bar{\lambda}^\alpha_L\gamma^\mu\lambda^\beta_R)$$

$$+ \frac{1}{2} f^i_{\alpha\beta}(-\bar{\chi}_{Li}\sigma^{\mu\nu}\lambda^\beta_L(F^{-\alpha}_{\mu\nu} - \frac{1}{2} \bar{\psi}_{\rho L}\sigma_{\mu\nu}\gamma^\rho\lambda^\alpha_R$$

$$+ \frac{1}{2} \bar{\psi}_R\cdot\gamma\sigma_{\mu\nu}\lambda^\alpha_L) - \frac{1}{2} \bar{\psi}_R\cdot\gamma\chi_{Li}\bar{\lambda}^\alpha_L\lambda^\beta_L)$$

$$+ \frac{1}{4} f^{ij}_{\alpha\beta}\bar{\chi}_{Li}\chi_{Lj}\bar{\lambda}^\alpha_L\lambda^\beta_L + \text{h.c.} \qquad (4.13)$$

The kinetic part of Eq.(4.13) can be further diagonalized through the redefinition

$$\psi_{\mu L} \to \psi_{\mu L} - \gamma_\mu \frac{\phi_i}{\Phi} \chi^i_R \qquad (4.14)$$

If, in addition, we perform the following chiral redefinitions on the fermion fields,

$$\psi_{\mu L} \to (\frac{g}{g^*})^{1/4}\psi_{\mu L},\ \chi_{Li} \to (\frac{g}{g^*})^{-1/4}\chi_{Li},\ \lambda^{\alpha}_{L} \to (\frac{g}{g^*})^{1/4}\lambda^{\alpha}_{L} \tag{4.15}$$

and define the function

$$G = J - \log \frac{1}{4}\,|g|^2 \tag{4.16}$$

so that

$$G^i = J^i - \frac{g^i}{g}\ ,\ G^i_j = J^i_j \tag{4.17}$$

the final Lagrangian becomes

$$L = L_B + L_{FK} + L_{FM} + L_{(4)F} \tag{4.18}$$

where

$$\begin{aligned} e^{-1}L_B = & -\frac{1}{2} R + G^i_j D_\mu z_i D^\mu z^{*j} - \frac{1}{4} \mathrm{Re} f_{\alpha\beta} F^\alpha_{\mu\nu} F^\beta_{\mu\nu} \\ & + \frac{1}{4} i \mathrm{Im} f_{\alpha\beta} F^\alpha_{\mu\nu} \tilde{F}^\beta_{\mu\nu} + e^{-G}(3 + G_k G^{-1k}_{\ell} G^\ell) \\ & + \frac{1}{2} \tilde{g}^2 \mathrm{Re} f^{-1}_{\alpha\beta} (G^i T^{\alpha j}_i z_j)(G^k T^{\beta\ell}_k z_\ell) \end{aligned} \tag{4.19}$$

$L_{F,K}$ is the curved generalization of the fermionic terms; $L_{F,M}$ is the quadratic part in the fermion fields without derivative terms; $L_{(4)F}$ is the part of the Lagrangian which contains four-fermions interaction terms [not coming from the spin 3/2 torsion contribution which are already contained in the Lorentz connection $\omega_{\mu mn}(e,\psi)$] and also other interactions of dimension greater than four containing covariant derivatives of boson fields:

$$\begin{aligned} e^{-1}L_{Fk} = & -\frac{1}{4} \mathrm{Re} f_{\alpha\beta} \bar{\lambda}^\alpha \not{D} \lambda^\beta - \frac{1}{4} e^{-1} \bar{\psi}_\mu \gamma_5 \gamma_\nu D_\rho \psi_\sigma \varepsilon^{\mu\nu\rho\sigma} \\ & + G^i_j \bar{\chi}_{Li} \not{D} \chi^j_R - \frac{i}{8} \mathrm{Im} f_{\alpha\beta} e^{-1} D_\mu (e \bar{\lambda}^\alpha \gamma_5 \gamma_\mu \lambda^\beta) + \text{h.c.} \end{aligned} \tag{4.20}$$

$$e^{-1}L_{F,M} = e^{-G/2}\bar{\psi}_{\mu R}\sigma^{\mu\nu}\psi_{\nu R} - \bar{\psi}_R \cdot \gamma (e^{-G/2} G^i \chi_{Li}$$

$$- \frac{i}{2} \tilde{g} G^i T^{\alpha j}_i z_j \lambda^\alpha_L) + e^{-G/2}[G^{ij} - G^i G^j$$

$$- G^{\ell}G_{\ell}^{-1k}G_{k}^{ij}]\bar{\chi}_{Li}\chi_{Lj} - 2i\tilde{g}G_{i}^{k}z^{*j}T_{j}^{\alpha i}\bar{\lambda}_{L}^{\alpha}\chi_{Lk}$$

$$+ \frac{1}{2} f_{\alpha\beta}^{k}\left\{\frac{1}{2} e^{-G/2}G_{\ell}G_{k}^{-1\ell}\bar{\lambda}_{L}^{\alpha} + i\tilde{g}\mathrm{Ref}_{\alpha\gamma}^{-1}G^{i}T_{i}^{\gamma j}z_{j}\bar{\chi}_{Lk}\right\}\lambda_{L}^{\beta} + \text{h.c.}$$

$$e^{-1}L_{(4)F} = \frac{1}{2}\,\mathrm{Ref}_{\alpha\beta}(\frac{i}{2}\,\bar{\lambda}^{\alpha}\gamma^{\mu}\sigma^{\rho\sigma}\psi_{\mu}F_{\rho\sigma}^{\beta} - \frac{1}{2}\,\bar{\lambda}_{L}^{\alpha}\gamma_{\mu}\lambda_{R}^{\beta}G^{i}D_{\mu}z_{i})$$

$$- \frac{1}{2} f_{\alpha\beta}^{i}\bar{\chi}_{Li}\sigma.F^{\alpha}\lambda_{L}^{\beta} + \frac{1}{8}\,e^{-1}\varepsilon^{\mu\nu\rho\sigma}\bar{\psi}_{\mu}\gamma_{\nu}\psi_{\rho}G^{i}D_{\sigma}z_{i}$$

$$- G_{i}^{j}\bar{\psi}_{\mu L}\not{D}z^{*i}\gamma^{\mu}\chi_{Lj} - \bar{\chi}_{Li}\not{D}z_{j}\chi_{R}^{k}(G_{k}^{ij} + \frac{1}{2}\,G_{k}^{i}G^{j})$$

$$+ \frac{1}{32}\,G_{\ell}^{-1k}f_{\alpha\beta}^{\ell}f_{\gamma\delta k}^{*}\bar{\lambda}_{L}^{\alpha}\lambda_{L}^{\beta}\bar{\lambda}_{R}^{\gamma}\lambda_{R}^{\delta} + \frac{3}{32}\,(\mathrm{Ref}_{\alpha\beta}\bar{\lambda}_{L}^{\alpha}\gamma_{m}\lambda_{R}^{\beta})^{2}$$

$$+ \frac{1}{8}\,\mathrm{Ref}_{\alpha\beta}\bar{\lambda}^{\alpha}\gamma^{\mu}\sigma^{\rho\sigma}\psi_{\mu}\bar{\psi}_{\rho}\gamma_{\sigma}\lambda^{\beta} + \frac{1}{2}\,f_{\alpha\beta}^{i}(\bar{\chi}_{Li}\sigma^{\mu\nu}\lambda_{L}^{\alpha}\cdot\bar{\psi}_{\nu L}\gamma_{\mu}\lambda_{R}^{\beta}$$

$$+ \frac{1}{4}\,\bar{\psi}_{R}\cdot\gamma\chi_{Li}\bar{\lambda}_{L}^{\alpha}\lambda_{L}^{\beta}) + \frac{1}{8}\,G_{i}^{j}\bar{\chi}_{k}^{i}\gamma_{d}\chi_{Lj}(\varepsilon^{abcd}\bar{\psi}_{a}\gamma_{b}\psi_{c}$$

$$- \bar{\psi}_{a}\gamma_{5}\gamma^{d}\psi_{a}) + \frac{1}{16}\,\bar{\chi}_{Li}\gamma^{\mu}\chi_{R}^{j}\bar{\lambda}_{R}^{\delta}\gamma_{\mu}\lambda_{L}^{\gamma}(-2G_{j}^{i}\mathrm{Ref}_{\gamma\delta}$$

$$+ \mathrm{Ref}_{\alpha\beta}^{-1}f_{\alpha\gamma}^{i}f_{\beta\delta j}^{*}) - \frac{1}{16}\,\bar{\chi}_{Li}\sigma_{\mu\nu}\chi_{Lj}\bar{\lambda}_{L}^{\gamma}\sigma^{\mu\nu}\lambda_{L}^{\delta}$$

$$\times\ \mathrm{Ref}_{\alpha\beta}^{-1}f_{\alpha\gamma}^{i}f_{\beta\delta}^{j} + \frac{1}{16}\,\bar{\chi}_{Li}\chi_{Lj}\bar{\lambda}_{L}^{\gamma}\lambda_{L}^{\delta}$$

$$\times\ (-4G_{k}^{ij}G_{\ell}^{-1k}f_{\gamma\delta}^{\ell} + 4f_{\gamma\delta}^{ij} - \mathrm{Ref}_{\alpha\beta}^{-1}f_{\alpha\gamma}^{i}f_{\beta\delta}^{j})$$

$$+ (-\frac{1}{2}\,G_{k\ell}^{ij} + \frac{1}{2}\,G_{m}^{ij}G_{n}^{-1m}G_{k\ell}^{n} - \frac{1}{4}\,G_{k}^{i}G_{\ell}^{j})$$

$$\times\ \bar{\chi}_{Li}\chi_{Lj}\bar{\chi}_{R}^{k}\chi_{R}^{\ell} + \text{h.c.} \qquad (4.22)$$

The final form of the Lagrangian, as given by Eq.(4.18), is invariant under the following local supersymmetry transformations:

$$\delta e_{\mu}^{m} = \frac{1}{2}\,\bar{\varepsilon}_{L}\gamma^{m}\psi_{\mu R} + \text{h.c.}$$

$$\delta\psi_{\mu L} = (\partial_{\mu} + \frac{1}{2}\,\omega_{\mu mn}(e,\psi)\sigma^{mn})\varepsilon_{L} + \frac{1}{2}\,\sigma_{\mu\nu}\varepsilon_{L}G_{i}^{j}\bar{\chi}_{R}^{i}\gamma^{\nu}\chi_{Lj}$$

$$+ \frac{1}{2}\,\gamma_{\mu}\varepsilon_{R}e^{-G/2} + \frac{1}{4}\,\psi_{\mu L}(G^{i}\bar{\varepsilon}_{L}\chi_{Li} - G_{i}\bar{\varepsilon}_{R}\chi_{R}^{i})$$

$$- \frac{1}{4}\,\varepsilon_{L}(G^{i}D_{\mu}z_{i} - G_{i}D_{\mu}z^{*i}) + \frac{1}{4}\,(\sigma_{\nu\mu} + g_{\mu\nu})\varepsilon_{L}\bar{\lambda}_{L}^{*}\gamma^{\nu}\lambda_{R}^{\beta}\mathrm{Ref}_{\alpha\beta}$$

$$\delta B_{\mu}^{\alpha} = -\frac{1}{2}\,\bar{\varepsilon}_{L}\gamma_{\mu}\lambda_{R}^{\alpha} + \text{h.c.}$$

$$\delta\lambda_R^\alpha = \frac{1}{2}\sigma^{\mu\nu}\hat{F}^\alpha_{\mu\nu}\varepsilon_R + \frac{1}{2} i\varepsilon_R \mathrm{Ref}^{-1}_{\alpha\beta}(-\tilde{g}G^i T_i^{\beta j} z_j + \frac{1}{2} if^i_{\beta\gamma}\bar{\chi}_{Li}\lambda_L^\gamma$$
$$- \frac{1}{2} if^*_{\beta\gamma i}\bar{\chi}^i_R\lambda^\gamma_R) - \frac{1}{4}\lambda^\alpha_R(G^i\bar{\varepsilon}_L\chi_{Li} - G_i\bar{\varepsilon}_R\chi_R)$$

$$\delta z_i = \bar{\varepsilon}_L\chi_{Li}$$

$$\delta\chi_{Li} = \frac{1}{2}\hat{\not{\partial}} z_i\varepsilon_R - \frac{1}{2}\varepsilon_L e^{-G/2}G_i^{-1j}G_j - \frac{1}{8}\varepsilon_L\bar{\lambda}^\alpha_R\lambda^\beta_R G_i^{-1k}f^*_{\alpha\beta k}$$
$$+ \frac{1}{2}\varepsilon_L G_i^{-1k}G_k^{j\ell}\bar{\chi}_{Lj}\chi_{L\ell} + \frac{1}{4}\chi_{Li}(G_j\bar{\varepsilon}_R\chi^j_R - G^j\bar{\varepsilon}_L\chi_{Lj}) \qquad (4.23)$$

The transformations (4.23) are directly obtained from the original transformations (3.25), with the substitution of Eq.(4.5) in the fermionic field variations and with the redefinitions (4.8), (4.14), and (4.15) for the vierbein field and for the fermionic fields. In addition, the local supersymmetry parameter in Eq.(4.23) has also been redefined, compared to Eq.(3.25), as follows:

$$\varepsilon_L \to e^{-\frac{1}{2}J}\left(\frac{g}{g^*}\right)^{1/4}\varepsilon_L \qquad (4.24)$$

We note that, by using conformal tensor calculus, we could avoid the redefinitions given by Eqs.(4.8), (4.9), (4.14), (4.15), and (4.24), by suitably choosing a special conformal gauge. This has been shown elsewhere [17].

Finally, we discuss the case in which the gauge group is G x $U_R(1)$, corresponding to the superspace action formulae given in Eqs.(3.11) and (3.12). As a consequence of Eq.(3.35), the presence of a gauged R-symmetry is most easily discussed in the final form of the Lagrangian and transformation laws as given by Eqs.(4.18) and (4.23).

In fact, the Lagrangian has precisely the same expression as in Eq.(4.18) with suitable gauge covariantization with respect to the $U_R(1)$ gauge field. The only difference now is the fact that the D-term of the R-multiplet, defined as

$$D_R = g_R G^i n_i z_i \qquad (4.25)$$

can have a constant term, due to the non-invariance of the superpotential under R-transformations:

$$g^i(z)n_i g_R z_i = 3g_R g(z) \qquad (4.26)$$

where $n_i g_R = q_i$ are the U(1) charges of the scalar fields z_i.

It is obvious that these considerations are no longer valid for a vanishing superpotential function $g(z) = 0$. In this case the final form of the Lagrangian and transformation laws comes directly from Eqs.(4.18) and (4.25) with the following substitution rules [11,18]:

i) $e^{-G}, e^{-G/2} \to 0$

$$G^i \nabla_m z_i \Rightarrow J^i \nabla_m z_i + \frac{3}{2} i g_R V_m^R$$

ii) $(G_i \nabla_m z^{*i} \Rightarrow J_i \nabla_m z^{*i} - \frac{3}{2} i g_R V_m^R)$

iii) $\tilde{g}_\alpha G^i T_i^{\alpha j} z_j \Rightarrow \tilde{g}_\alpha J^i T_i^{\alpha j} z_j - \frac{3}{2} g_R \delta_R^\alpha$

iv) for other derivatives of G not in the form given by (ii) and (iii), replace $G \to J$.

(4.27)

We note, in particular, that the substitution rule (ii) produces new minimal gauge couplings of the R-gauge field to the gravitino, gauginos, and chiral fermions (but not to the scalars) which are usually eliminated through the chiral redefinitions (4.15) when $g(z) \neq 0$.

It is important to notice that some models discussed in the literature fall within this class [19]. However, these models can only have spontaneously broken supersymmetry with a non-vanishing cosmological constant.

In the more interesting physical situation of spontaneously broken R-symmetry and supersymmetry with a vanishing cosmological constant, the superpotential cannot vanish and the final form of the Lagrangian as given by Eq.(4.18) applies.

5. THE SUPER-HIGGS EFFECT IN N=1 SUPERGRAVITY

Spontaneous breaking of local supersymmetry implies the super-Higgs effect, i.e. the gravitino mass generation and the disappearance of the massless goldstino which is absorbed by the "massive" spin 3/2 gravitino.

This phenomenon may or may not occur with a vanishing cosmological constant, i.e. in a Minkowski or de Sitter space-time.

The goldstino field is uniquely defined by the spin 1/2 fermion which couples to the spin 3/2 gravitino in $L_{F,M}$ given by Eq.(4.21):

$$-\bar{\psi}_R \cdot \gamma \eta_L, \quad \eta_L = e^{-G/2} G^i \chi_{Li} - \frac{i}{2} D^\alpha \lambda_L^\alpha \tag{5.1}$$

where

$$D^{\alpha} = \tilde{g}G^{i}T_{i}^{\alpha j}z_{j} \tag{5.2}$$

A necessary and sufficient condition for spontaneous supersymmetry breaking therefore implies that one of the quantities [6],

$$e^{-G/2}G^{i}, \ D^{\alpha} \tag{5.3}$$

is different from zero at the minimum of the potential given in Eq. (4.19):

$$V(z,z^{*}) = -e^{-G}(3+G^{i}G_{i}^{-1j}G_{j}) + \frac{1}{2}\,\mathrm{Re}f_{\alpha\beta}^{-1}D^{\alpha}D^{\beta} \tag{5.4}$$

In the more interesting situation of a Minkowski space-time, we also demand the vanishing of the cosmological constant

$$V(z_{o},z_{o}^{*}) = 0 \quad \text{at} \quad \left.\frac{\partial V}{\partial z}\right|_{z=z_{o}} = 0 \tag{5.5}$$

Under condition (5.5), the gravitino mass has a precise meaning, and this is given by the universal formula [5]

$$m_{3/2} = e^{-G/2} \tag{5.6}$$

We now discuss the particularly interesting case of "minimal" coupling of the Yang-Mills system to supergravity. This is defined by the two conditions

$$G_{j}^{i} = -\frac{1}{2}\delta_{j}^{i}\,, \quad f_{\alpha\beta} = \delta_{\alpha\beta} \tag{5.7}$$

in which case all the kinetic terms are canonical and the Lagrangian depends only on the superpotential $g(z)$. In this case we have

$$G(z,z^{*}) = -\frac{1}{2}\,|z|^{2} - \log\frac{1}{4}\,|g(z)|^{2} \tag{5.8}$$

The scalar potential (5.4) becomes

$$V(z,z^{*}) = e^{-G}[2G^{i}G_{i}-3] + \frac{1}{2}\,(D^{\alpha})^{2} \tag{5.9}$$

with

$$D^{\alpha} = -\frac{1}{2}\tilde{g}_{\alpha} z^{*i} T_i^{\alpha j} z_j + \xi \delta_R^{\alpha}$$

The term $L_{F,M}$ given by Eq. (4.21), which determines the fermion mass matrix, is

$$e^{-1} L_{F,M} = e^{-G/2}[\bar{\psi}_{\mu R}\sigma^{\mu\nu}\psi_{\nu R} - \bar{\psi}_R \cdot \gamma \tilde{\eta}_L - \frac{2}{3}\bar{\tilde{\eta}}_L \tilde{\eta}_L]$$

$$+ \bar{\chi}_{Li} M^{ij} \chi_{Lj} + 2\bar{\chi}_{Li} M^{i\alpha}\lambda_L^{\alpha} + \bar{\lambda}_L^{\alpha} M^{\alpha\beta}\lambda_L^{\beta} + \text{h.c.} \tag{5.10}$$

where

$$\tilde{\eta}_L = G^i \chi_{Li} - \frac{i}{2} e^{G/2} D^{\alpha}\lambda_L^{\alpha} \tag{5.11}$$

The spin 1/2 fermion mass matrix has the form

$$M^{ij} = e^{-G/2}[G^{ij} - \frac{1}{3} G^i G^j]$$

$$M^{i\alpha} = -\frac{1}{3} iG^i D^{\alpha} - iD^{\alpha i}$$

$$M^{\alpha\beta} = -\frac{1}{6} e^{G/2} D^{\alpha} D^{\beta} \tag{5.12}$$

We can now compute the quadratic mass relation which generalizes the result of spontaneously broken globally supersymmetric Yang-Mills theories. The scalar contribution to the trace of the square mass is

$$\text{Tr}M_0^2 = -(N+4)D^{\alpha}D^{\alpha} + 2G^i G_i D^{\alpha} D^{\alpha} + \eta e^{-G}(4G_{ij}G^{ij} + N)$$

$$+ 4D^{\alpha i} D_i^{\alpha} + 4D^{\alpha} D_i^{\alpha i} \tag{5.13}$$

The spin-1 contribution is

$$\text{Tr}M_1^2 = 12 D_i^{\alpha} D^{i\alpha} \tag{5.14}$$

The spin-1/2 contribution is

$$\text{Tr}M_{1/2}^2 = 8e^{-G}(G^{ij} - \frac{1}{3} G^i G^j)(G_{ij} - \frac{1}{3} G_i G_j) + 16(D^{\alpha i}$$

$$+ \frac{1}{3} G^i D^{\alpha})(D_i^{\alpha} + \frac{1}{3} G_i D^{\alpha}) + \frac{2}{g} e^G (D^{\alpha})^2 (D^{\beta})^2 \tag{5.15}$$

and the gravitino contribution is

$$\mathrm{Tr}M_{3/2}^2 = 4e^{-G} \tag{5.16}$$

Therefore we finally get the mass formula

$$\text{Super trace } M^2 = \sum_{J=0}^{3/2} (-)^{2J}(2J+1)m_j^2 = (N-1)(2m_{3/2}^2 - K^2 D^\alpha D^\alpha) - 2\tilde{g}_\alpha D^\alpha \mathrm{Tr}T^\alpha \tag{5.17}$$

Equation (5.17) gives back the result of global supersymmetry [20] if we set $m_{3/2} = 0$, $k = 0$. It also reproduces the result of Ref.11 for N=1 and $\tilde{g} = 0$. We also note that the last term in Eq.(5.17) is only possible for Abelian U(1) factors of G with Tr $T^\alpha \neq 0$. Equation (5.17) summarizes the mass relations of spontaneously broken supersymmetry when the simultaneous occurrence of the Higgs and super-Higgs effects takes place.

6. LOW-ENERGY LIMIT OF SPONTANEOUSLY BROKEN LOCALLY SUPERSYMMETRIC LAGRANGIANS

In the practical current applications of spontaneously broken supergravity models to particle interactions, the Lagrangians described in the previous sections will contain, beyond the Planck scale M_P, some lower energy scales.

In particular they will contain the weak interaction scale M_W and other possible fundamental scales, such as the GUT scale $M_X \simeq 0(10^{15}$-10^{16} GeV).

Although our previous formalism is valid for any value of the gravitino mass, which is related to the supersymmetry-breaking scale M_{SB} as follows [4,5]:

$$m_{3/2} = \sqrt{\frac{8\pi}{3}}\,\frac{M_{SB}^2}{M_P} \tag{6.1}$$

we will consider the particular situation of an intermediate supersymmetry breaking scale

$$M_{SB} = \sqrt{\mu M_P} \tag{6.2}$$

where μ is a scale parameter $0(M_W)$. Under this circumstance

$$m_{3/2} = \sqrt{\frac{8\pi}{3}} = 0(M_W) \tag{6.3}$$

and supergravity corrections to low-energy physics play a crucial role. This is already implied by the modification of the mass spectrum through the quadratic mass relation given by Eq.(5.17).

The purpose of the present section is to derive all possible supergravity-induced breaking terms which survive in the low-energy limit defined by letting $M_P \to \infty$ with $m_{3/2}$ kept fixed.

The final result will be that in the limit $m_{3/2}/M_P << 1$, spontaneously broken supergravity will manifest itself as a softly explicitly broken global supersymmetric theory with well-defined renormalizable interactions [21].

In particular, the soft breaking terms, at most of dimension three in the particle fields, are such that no quadratic divergence arises through radiative corrections if gravitational interactions are neglected [22]. This circumstance has the resulting effect of not destroying the characteristic feature of supersymmetric theories, i.e. of keeping radiative corrections of light particles, including Higgs, generally small, compared to some other large scale of the theory such as the GUT scale M_X or the Planck scale M_P.

The important point, in obtaining the low-energy limit of a spontaneously broken supergravity model, is to realize that the cancellation of the cosmological constant [see Eq.(5.5)], implies the presence in the theory of some scalar fields x_i acquiring v.e.v.'s of order M_P which break supersymmetry [5]. It is natural, although not compulsory, that these fields are singlets under the Yang-Mills group G introduced in the previous section, especially if we do not want the GUT scale $M_X < M_P$ to be related to the Planck scale. The full set of scalar fields z_i belonging to the chiral multiplets S^i of the theory therefore split into

$$z_i = (x_i, y_i) \tag{6.4}$$

where $<x_i> = O(M_P)$ and $<y_i>/<x_j> << 1$.

The remaining fields y_i are called low-energy fields and generally transform non-trivially under G.

The requirement that the low-energy effective Lagrangian indeed describes light fields with renormalizable interactions can be fulfilled under the condition that a decoupling occurs between x-type and y-type multiplets as well as gauge vector multiplets. Furthermore, the gravitino mass $m_{3/2}$ must be finite for $M_P \to 0$ [21].

We define the dimensionful G function to be (dim G = 2):

$$G(z,z^*) = J(z,z^*) - \frac{1}{k^2} \log |k^3 g(z)|^2 \tag{6.5}$$

The y-dependent part of the Kahler potential is

$$J(\tfrac{1}{k} S, \tfrac{1}{k} S^*; y,y^*) = -\frac{1}{2} y_i \Lambda^i_j(S,S^*) y^{*j}$$

$$+ B^i(S,S^*) y_i + B^*_i(S,S^*) y^{*i} \tag{6.6}$$

where $\zeta_i = kx_i$. Analogously, the y-dependent part of the superpotential $g(z) = g(x,y)$ must be a finite polynomial, at most of degree three in the y-variables with ζ-dependent coefficients.

Equation (3.5) defines the Kahler potential $J(z,z^*)$ up to a Kahler transformation

$$J(z,z^*) \to J(z,z^*) + f(z) + f^*(z^*)$$

$$g(z) \to g(z)\exp k^2 f(z) \tag{6.7}$$

The behaviour of the y-dependent part of the superpotential $g(z)$ for $M_P \to \infty$, consistent with Eq. (6.6), must be

$$g(S,0) = O(M_P^2)$$

$$k \to 0 \tag{6.8}$$

In addition, the dimensionless function $f_{\alpha\rho}(z)$ must behave as

$$f_{\alpha\beta}(z) = \delta_{\alpha\beta}(1 + i\theta_\alpha)$$

$$k \to 0 \tag{6.9}$$

for $M_P \to \infty$ and x_i, y_i fixed.

Let us now consider the general form of the Lagrangian (4.18) in the limit $M_P \to \infty$ ($k \to 0$) in the fields y_i, after the x_i fields have acquired v.e.v.'s of $O(M_P)$ which break supersymmetry.

Equation (6.8) implies a finite gravitino mass

$$m_{3/2} = \lim_{k\to 0} k^2 |\langle g\rangle| \exp - \frac{k^2}{2} \langle J\rangle \tag{6.10}$$

The low-energy fields also include, out of the x-type fields, the goldstino which provides the missing degrees of freedom of the massive spin 3/2 gravitino and two G-singlet scalar fields, the scalar partners of the would-be goldstino, with masses $O(m_{3/2})$ [6]. However, these fields are not included in the low-energy Lagrangian of the y fields, because in the limit $M_P \to \infty$ they decouple from the y-type multiplets.

The gravitino mass defined by Eq.(6.10) provides the scale of all explicit supersymmetry-breaking terms appearing in the limiting Lagrangian for the low-energy fields y_i. This has the general form [23]

$$\begin{aligned} L &= L(\mathrm{susy};f) + m_{3/2}(c_\alpha \bar{\lambda}^\alpha_L \lambda^\alpha_L + \mathrm{h.c.}) \\ &\quad - m_{3/2}[h(y) + h^*(y^*)] - \frac{1}{2} m^2_{3/2} y^{*j} \Lambda^i_j y_i \end{aligned} \tag{6.11}$$

The first term is a globally supersymmetric Lagrangian with superpotential

$$f(y) = \lim_{k\to 0} g(<x>,y) \exp - (\frac{1}{2} k^2 <J>) \tag{6.12}$$

The function h(y) is another analytic polynomial, invariant under G, of (at most) degree three in the y fields, the $c_{\alpha,s}$ in the gaugino mass terms are complex dimensionless coefficients of order 1, and Λ^j_i is a numerical Hermitian matrix which depends on the Kähler potential J defined by Eq.(6.6).

In addition, the D-term of the supersymmetric part of Eq.(6.11) can contain dimensionful constants (Fayet-Iliopoulos terms), corresponding to U(1) factors of G:

$$D^2_{U(1)} = (\tilde{g} y_i Q^i_j y^{*j} + \xi)^2 \tag{6.13}$$

this even if the original Lagrangian given by Eq.(4.18) does not contain any gauged R-symmetry, as defined in Section 3. The reason is that if the scalar fields x_i have v.e.v.'s $<x_i> = O(M_P)$, then we can get a purely gravitationally induced effective D-term coming from terms linear in the U(1) gauge multiplet which couples to the x-fields[F2]:

$$k^2 D^\alpha W_\alpha \Psi(S,\bar{S})|_D \tag{6.14}$$

producing an effective interaction

$$D_{U(1)}\psi^j_i h^i h^*_j \rightarrow cD_{U(1)} m^2_{3/2} \tag{6.15}$$

for $\psi^j_i \sim \delta^j_i$ and $|h_x| = 0(m_{3/2})$.

This term should not be confused with a possible intrinsic ξ term in the original Lagrangian (4.18) which could be introduced by the gauging of R-symmetry as explained in Section 3. The term in Eq.(6.13) rather produces a soft-breaking term of dimension two in the scalar fields, of the type

$$L_{SOFT} \simeq cm^2_{3/2}\ y_i(Q)^i_j y^{*j} \tag{6.16}$$

where Q_i is the U(1) charge matrix of the y fields and which looks like a Fayet-Iliopoulos term since it can be rewritten in the form (6.13).

It is interesting to note that the general form (6.11) can also be obtained by the coupling of a locally supersymmetric Lagrangian for the matter fields y_i, to a Volkov-Akulov field, i.e. the non-linearly realized goldstino field. This has recently been shown by Samuel and Wess [24].

The reason why this approach gives the same result of the linear realization is due to the decoupling which occurs, for $M_P \rightarrow \infty$, between the goldstino multiplet and the light-matter fields y_i. In particular, the scalar degrees of freedom of the goldstino multiplet, which acquire v.e.v.'s of $O(M_P)$, do not play any role in the final softly broken Lagrangian we have considered. Then the entire goldstino multiplet can be simply replaced by a Volkov-Akulov field θ transforming, for constant ε, as [24]:

$$\delta_\varepsilon \theta(x) = M^2_{SB}\varepsilon + \frac{1}{M^2_{SB}}\ \xi^\mu(x)\partial_\mu\theta(x) \tag{6.17}$$

with

$$\xi^\mu(x) = \frac{1}{2}\ \bar{\varepsilon}_L\gamma^\mu\theta_R(x) - \frac{1}{2}\ \bar{\theta}_L(x)\gamma^\mu\varepsilon_R \tag{6.18}$$

The Lagrangian (6.11) shows that the low-energy effects due to spontaneously broken supergravity, with $m_{3/2} = O(M_W)$, result in explicit soft-breaking terms in the global limit which do not alter the ultra-violet behaviour of the purely globally supersymmetric part of the Lagrangian. In particular, they do not introduce quadratic divergences since these terms are soft in a supersymmetric sense, according to the analysis of

Ref.22.

To understand the origin of the various soft-breaking terms in Eq.(6.11), it is useful to consider special forms for the functions $G(z,z^*)$ and $f_{\alpha\beta}(z)$ which appear in the general form of the matter-Yang-Mills supergravity action given by Eq.(4.18).

In the case of minimal coupling, defined by Eq.(5.7), we get [21, 25]

$$h(y) = (A-3)f(y) + \sum_I y_i \frac{\partial f}{\partial y_i} \tag{6.19}$$

$$c_\alpha = 0$$

$$\Lambda_i^j = \delta_i^j$$

where A is a parameter depending on the particular way the x fields enter into the superpotential $g(z) = g(x,y)$.

For example, if

$$g(x,y) = g(x) + g(y) \tag{6.20}$$

with

$$g(x) = ax + b \tag{6.21}$$

then

$$A = 3 - \sqrt{3} \tag{6.22}$$

In the minimal case we may obtain an arbitrary function h(y) for the soft-breaking term in Eq.(6.11) which is linear in $m_{3/2}$ by relaxing the property (6.19)[F3].

If we introduce general mixing terms of the fields y to the fields x, then h(y) receives an additional contribution given by

$$\Delta h(y) = \lim_{k\to 0} \frac{\sqrt{3}}{k} \left.\frac{\langle\partial g(x,y)\rangle}{\partial x}\right|_{x=x_o} \exp\left(-\frac{k^2}{2}\langle J\rangle\right) \tag{6.23}$$

where $\partial g/\partial x$ is the derivative of g in the direction of the would-be goldstino.

Note that in order to get Eq.(6.22), what is needed is to introduce a superpotential of the form[F4]

$$g(x,y) = g(x) + g(y) + xg'(y) + O(\frac{1}{M_P}) + \ldots \tag{6.24}$$

The reason for introducing non-vanishing gaugino mass terms in Eq. (6.11) is because we can have an x-dependence in the function $f_{\alpha\beta}(z)$ which appears in the Yang-Mills part of the original Lagrangian [6]. The coefficients c_α in Eq.(6.11) are related to the first derivative of the function $f_{\alpha\beta}$ in the x-direction:

$$\begin{aligned} m_{3/2} c_\alpha \delta_{\alpha\beta} &= \lim_{k\to 0} \frac{1}{k} \; f^i_{\alpha\beta} G_j (-J^{-1})^j_i \exp{-k^{2<G>}}/2 \\ &= \lim_{k\to 0} \frac{1}{k} <f^*_{\alpha\beta} G_x (-J^{-1})^x_x \exp - \frac{k^2}{2} <G> \end{aligned} \tag{6.25}$$

Finally, to get mass terms for the y fields of the form $y_i \Lambda^i_j y^{*j}$ with $\Lambda^i_j \neq \delta^i_j$, we need a non-minimal Kähler potential [23]

$$J(z,z^*) = -\frac{1}{2}|x_i|^2 - \frac{1}{2}|y_i|^2 - \frac{1}{2}k^2|x_i|^2 \; y^{*\ell} A^m_\ell y_m \tag{6.26}$$

where A^m_ℓ is a numerical Hermitian matrix.

Then one gets in this case the matrix relation

$$\Lambda = \mathbb{1} + 3A \tag{6.27}$$

It turns out that Eqs.(4.24) and (4.26) give a correction to the general y-independent mass sum rule for minimal coupling regarding the y multiplets [6]:

$$\sum_{J=0}^{1} (-)^{2J}(2J+1)m_J^2 = 2\dim T^\alpha m^2_{3/2} - 2\tilde{g}^\alpha D^\alpha \mathrm{Tr} T^\alpha \tag{6.28}$$

This is related to the modification of the general quadratic mass sum rule in the case of non-minimal coupling [16].

For instance, a non-minimal Kähler potential gives an additional contribution to Eq.(6.27) of the form [16]

$$e^{-G} G_a (G^{-1})^a_b G^c (G^{-1})^d_c R^b_d, \quad R^b_d = \partial^b \partial_d \log\mathrm{Det} G^m_n \tag{6.29}$$

When R^ℓ_i is the contracted Riemann tensor (Ricci tensor) constructed out of the Kahler metric G^j_i.

For the particular example given by Eq.(6.25), the term in Eq. (6.28) gives a contribution to the quadratic mass sum rule proportional to $3m^2_{3/2}$ Tr A.

Further examples of a super-Higgs sector of the theory with non-trivial Kähler curvature and non-trivial mass relations will be given in the next section.

The Lagrangian given by Eq.(6.11) is a tree-level result. If we start with a minimal coupling of matter to supergravity, then $c_\alpha = 0$ and $\Lambda_i^j = \delta_i^j$. It is clear that for general values of Λ_i^j and c_α the low-energy Lagrangian (6.11) is stable under radiative corrections. However, if we start with the minimal coupling [Eq.(5.7)] then radiative corrections, including gravitational ones, will induce corrections which at most will be reabsorbed in definitions of the coefficients c_α and the matrix Λ_i^j as well as in a possible Fayet-Iliopoulos term for Abelian factors of G. These corrections, depending on the particular model under consideration, can turn out to be small. In this situation the low-energy spectrum can be described by Eq.(6.18), in terms of a few parameters, and we can build simple extensions of the standard model for weak interactions as well as GUT models with well-defined predictions for particle phenomenology in the 100 GeV-1 TeV energy range.

Simple examples of the generalization of the Standard Model, based on low-energy supergravity, will be given in Section 8.

7. FURTHER EXAMPLES OF A SUPER-HIGGS SECTOR: FLAT POTENTIALS AND GAUGED R-SYMMETRY

In this section we will consider the super-Higgs effect in situations that are different from the flat Kähler potential situation

$$G_i^j = -\frac{1}{2}\delta_i^j \tag{7.1}$$

as considered in Sections 5 and 6.

We again consider the situation of a single chiral multiplet (z,χ_L). The z-potential is given, in general, by the first term of Eq.(5.4) for i = 1:

$$V(z,z^*) = -e^{-G}[G_z G_{z^*} G^{-1}_{zz^*} + 3] \tag{7.2}$$

The gravitino mass is given by Eq.(5.6), provided $V(z,z^*)$ vanishes at the point $V_{,z} = 0$. The Kähler curvature, for a single scalar field, is given by [16]

$$R_{zz^*} = \partial_z \partial_{z^*} \log G_{zz^*} \tag{7.3}$$

and it vanishes for the "minimal kinetic term" (7.1). This corresponds to the super-Higgs sector described by a Polony field [28]. For a non-trivial, non-constant metric G_{zz^*}, the mass-sum rule of the super-Higgs sector becomes [16]:

$$m_A^2 + m_B^2 - 4m_{3/2}^2 = -2e^{-G} \frac{G_z G_{z^*}}{(G_{zz^*})^2} R_{zz^*} \tag{7.4}$$

There are interesting cases where the scalar potential given by Eq.(7.2) is identically equal to zero, although the supersymmetry may be broken. Rewriting the scalar potential of Eq.(7.2) as [5]

$$V = -ge^{-\frac{4}{3}G} G_{zz^*}^{-1} \partial_z \partial_{z^*} e^{\frac{1}{3}G} \tag{7.5}$$

it follows immediately that the potential is identically equal to zero [29] if[F5)]

$$G = \frac{3}{2} \log(\phi(z) + \phi^*(z^*))^2 \tag{7.6}$$

Contrary to the minimal kinetic term case, the Kahler manifold is not flat ($R_{zz^*} = 0$), since using Eq.(7.3) is found that

$$R_{zz^*} = - \frac{2}{3} G_{zz^*} \tag{7.7}$$

Therefore, this defines an Einstein-Kähler manifold. However, relation (7.7) is not sufficient to ensure the vanishing of the potential.

Note that in the positive kinetic energy domain defined by

$$\mathcal{D} : G_{zz^*} < 0 \tag{7.8}$$

supersymmetry is spontaneously broken and the gravitino mass is different from zero:

$$m_{3/2}^2 = e^{-G} = \frac{1}{|\phi(z)+\phi^*(z^*)|^3} \text{, for all } z \in \mathcal{D} \tag{7.9}$$

From the explicit expression for G_{zz^*}, it follows that

$$\mathcal{D} : |\phi_z| \neq 0, \ (\phi+\phi^*)^2 < \infty \tag{7.10}$$

So the "gravitino mass" is zero only at the boundary points $(\phi+\phi^*) = \infty$. The fact that the scalar potential is degenerate ($V \equiv 0$) reflects an arbitrariness for the value of the gravitino mass. This ambiguity can easily be removed if we demand the flatness condition ($V \equiv 0$) only locally, around a point $z_o \in \mathcal{D}$.

$$V_{z \to z_o} \simeq C|z-z_o|^{2n}, \quad n > 2 \tag{7.11}$$

In fact we can build a large class of positive definite potentials demanding the relation

$$\partial_z \partial_{z^*} e^{\frac{1}{3}G} = \Phi_{zz^*}(z,z^*) \tag{7.12}$$

where the real function $\Phi(z,z^*)$ satisfies the conditions

$$\Phi_{zz^*}(z,z^*) > 0, \quad \text{for all } z \in \mathcal{D} \tag{7.13}$$

with $\mathcal{D}$ the positive kinetic energy domain ($G_{zz^+} < 0$), and

$$\Phi_{zz^*}(z_o,z_o^*) = 0 \tag{7.14}$$

The general solution of Eq.(7.12) is then

$$G = \frac{3}{2} \log(\phi+\phi^*+\Phi)^2 \tag{7.15}$$

with

$$G_{zz^*} = 3 \frac{\Phi_{zz^*}(\phi+\phi^*+\Phi) - |\phi_z+\Phi_z|^2}{(\phi+\phi^*+\Phi)^2} < 0 \tag{7.16}$$

Equation (7.16) defines the positive kinetic energy domain $\mathcal{D}$.

The corresponding scalar potential is positive definite in $\mathcal{D}$, provided that $\Phi_{zz^*} > 0$ and $\phi+\phi^++\Phi > 0$, as can easily be seen from its analytic expression

$$V_o = 3 \frac{\Phi_{zz^*}(\phi+\phi^*+\Phi)}{|\phi+\phi^*+\Phi|^3[|\phi_z+\Phi_z|^2 - \Phi_{zz^*}(\phi+\phi^*+\Phi)]} \tag{7.17}$$

At the minimum ($z = z_o$), the potential vanishes identically and the gravitino mass is well defined. If the potential is locally flat at z_o, the fourth derivative of Φ at $z = z_o$ must also vanish:

$$\Phi_{zzz^*z^*}(z_o,z_o^*) = 0 \tag{7.18}$$

In that case, the relation (7.7) between the curvature R_{zz^*} and the Kähler metric G_{zz^*} is still valid locally ($z \sim z_o$), and the curvature contribution to the $(\text{mass})^2$ sum rule [16] is minus four times the gravitino mass squared:

$$\Delta m^2 = -2e^{-G} \frac{G_z G_{z^*}}{(G_{zz^*})^2} R_{zz^*}\Bigg|_{z=z_o} = -4m^2_{3/2} \tag{7.19}$$

We may then understand how the $(\text{mass})^2$ sum rule can be satisfied with a non-vanishing gravitino mass (signal of supersymmetry breaking) and vanishing scalar masses, as is obvious from the local flatness of the potential at the minimum.

Let us now examine the interesting physical case where the usual matter fields are also coupled to supergravity. In general, the positivity properties of the potential are destroyed when the matter fields are coupled in an arbitrary way. The solution to this problem is to assume that G is the sum of two uncoupled functions $G(z,z^*)$, as in Eq.(7.15), and $\bar{G}(y^i, y_i^*)$, so that the potential becomes

$$V = -e^{-G_T}\left(\frac{G_z G_{z^*}}{G_{zz^*}} + 3\right) - e^{-G_T}\,\bar{G}_i(\bar{G}^{-1})^i_j\bar{G}^j + \frac{1}{2}D^\alpha D^\alpha \tag{7.20}$$

with

$$G_T = G(z,z^*) + \bar{G}(y_i, y^{*i}) \tag{7.21}$$

and, therefore, remains positive definite provided that $-\bar{G}^i_j$ is a Hermitian positive matrix, the latter being a necessary condition for ensuring a meaningful kinetic term.

The vanishing minimum arises for $z = z_o$ as before, and

$$\bar{G}_i(\bar{G}^{-1})^i_j\bar{G}^j = 0$$

$$D^\alpha = 0 \tag{7.22}$$

The choice of minimal kinetic terms for the matter sector (y^i) leads to the following positive definite potential:

$$V(z,y_i) = e^{-\bar{G}}V_o(z,z^*) + e^{-G_T}\left(\sum_i |h_i^* + y_i|^2\right) + \frac{1}{2}(D^\alpha)^2 \tag{7.23}$$

where $V_o(z,z^*)$ is given by Eq.(7.17). The total G_T function reads

$$G_T = \frac{3}{2}\log(\phi+\phi^*+\Phi)^2 - y_i y^{i*} - h(y_i) - h^*(y^{*i}) \tag{7.24}$$

The absolute minimum of the potential (7.23) is zero and occurs at $z = z_o$ and $y_i = -h_i^*$ (if there are solutions). Note that the condition $y_i = -h_i^*$ automatically implies the vanishing of D terms because of the invariance of h under the internal group [6]. The vanishing of the cosmological constant is thus automatically satisfied without any unnatural fine tuning. The gravitino mass, in that case, is given by

$$m_{3/2}^2 = e^{-G_T(z,z^*,y_i,y^{i*})}\Big|_{\text{at the minimum}} \qquad (7.25)$$

In the flat limit, the hidden sector decouples and the scalar particle z remains massless for locally flat potentials. The potential for the matter fields becomes

$$V(y_i,y^{*i}) = \sum_i \left|\frac{\partial g(y_i)}{\partial y_i} + m_{3/2}y^{*i}\right|^2 + \frac{1}{2} D^\alpha D^\alpha \qquad (7.26)$$

where we have rescaled $h(y^i) = (1/M^2 m_{3/2})g(y^i)$ in order to have a residual effect. The D term in that case is given by

$$D^\alpha = g^\alpha y^{*i} T_i^{\alpha j} y_j \qquad (7.27)$$

The form of the potential presented in Eq.(7.26) no longer depends on the specific form of the hidden sector, and gives rise to soft global supersymmetry-breaking terms of the same form as in the A = 3 case or the case of a factorized form of the superpotential [27], and therefore the supertrace formula for the matter fields remains the same as in the minimal case [6,27]

$$\sum_B m_B^2 - 2\sum_F m_F^2 = 2N_y m_{3/2}^2 - 2r_i m_{\lambda_i}^2 \qquad (7.28)$$

where N_y is the number of chiral multiplets in the y sector, r_i is the dimension of the internal gauge group, and m_{λ_i} are the gaugino masses coming from a non-minimal choice of f_{AB} [6,27]. In the gravity hidden sector, the bosonic degrees of freedom are massless, although their fermionic partners obtain masses equal to the gravitino mass ($m_{3/2}$). To appreciate this fact better, let us generalize our results to the case of N chiral superfields (z_a,χ_a) in the hidden sector. In that case, the corresponding scalar potential of Eq.(7.2) becomes

$$V = -e^{-G}[G_a(G^{-1})_b^a G^b + 3] \qquad (7.29)$$

and may be rewritten as

$$V = -\frac{g}{N^2} e^{-\frac{N+3}{N}G} (G^{-1})^a_b \partial_a \partial^b e^{\frac{N}{3}G} \tag{7.30}$$

The flatness of the potential in this case implies a particular G, so that

$$(G^{-1})^a_b \partial_a \partial^b e^{\frac{N}{3}G} = 0 \tag{7.31}$$

For simplicity, we only examine the case of N chiral superfields which are singlets under the internal group. An obvious particular solution of Eq.(7.31) is

$$G = \sum_{a=1}^{N} \frac{3}{2N} \log(\phi_a(z_a) + \phi_a^*(z_a^*)) \tag{7.32}$$

where $\phi_a(z^a)$ is a function of the z^a field only. The potential is identically zero, as in the one-superfield case. Here also, we can realize the locally flat potential requirement ($z_a \sim z_a^o$) by modifying G to

$$G = \sum_{a=1}^{N} \frac{3}{2N} \log(\phi_a(z_a) + \phi_a^*(z_a^*) + \Phi_a(z_a, z_a^*)) \tag{7.33}$$

with

$$\partial_a \partial^a \Phi_a > 0 \qquad \text{for all } z_a \in \mathcal{D}$$

$$\partial_a \partial^a \partial_a \partial^a \Phi_a = 0 \qquad \text{at } z_a^o \tag{7.34}$$

All scalar fields are then massless, contrary to their fermionic partners and the gravitino whose masses satisfy the following $(\text{mass})^2$ formula [16]:

$$-2\sum_{F=1}^{N-1} m_F^2 - 4m_{3/2}^2 = -2e^{-G} G_a (G^{-1})^a_b G^c (G^{-1})^d_c R^b_d + 2(N-1)m_{3/2}^2 \tag{7.35}$$

where the curvature tensor R^b_a is given by [16]

$$R^b_a = \partial^b \partial_a \log \text{Det}[G^m_n] \tag{7.36}$$

The particular structure of the function G in Eq.(7.33) implies the following non-trivial R^b_a:

$$R^b_a = -\frac{2N}{3} G^b_a \tag{7.37}$$

Here also the Kähler space is an Einstein manifold.

Using Eq.(7.37) we find

$$e^{-G}G_a(G^{-1})^a_bG^c(G^{-1})^d_cR^b_d = 2Nm^2_{3/2} \tag{7.38}$$

and, using Eqs.(7.5) and (7.7), finally obtain the following mass formula:

$$2\sum_{F=1}^{N-1} m_F^2 = 2(N-1)m^2_{3/2} \tag{7.39}$$

which means that every fermionic degree of freedom acquires a mass equal to the gravitino mass. Their bosonic partners are massless. Note the reverse role of bosonic and fermionic degrees of freedom. In fact, contrary to the minimal kinetic term case [6], the extra mass contribution originating in the supersymmetry breaking is distributed among the fermionic degrees of freedom, whilst their bosonic partners remain massless.

Just as in the case of one chiral superfield in the hidden sector, the matter fields must be coupled in such a way that the properties of the hidden sector are not destroyed. We must choose the total G Function as the sum of the hidden sector $G(z^a,z^*_a)$ and the matter one $G(y^i,y^*_i)$. The resulting potential for the y sector is the same as before.

In the second part of this section we consider a super-Higgs sector implemented by local R-symmetry [15]. The introduction of local R-symmetry in supergravity has been extensively explained in Section 6, and it allows us to generalize the Fayet-Iliopoulos mechanism [13] for spontaneous supersymmetry breaking to local supersymmetry.

We consider a class of "minimal" models in which the super-Higgs sector is due to spin $(0^\pm, 1/2)$ chiral multiplet carrying non-trivial R-charge and a spin (1, 1/2) vector multiplet gauging R-symmetry [32]. This is the minimal multiplet content for a super-Higgs sector with local R-symmetry if we demand the vanishing of the cosmological constant with positive definite potential. We exhibit a simple model in which the supersymmetry breaking receives equal contributions from the D-term due to the vector multiplet and the f-term due to the chiral multiplet.

The final mass spectrum consists of a real scalar, a chiral spinor, and a massive vector, all with the same mass $m = 2m_{3/2}$. In an R-symmetric theory with a single chiral multiplet coupled to N=1 supergravity the most general form of the scalar potential is [11]

$$V(z,z^*) = -e^{-G}[G_z G_{z^*} G_{zz^*}^{-1} + 3] + \frac{1}{2} g^2 D^2 \tag{7.40}$$

where $G = G(z,z^*)$ and $D = -G_z z$; g is the R-gauge coupling constant. Using the identity

$$G_z = z^* G'(\rho),\ G_{z^*} = zG'(\rho),\ G_{zz^*} = G'(\rho) + \rho G''(\rho) = -D'(\rho)$$
$$= -\frac{1}{\rho} D'(\omega) \tag{7.41}$$

$$\rho = zz^*,\ \omega = \log\rho$$

we can rewrite (7.40) as follows:

$$V(\omega) = e^{-G}\left(\frac{D^2(\omega)}{D'(\omega)} - 3\right) + \frac{1}{2} g^2 D^2(\omega) \tag{7.42}$$

with $D' = f(D)$, $dD/f(D) = d\omega$.

The G function is then given by the following integral:

$$G(\omega) = -\int d\omega D(\omega) = -\int dD \frac{D}{f(D)} \tag{7.43}$$

If we now define

$$e^{-G} = \phi(D) \tag{7.44}$$

then

$$-G'e^{-G} = \phi'(D)D' = \phi(D)D;\ \frac{1}{D'} = \frac{\phi'(D)}{\phi(D)D} \tag{7.45}$$

so we finally get

$$\omega = \int \frac{dD}{D} \frac{\phi'(D)}{\phi(D)} \tag{7.46}$$

and the potential can be written in the D-variable as follows [32]:

$$V = \phi(D)\left(\frac{D^2\phi'(D)}{D\phi(D)} - 3\right) + \frac{1}{2} g^2 D^2 = -3\phi(D) + D\phi'(D) + \frac{1}{2} g^2 D^2 \tag{7.47}$$

Equation (7.47) defines the most general R-symmetric potential in terms of the function $\phi(D)$, which is related to G and to D through Eqs.(7.44) and (7.45).

Positive definite potentials correspond to different choices of $\phi(D)$, so that Eq.(7.47) is semipositive definite.

The simplest case is to take $\phi(D)$ linear in D [32]:

$$\phi(D) = \alpha D - \beta/3 \tag{7.48}$$

Then Eq.(7.47) becomes

$$V = \frac{1}{2} g^2 (D^2 - \frac{4\alpha}{g^2} D + \frac{2\beta}{g^2}) \tag{7.49}$$

Positivity means

$$V = \frac{1}{2} g^2 (D-\xi)^2 \tag{7.50}$$

which demands

$$\frac{2\alpha}{g^2} = \xi, \ \frac{2\beta}{g^2} = \xi^2; \ \phi(D) = \frac{1}{2} g^2 \xi D - \frac{1}{6} g^2 \xi^2 \tag{7.51}$$

At the minimum

$$D = \xi, \ \phi(\xi) = \frac{1}{3} g^2 \xi^2, \ m^2_{3/2} = \frac{1}{3} g^2 \xi^2 \tag{7.52}$$

Using Eq.(7.46) we can rewrite the potential in terms of ρ:

$$D = \frac{\xi}{3} \frac{1}{1-(\rho/\rho_o)^{\xi/3}} \tag{7.53}$$

So we finally get

$$V(\rho) = \frac{2}{3} m^2_{3/2} \frac{\left[1 - (\frac{\rho}{\rho_{min}})^{\xi/3}\right]^2}{\left[1 - \frac{2}{3} (\frac{\rho}{\rho_{min}})^{\xi/3}\right]^2}, \ \rho_{min} = \frac{2m^3_{3/2}}{g^2} \tag{7.54}$$

where the following conditions have been used to compute ρ_{min} and ρ_o:

$$D(\rho_{min}) = \xi, \ D'(\rho_{min}) = 1 \tag{7.55}$$

From Eqs.(7.55) we get

$$\rho_{min} = \frac{2}{3} \xi^2, \ (\rho_{min}/\rho_o)^{\xi/3} = \frac{2}{3} \tag{7.56}$$

Note that the potential (7.54) becomes singular for $\rho/\rho_{min} = (3/2)^{3/\xi}$, so the variable ρ is constrained in the region $0 < \rho < \rho_{min}(3/2)^{3/\xi}$.

From Eq.(7.47) we also get, at the minimum,

$$-e^{-G}G_zG_{z^*}G^{-1}_{zz^*} = \frac{1}{2}\, g^2D^2 = \frac{1}{2}\, g^2\xi^2 = \frac{3}{2}\, m^2_{3/2}$$

which means that the gravitino mass gets an equal contribution from the "D"-breaking and the "f"-breaking terms. As we will see below, this means that the would-be goldstino is an equal mixture of the χ and λ spinors of the chiral and vector multiplets. It is now straightforward to compute the tree-level particle masses of the theory.

The scalar square-mass matrix is

$$M^2_B = g^2\rho_{min}\begin{pmatrix}1 & 1\\ 1 & 1\end{pmatrix} \text{ i.e. } M^2_1 = 0,\ M^2_2 = 2g^2\rho_{min} = 4m^2_{3/2} \tag{7.57}$$

The massless mode is the would-be Goldstone boson of spontaneously broken R-symmetry; the physical massive mode has mas $M_2 = 2m_{3/2}$.

The vector boson mass is

$$m^2_V = 2g^2\rho_{min} = 4m^2_{3/2} \tag{7.58}$$

For the fermion mass matrix we notice that the $\chi\chi$ term of the N=1 supergravity Lagrangian [6]

$$+\ \bar{\chi}_L\chi_L[G_{zz}-G_zG_z-G_zG_{zzz^*}G^{-1}_{zz^*}] \tag{7.59}$$

is absent owing to the vanishing of the expression in brackets in our model.

The remaining terms of the spin 1/2 mass matrix is

$$\bar{\lambda}_L\chi_L(2igz^*) + \text{h.c.} \tag{7.60}$$

The goldstino mode is defined by the term coupled to the spin 1/2 gauge field in the supergravity Lagrangian [6]

$$\bar{\psi}_R\cdot\gamma\eta_L + \text{h.c.} \tag{7.61}$$

$$\eta_L = +m_{3/2}G_z\chi_L - \frac{i}{2}\, gzG_z\lambda_L = i\sqrt{\frac{3}{2}}\, m_{3/2}\left(\frac{i\sqrt{2}\chi_L+\lambda_L}{\sqrt{2}}\right) \tag{7.62}$$

Therefore, Eq.(7.60) can be rewritten as:

$$\bar{\lambda}_L(i\sqrt{2}\chi_L)\sqrt{2}\ gz^* + \text{h.c.},\ \sqrt{2}\ gz^*_{min} = 2m_{3/2} \tag{7.63}$$

Equation (7.63) shows that the orthogonal combination to the would-be Goldstone fermion

$$\psi_L = \frac{i\sqrt{2}.\chi_L - \lambda_L}{\sqrt{2}} \tag{7.64}$$

has mass $m_\psi = 2m_{3/2}$. Therefore, the physical spectrum consists of a massless spin 2 graviton, a massive spin 3/2 gravitino, and a scalar, a spinor, and a vector, all with the same mass:

$$m_S = m_V = m_\psi = 2m_{3/2} \tag{7.65}$$

The square-mass sum rule gives in this case

$$\mathrm{St}M^2 = \sum_{J=0}^{3/2} (-)^{2J} m_J^2 = 4m_{3/2}^2 \tag{7.66}$$

To understand this result, we have to apply the general mass formula in N=1 supergravity coupled to matter in the presence of non-vanishing Kahler curvature and the non-vanishing D-term [16]:

$$\mathrm{St}M^2 = 2(g^2D+g^2z\Gamma_z D) - 2R_{zz^*} \frac{G_z G_{z^*}}{(G_{zz^*})^2} e^{-G} \tag{7.67}$$

where $z\Gamma_z$ is the Kahler connection,

$$z\Gamma_z = z\partial_z \log G_{zz^*} \tag{7.68}$$

In our case,

$$z\Gamma_z\Big|_{min} = z\partial_z \log D'(\rho)\Big|_{min} = (-1 + \tfrac{5}{3}\xi) \tag{7.69}$$

$$R_{zz^*} = -2G_{zz^*} \tag{7.70}$$

which at the minimum gives

$$R_{zz^*}\Big|_{min} = 2 \tag{7.70'}$$

Therefore we get

$$\begin{aligned} \mathrm{St}M^2 &= 2[g^2\xi+g^2\xi(-1 + \tfrac{5}{3}\xi)] + 4m_{3/2}^2 \frac{G_z G_{z^*}}{G_{zz^*}} \\ &= \frac{10}{3} g^2\xi^2 - 4m_{3/2}^2 \frac{\xi^2}{\rho} = 10m_{3/2}^2 - 4 \times \tfrac{3}{2} m_{3/2}^2 = 4m_{3/2}^2 \end{aligned} \tag{7.71}$$

which agrees with the left-hand side given by Eq.(7.66). Incidentally, we note that Eq.(7.71) shows that the Kähler manifold is an Einstein space.

We now consider the potential given by Eq.(7.54) in the low-energy limit $M_P \to \infty$ with $m_{3/2}$ fixed [32]. By means of Eq.(7.52) we can express the dimensionless variable ξ as

$$\xi = \sqrt{24\pi}\,\frac{m_{3/2}}{gM_P} \tag{7.72}$$

so that two different limits are possible.

If $g \to 0$ with $M_P \to \infty$, then it is easy to see that

$$V_{M_P\to\infty,\, g\to 0} = 2m_{3/2}^2\,\phi^2 + 0\left(\frac{1}{M_P}\right) \qquad (7.73$$

where ϕ is the physical scalar degree of freedom. The vector, spinor, and scalar degrees of freedom decouple in this limit and we get a free theory of massive particles of spin 0, 1/2, and 1, respectively. This situation is entirely analogous to the normal super-Higgs effect without local chiral symmetry when the scalar field of the hidden sector just decouples in the limit $M_P \to \infty$, $m_{3/2}$ fixed [21].

However, in the presence of a gauged R-symmetry a second nontrivial limit exists for the supergravity Lagrangian with $M_P \to \infty$, $m_{3/2}$ fixed, and g fixed [32]. In this case the gauge-R interaction is nongravitational in the sense that g does not vanish with $M_P \to \infty$.

From a physical point of view, this second limit is less interesting: it corresponds to non-renormalizable interactions at low-energy, since non-renormalizable terms in the effective Lagrangians appear, which are scaled by inverse powers of $m_{3/2}$.

8. GENERALIZATION OF THE STANDARD MODEL TO LOCAL SUPERSYMMETRY: MINIMAL MODELS

"Minimal" models for the low-energy gauge theory $SU(3)_C \times SU(2)_L \times U(1)$ based on low-energy SUGRA with tree-level breaking of $SU(2)_L \times U(1)$ have been constructed [21,25,27,33-35].

The ordinary quarks and leptons are embedded in chiral spin (1/2, 0) multiplets with the following $SU(3) \times SU(2)_L \times U(1)$ quantum numbers:

$$Q = (3,2,1/3),\quad u^c = (\bar{3},1,-4/3),\quad d^c = (\bar{3},1,2/3)$$

$$L = (1,2,-1),\quad e^c = (1,1,2) \tag{8.1}$$

The Higgs sector contains three chiral multiplets, two SU(2) doublets,

$$H = (1,2,-1),\ H^c = (1,2,1) \qquad (8.2)$$

and an over-all singlet Y

$$Y = (1,1,0) \qquad (8.3)$$

After $SU(2)_L \times U(1)$ breaking down to $U_{em}(1)$, the scalar Higgs sector contains two charged and five neutral scalars.

Equation (8.1) defines an ordinary fermion family. The present fermion spectrum, with the inclusion of the t-quark, is obtained by repeating the chiral multiplets three times, as defined by Eq.(8.1). Extension of these models to SU(5) with the same set of particle states has been considered [36-38]. This requires the following structure of SU(5) representations for the basic spin (1/2,0) chiral multiplets:

$3(10 + \bar{5})$	Quarks and Leptons	
$5,\bar{5},1,24$	Higgs sector	(8.4)
24	Gauge multiplet	

Models have been constructed with a tree-level breaking $SU(5) \to SU_L(3) \times SU(2) \times U(1) \to SU(3)_C \times U(1)_{em}$ with the hierarchy $M_W[O(m_{3/2})] < M_X < M_P$. Unfortunately, unless there are further modifications, this class of models has an unstable hierarchy. In fact the very existence of the Y-SU(5) singlet spoils, at the one-loop level, the tree-level hierarchy $M_W << M_X$. To understand this point, it is sufficient to observe that because of the $YH\bar{H}$ coupling and of the heavy triplet mass splitting $\delta m^2_{3/2} = O(M_X\, m_{3/2})$, Fig.1 destabilizes the tree-level hierarchy; it induces Higgs doublet radiative square masses $\delta m_2^2 \simeq O(m_{3/2}\, M_X)$ [39]. This is the sliding singlet problem and it was first observed, in the context of global SUSY GUTs, by Polchinski and Susskind [3]. The sliding singlet problem in local SUSY GUTs can be partially cured by introducing additional 50, $\overline{50}$, and 75 representations for the chiral multiplets [38]. The missing partner scenario [40] can then avoid $H_3H_3^c$ direct mass terms as well as triplet direct mass terms contained in the 50 and $\overline{50}$ representations. Upon

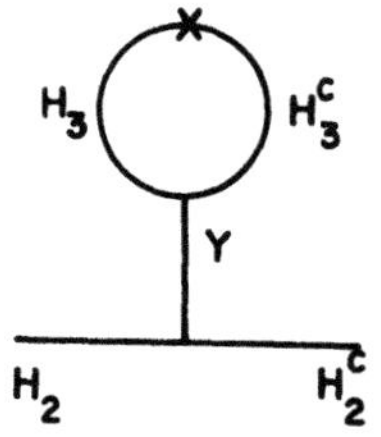

Fig. 1

integration on the superheavy fields, the low-energy SUGRA theory reduces to the form given by the first two lines of Eq.(6.11) and involves

i) the chiral multiplets given by Eqs.(8.1) and (8.2);

ii) the spin (1,1/2) gauge multiplets of SU(3) x SU(2) x U(1).

The complete particle spectrum of the minimal low-energy SUGRA theory with tree level breaking is given by Table 1.

Table 1

Ordinary particles		Superpartners
12 x 3 $(\text{quarks})_{\text{LEFT}}$	→	24 x 3 (squarks)
3 x 3 $(\text{leptons})_{\text{LEFT}}$	→	6 x 3 (sleptons)
$H^{\pm}$, 3 neutral (Higgs)	→	H_1^o (shiggs)
2 neutral (Y-Higgs)	→	λ_Y (Y-Majorana)
3 $(W^{\pm}, Z^o)$	→	$\lambda_1^{\pm}, \lambda_2^{\pm}, \lambda_1^o, \lambda_2^o$ $\begin{Bmatrix}\text{W-inos}\\ \text{Z-inos}\end{Bmatrix}$
1 (γ)	→	λ_γ (photino)
8 (g)	→	(8) λ_g (gluinos)
Gravitational sector		
1 (graviton)	→	(1) gravitino
	→	2 scalars (goldstino partners)

An example of tree-level mass spectrum for the superpartners of Table 1 is given above [38]: it corresponds to the choice A = 3 of Eq.(6.19) and the following Higgs superpotential:

$$h(Y)_{\text{Higgs}} = \alpha Y H H^c + \frac{1}{3}\beta Y^3 \tag{8.5}$$

Fermions (gauginos and Higgses)

$$m^2_{W(1,2)} = M_W^2 + \frac{1}{2} m^2_{3/2}\left[1 \pm \left(1 + \frac{4M_W^2}{m^2_{3/2}}\right)^{1/2}\right]$$

$$m^2_{Z(1,2)} = M_Z^2 + \frac{1}{2} m^2_{3/2}\left[1 \pm \left(1 + \frac{4M_Z^2}{m^2_{3/2}}\right)^{1/2}\right]$$

$$m^2_{0(3)} = m^2_{3/2}, \; m^2_{0(4)} = m^2_{3/2}\,(2-x)^2$$

$$m_g, \; m_\gamma \;\; \text{arbitrary} \tag{8.6}$$

Scalar Higgs sector

$$M^2_\pm = M^2_W + 4m^2_{3/2}, \; M^2_o = M^2_z + 4m^2_{3/2}$$

$$M^2_{o(2,3)} = \frac{1}{2}\,(1-x)m^2_{3/2}[5-x \pm (9-10x+x^2)^{1/2}]$$

$$M^2_{o(4,5)} = \frac{1}{2}\,m^2_{3/2}[9-2x+x^2 \pm (3+x)(9-10x+x^2)^{1/2}] \tag{8.7}$$

Squarks and sleptons

$$M^\pm_{sq(sl)} = |m_{3/2} \pm m_{q(\ell)}| \tag{8.8}$$

The entire spectrum of the superpartners of quarks, leptons, and gauge vector bosons is predicted in terms of two parameters: the gravitino mass $m_{3/2}$ and a dimensionless (Higgs coupling) constant $x = \beta/\alpha$.

The particle spectrum satisfies the tree-level mass sum rules [38]:

$$\sum_{J=0}^{1} (-)^{2J}(2J+1)m^2_J(\text{w sector}) = 4m^2_{3/2}$$

$$\sum_{J=0}^{1} (-)^{2J}(2J+1)m^2_J(\text{z sector}) = 2m^2_{3/2}$$

$$\sum_{0}^{1/2} (-)^{2J}(2J+1)m^2_J(\text{quark, lepton}) = 2m^2_{3/2}$$

$$\sum_{J=0}^{1/2} (-)^{2J}(2J+1)m^2_J(\text{4 neutral}) = 4m^2_{3/2} \tag{8.9}$$

There is a remarkable class of tree-level models considered in Refs.27 and 34, in which the tree-level mass spectrum is further simplified. This is obtained when the bare gaugino masses in Eq.(6.11) are set equal to the gravitino masses, and the $W^\pm$-ino and Z^o-ino masses are found to be

$$m_{w\pm(1,2)} = |m_{3/2} \pm M_w|, \; m_{z(1,2)} = |m_{3/2} \pm M_z|$$

whilst for the associated Higgs sector the same mass relations as given by the first line of Eq.(8.7) hold. Also, the squark and slepton masses are given by Eq.(8.8). In this situation the first two mass formulae of Eq.(8.9) are modified accordingly:

$$\sum_{J=0}^{1} (-)^{2J}(2J+1)m_J^2(\text{w sector}) = 0$$
$$\sum_{J=0}^{1} (-)^{2J}(2J+1)m_J^2(\text{z sector}) = 0 \qquad (8.10)$$

The (s) particle mass spectrum (superpartners of known particles) can be plotted in this case as a function of $m_{3/2}$ and is given by Fig.2. (For definiteness, a value of the top-quark mass $m_t \simeq 20$ GeV has been assumed.)

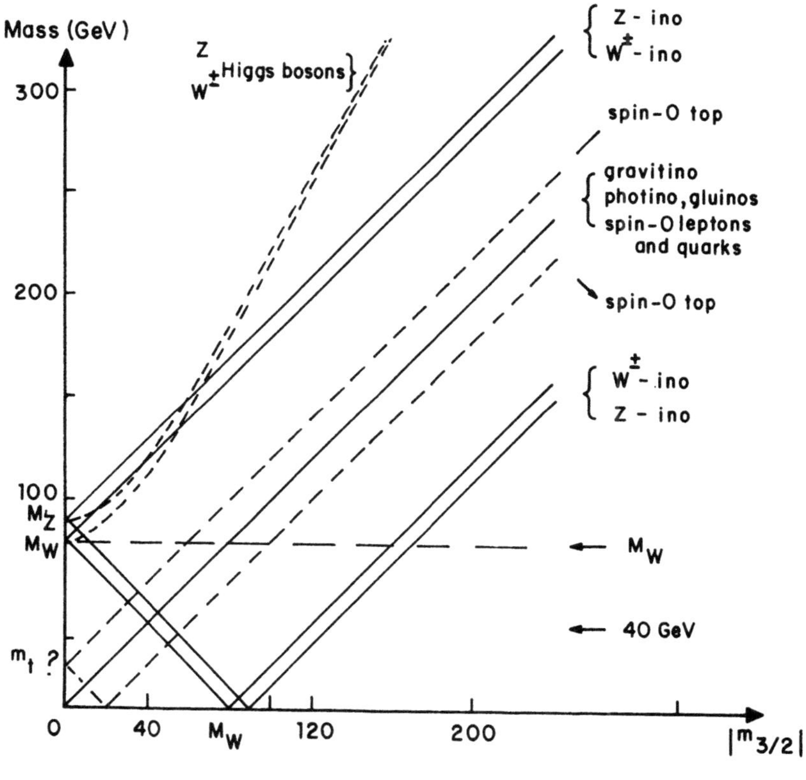

Fig. 2

Recently, a class of SUGRA GUT models based on radiative $SU(2)_L \times U(1)$ breaking have been constructed without the sliding singlet problem [41].

These models are somewhat similar to the globally supersymmetric models with induced radiative breaking, with the only (but important) difference that the radiative breaking is now induced by the soft breaking terms due to the supergravity rather than to the Fayet-O'Raifeartaigh hidden sector.

These models fall into two classes [41]:

i) very large top-quark mass: $m_t \geqslant 60$ GeV and $m_{3/2} \simeq O(100 \text{ GeV})$;

ii) small top-quark mass: $m_t \geqslant 20$-40 GeV, and a light gravitino: $m_{3/2} \geqslant O(20 \text{ GeV})$.

This second class of model is particularly interesting because it can have a rather light mass spectrum for the superpartners of ordinary particles (near the experimental bound). Moreover, they predict a light neutral scalar with mass of a few GeV. A typical mass spectrum (in GeV units) for the superpartners (Table 2) is reported here for $m_{3/2} = 20$ GeV and different values of the top-quark mass. The (s)particle mass spectrum in GeV unit is from Kounnas et al. in Ref. 41.

Table 2

Top		25	35	50
All families	$(\text{sleptons})_L$	27	36	25
	$(\text{sleptons})_R$	20	23	19
1st and 2nd family	$(\text{squark})_L$	77	108	67
	$(\text{squark})_R$	74	104	65
		74	103	65
3rd family	$(\text{sbottom})_L$	76	106	66
	$(\text{sbottom})_R$	74	104	65
	$(\text{stop})_L$	96	132	106
	$(\text{stop})_R$	54	78	37
Charged Higgses		93	94	83
Neutral Higgses		104	105	95
		3	5.3	5
		46	48	19

Table contd.

Table contd.

Gluinos	84	118	72
Photino	4.6	7.3	3
W-inos	87	94	90
	82	79	75
Z-inos	108	116	106
	85	80	83
Neutral Majorana	23	24	9

9. N=2 SUPERGRAVITY COUPLED TO VECTOR MULTIPLETS: GENERAL COUPLINGS AND THE SCALAR POTENTIAL

Lagrangians based on spontaneously broken N=1 supergravity provide phenomenologically acceptable models for particle physics with interesting new predictions in the 100-1000 GeV energy range where new particles, the superpartners of the known elementary particles, should be soon discovered [2,41]. However, from a more fundamental point of view, N=1 supergravity, being non-renormalizable, cannot be the final answer for the unification of fundamental forces, and new symmetries, perhaps extended supersymmetry, should possibly manifest themselves at higher energies.

In the present section, we focus on the simplest extension of N=1 local supersymmetry, namely N=2 local supersymmetry [42]. For computational reasons, it is important to have the fundamental Lagrangians written in the simplest possible way. In particular, if a possible scenario is envisaged, in which N=2 supergravity is broken down to N=1 or N=0 supersymmetry, it is relevant to have a formulation of N=2 Lagrangians in terms of N=1 variables [44].

Our formulation is particularly useful in connection with the σ-model structure of $N > 4$ extended supergravities. In the N=2 language, the scalar sector of these theories is given by an interaction of a Kählerian σ-model related to N=2 vector multiplets [43] and a quaternionic σ-model related to N=2 hypermultiplets [45,46], these multiplets having a unique embedding in the $N > 4$ supergravity multiplets. We claim that in the case in which the quaternionic manifold of the hypermultiplet sector is also Kählerian these theories have a simple description in terms of N=1 invariant functions. This is the case for N=4,5, and 6 extended supergravity. Remarkably enough, the general form of the scalar potential given in this section, as shown below, does indeed apply to these theories.

Our task is to formulate the most general N=2 coupling of matter and Yang-Mills multiplets to supergravity in terms of the N=1 invariant functions [5,6]:

$$G(z,z^*) = + J(z,z^*) - \ln|h(z)|^2 \tag{9.1}$$

and

$$\tilde{f}_{ab}(z) = \tilde{f}_{ba}(z) \tag{9.2}$$

which determine the most general coupling of a Yang-Mills-matter system to N=1 supergravity. The z_i variables in Eqs.(9.1) and (9.2) denote the complex scalar fields of chiral supermultiplets, which in N=1 supergravity can belong to arbitrary (generally complex) representations of the gauge group G.

Let us recall the properties of the functions $G(z,z^*)$ and $\tilde{f}_{ab}(z)$. The function $G(z,z^*)$ is a G-invariant function[F6)]

$$G^i T^{aj}_i z_j = z^{*i} T^{aj}_i G_j \tag{9.3}$$

containing the Kähler potential $J(z,z^*)$ and the superpotential $h(z)$. The scalar-kinetic terms are [6]:

$$G^j_i \partial_\mu z_j \partial_\mu z^{*i}, \quad G^j_i = + J^j_i \tag{9.4}$$

whilst the chiral part of the scalar potential is given by

$$-e^{-G}[G^i (G^{-1})^j_i G_j + 3] \tag{9.5}$$

The function $\tilde{f}_{ab}(z)$ is related to the Yang-Mills part of the Lagrangian [6]. For instance, the part of the Lagrangian quadratic in the Yang-Mills field strengths is:

$$-\frac{1}{4} \mathrm{Re}\, \tilde{f}_{ab} F^a_{\mu\nu} F^b_{\mu\nu} + \frac{i}{4} \mathrm{Im}\, \tilde{f}_{ab} F^a_{\mu\nu} \tilde{F}^b_{\mu\nu} \tag{9.6}$$

whilst the gauge contribution to the scalar potential is

$$\frac{1}{2} (\mathrm{Re}\tilde{f})^{-1ab} D_a D_b$$

$$D_a = gG^i T^j_{ai} z_j = -gJ^i T^j_{ai} z_j \tag{9.7}$$

g denoting the Yang-Mills gauge coupling constant. Spontaneous symmetry-breaking in local supersymmetry occurs whenever the vacuum state corresponds to [5,6]

$$\left\{ < e^{-G/2} G_i >, < D_a > \right\} \neq 0 \tag{9.8}$$

We now consider the case of N=2 supergravity coupled to matter. We use three different multiplets of physical fields:

i) the supergravity multiplet [42]

$$(e_{m\mu}, \Psi_{\mu\pm}, B_\mu) \tag{9.9}$$

containing the vierbein field, two Majorana gravitinos, and a U(1) gauge field (graviphoton);

ii) the Yang-Mills multiplet [47]

$$(A^a_\mu, \Omega^a_\pm, z^a) \quad a=1,\ldots,n=\dim G \tag{9.10}$$

containing the Yang-Mills gauge bosons, two gauginos and a complex scalar field;

iii) the hypermultiplet [47]

$$(\xi^i, A^i_\pm) \quad i=1,\ldots,2r \tag{9.11}$$

transforming as a 2r-dimensional representation of G and containing 2r Majorana spinors and 4r scalars satisfying a reality property which should be respected by the representation of the gauge group G.

The N=2 system has an over-all SO(2) x G gauge invariance. We identify the SO(2) with the gauge invariance of the B_μ field.[F7)] We denote by g' its gauge coupling constant. This gauge invariance acts on the ± indices in formulae (9.9) to (9.11), and the only scalars which are not SO(2) inert are thus the scalar fields of the hypermultiplet (9.11).

Couplings of Abelian vector multiplets [50] or hypermultiplets [50] to N=2 supergravity have been constructed first using the Noether method. The development of N=2 tensor calculus [43,45,51,53] allowed more general actions to be obtained. In particular, for the coupling of vector multiplets to supergravity, it was shown by de Wit et al.[43] that there exist more general actions, whose scalar sector describes a σ-model with a Kähler structure. The Kähler potential is in this case

$$J = \ell n\ y;\ y = z^{*I} N_{IJ} z^J \tag{9.12}$$

where the index I runs over the values 0 to $n = \dim G(z^o = 1)$. The action is constructed from a function $f(z^a)$ which is a G-invariant analytic function in the complex variables z^a. N_{IJ} and y in Eq.(9.12) are then given by

$$N_{oo} = \frac{1}{2}\left(f - f_a z^a + \frac{1}{2} f_{ab} z^a z^b\right) + \text{c.c.},$$

$$N_{oa} = \frac{1}{4}\left(f_a - f_{ab} z^b\right) + \text{c.c.},$$

$$N_{ab} = \frac{1}{4} f_{ab} + \text{c.c.},$$

$$y = \frac{1}{2}\left[f + f^* + \frac{1}{2}\left(f_a - f_a^*\right)\left(z^{*a} - z^a\right)\right]$$

$$\left(f_a = \partial f/\partial z^a,\ f_{ab} = \partial^2 f/\partial z^a \partial z^b\right) \tag{9.13}$$

If we neglect the vector B_μ [as we omit a multiplet of spins (3/2, 1) in the comparison between N=2 and N=1], we obtain kinetic terms as in formula (9.6) with

$$\tilde{f}_{ab} = -\frac{1}{4} f_{ab} + \frac{y_a y_b}{z^{*I} y_I},$$

$$y_a = \frac{\partial y}{\partial z^a} = N_{aI} z^{*I},\ y_o = N_{oI} z^{*I} \tag{9.14}$$

From Eq.(9.14) we notice that the N=2 Yang-Mills metric $\tilde{f}_{ab}$ has a non-analytic part $y_a y_b/(z^{*I} y_I)$. This term comes from the elimination of an (antisymmetric tensor) auxiliary field [49,53] of the spin (3/2,1) multiplet so it cannot be directly derived from the N=1 superspace Lagrangian of Ref.6.

The "minimal coupling" is the case where

$$f = 1 - z^a z^a \tag{9.15}$$

leading to

$$N_{oo} = 1,\ N_{oa} = 0,\ N_{ab} = -\delta_{ab} \tag{9.16}$$

The linear invariance of f is now extended to $SO(n,\mathbb{C})$. The complex form of G, $G_{\mathbb{C}}$, is a maximal subgroup of $SO(n,\mathbb{C})$. The corresponding function y is:

$$y = 1 - z^a z^{*a}, \tag{9.17}$$

whose linear invariance is now U(n). The scalar kinetic terms derived from the Kahler potential

$$J = \ell n(1 - z^a z^{*a}) \tag{9.18}$$

have the structure of a CP^n sigma model with scalar fields living on a Grassmannian manifold $U(n,1)/U(n) \times U(1)$.

For the couplings of hypermultiplets to N=2 supergravity, Bagger and Witten [46] used the Noether method to show that the scalar sector always describes a σ-model with a quaternionic structure. Some of these models were already obtained by the tensor calculus [43,45,52]. Combined general couplings of hypermultiplets and vector multiplets have been studied by de Wit et al.[43].

We temporarily neglect hypermultiplets and confine ourselves to the general local coupling of vector multiplets described by the G-invariant analytic function f(z),

$$f_b T_a{}^b{}_c z^c = 0 \tag{9.19}$$

where $iT_a{}^b{}_c$ are the (real) structure constants of G. The over-all scalar potential of the coupled N=2 Yang-Mills supergravity system has been established by de Wit et al. [43,49,53], and it reads:

$$V = -8g'^2(N^{-1})^{oo} - 16g'^2 y^{-1} - g^2 y^{-2} N_{ab}(z^{*c} T_c{}^a{}_d z^d)(z^{*e} T_e{}^b{}_m z^m) \tag{9.20}$$

where g is the coupling constant of the gauge group G, whilst g' is the coupling constant of the SO(2) local symmetry of gauged N=2 supergravity [54]. It is apparent that the gauge part of V has the D^2 structure which is familiar from N=1 supergravity.

In the case of "minimal coupling" [Eqs.(9.16) to (9.18)], one finds for the potential

$$V_{min} = -8g'^2\left(1 + \frac{2}{1-z^a z^{*a}}\right) + \frac{g^2}{(1-z^a z^{*a})^2} \left|z^{*c} T_c{}^b{}_d z^d\right|^2 \tag{9.21}$$

All stationary points, supersymmetry, and symmetry-breaking patterns for this potential have been studied in Ref.55.

Our aim is to recast formula (9.20) in the form given by formulae (9.5) and (9.7), namely to find the gauge-invariant function G given by formula (1). The answer turns out to be extremely simple: we claim that the function G which reproduces the Yang-Mills sector of N=2 supergravity coupled to gauge multiplets is simply given by:

$$G = - \ln(\frac{8g'^2}{y}) = - \ln(8g'^2) + \ln y \tag{9.22}$$

This means that the superpotential is fixed to be a constant

$$h(z) = 2\sqrt{2}g' \tag{9.23}$$

proportional to the SO(2) gauge coupling constant g'. As we will show below, the same expression for G will give the potential for N=4 and N=5 supergravities.

To derive this result, we need to calculate the potential of the N=1 theory specified by the two functions $\tilde{f}_{ab}$ [Eq.(9.14)] for the coupling of the N=1 Yang-Mills multiplet and G [Eq.(9.22)] for the chiral multiplet in the adjoint representation. The chiral part of this potential is given by Eq.(9.5). From the logarithmic form (9.22) for G we derive[F8)]

$$\begin{aligned} G_{\bar{a}b} &= + \frac{y_{\bar{a}b}}{y} - \frac{y_{\bar{a}}y_b}{y^2} \\ (G^{-1})^{a\bar{b}} &= + y(y^{-1})^{a\bar{b}} + \frac{(y^{-1})^{a\bar{c}}}{1-w} y_{\bar{c}}y_d (y^{-1})^{d\bar{b}} \end{aligned} \tag{9.24}$$

where

$$w = \frac{1}{y} y_a (y^{-1})^{a\bar{b}} y_{\bar{b}} \tag{9.25}$$

The chiral potential is then

$$V = 8g'^2 \frac{1}{y} (\frac{w}{w-1} - 3) \tag{9.26}$$

This form of the N=1 potential is valid without any assumption for y. Notice, however, that in our N=2 case:

$$\frac{\partial^2 y}{\partial z^a \partial z^{*b}} = N_{ab} \tag{9.27}$$

The expression for W [Eq.(9.25)] in N=2 is

$$y(1-w) = z^{*I}N_{IJ}z^J - z^{*I}N_{Ia}(y^{-1})^{ab}N_{bJ}z^J \qquad (9.28)$$

The only non-vanishing contribution comes from I = J = 0. The identity

$$(N_{oo} - N_{oa}y^{-1ab}N_{bo})(N^{-1})^{oo} = 1 \qquad (9.29)$$

which can be proved using

$$N_{oo}(N^{-1})^{oo} + N_{oa}(N^{-1})^{ao} = 1,$$

$$N_{bo}(N^{-1})^{oo} + N_{bc}(N^{-1})^{co} = 0,$$

$$y^{-1ab}N_{bc} = \delta^a_c, \qquad (9.30)$$

then implies

$$y(1-w) = \frac{1}{(N^{-1})^{oo}} \qquad (9.31)$$

The form (9.26) is now the first line of the N=2 potential (9.20).

We turn now to the gauge part, to identify the second line of Eq. (9.20) with Eq. (9.7). First, we note that the equation for the gauge invariance of f [Eq. (9.19)] can be derived with respect to z^d, giving:

$$f_{db}T_a{}^b{}_c z^c + f_b T^b_{ad} = 0 \qquad (9.32)$$

The expression for G [Eq. (9.22)] yields

$$D_a = - g\,\frac{1}{y}\,y_b T_a{}^b{}_c z^c \qquad (9.33)$$

We then use Eq. (9.14) for y_b, the definition of N [Eq. (9.13)], and Eq. (9.32) to obtain

$$D_a = - \frac{g}{4y}\,(z^{*d}f_{db}T_a{}^b{}_c z^c - f^*_{cb}T_a{}^b{}_d z^{*d}z^c) \qquad (9.34)$$

Up to now, the fact that the scalar fields span the adjoint representation has not been used: the matrix $(T_a)^b{}_c$ in Eq. (9.34) could be replaced by any representation. For the case of the adjoint representation, however, these matrices are antisymmetric in their lower indices. Hence, from Eq. (9.32), the quantity

$$f_{db}T_a{}^b{}_c z^c \tag{9.35}$$

is antisymmetric in the a, d indices. We then obtain

$$D_a = \frac{g}{y} N_{ab}(z^{*d}T_d{}^b{}_c z^c) \tag{9.36}$$

Using analogous manipulation we can prove that

$$y_a z^{*b}T_b{}^a{}_c z^c = 0 \tag{9.37}$$

which allows us to replace N_{ab} in Eq.(9.36) by -2 Re $\tilde{f}_{ab}$ [see Eq.(9.14)]. Equation (9.7) then yields

$$V_g = 2\,\frac{g^2}{y^2}\,\mathrm{Re}\,\tilde{f}_{ab}(z^{*d}T_d{}^a{}_c z^c)(z^{*e}T_e{}^b{}_m z^m) \tag{9.38}$$

which can indeed be identified with the second line in Eq.(9.20) using once more Eqs.(9.14) and (9.37).

To summarize, we have shown that the complete bosonic Lagrangian for N=2 gauge multiplets coupled to N=2 supergravity is completely specified in N=1 formulation by the analytic function f(z). Of course, we omitted the field B_μ in the N=1 formulation since this field belongs to the spin (3/2,1) multiplet which we neglected. As mentioned before, the elimination of an auxiliary field of this same multiplet is also responsible for the non-analytic term in the metric $\tilde{f}_{ab}$ as given in Eq.(9.14). This term does not allow the complete Yang-Mills part of the Lagrangian to be recast in an N=1 superfield form as given in Ref. [9.6].

When the SO(2) symmetry of supergravity is gauged, a chiral potential is induced, deriving from a fixed constant N=1 superpotential proportional to the SO(2) gauge coupling constant g'. Notice that there is no SO(2) gauge potential, simply because the scalar fields of the Yang-Mills multiplet are neutral under SO(2). On the contrary, an SO(2) gauge potential would appear for the scalars of a hypermultiplet. The latter situation is explicitly realized in N=5 supergravity.

Let us now consider some implications of our results for extended supergravities with N > 4. All these theories, in terms of N=2 supersymmetry representations, contain vector multiplets, and thus, scalar fields whose potential should fit in with our description. For N > 5, these theories will also contain hypermultiplets. But we will show that, at least for N=5, the scalars in these hypermultiplets also follow our prescription owing to the largest invariance of the manifold which contains all the scalar fields of the theory.

If we decompose N > 4 extended supergravity multiplets into N=2 supermultiplets, we can regard the scalar sector of these theories, which describes a σ-model over the manifold G/H, as a particular interaction of a Kählerian σ-model [55] (related to the scalar fields of the N=2 vector multiplets) with a quaternionic σ-model (related to the scalar fields of the N=2 hypermultiplets). The same procedure also applies to N=4 supergravity coupled to N=4 Yang-Mills multiplets.

We call coset disintegration the decomposition of a σ-model associated with G/H in terms of σ-models associated with G_1/H_1 and G_2/H_2, with $G_1, G_2 \subset G$ and $H_1, H_2 \subset H$.

Consider first the gravity multiplet of N=4 supergravity. It contains a complex scalar field, obviously inert under both SO(4) and the gauge group. This multiplet, when considered in terms of N=2 supersymmetry, splits into the N=2 gravity multiplet, an SU(2) doublet of gravitino multiplets and one vector multiplet which contains the scalar field. Thus, the scalar potential of the N=4 theory has to follow from our results. Moreover, it is known that this scalar field is a coordinate of the coset space SU(1,1)/U(1). The corresponding Kähler potential (up to a field redefinition) is then exactly the same as in the "minimal" coupling case, $f(z) = 1 - z^2$ [Eq.(9.16)], and the potential is then obtained from

$$G = -\ln(8g'^2(1-zz^*)^{-1}) \qquad (9.39)$$

which gives

$$V = -8g'^2\left(\frac{2zz^*}{1-zz^*} + 3\right) \qquad (9.40)$$

with vanishing gauge contributions for an SO(4)-singlet complex field z. This result agrees with the result of Cremmer et al. [57] for the gauging of SO(4) supergravity[F9)].

The N=5 supergravity with local SO(5) has been studied by de Wit and Nicolai [58]. The scalar fields, transforming like a complex $\underset{\sim}{5}$ of SO(5), have kinetic terms given by a non-linear σ-model based on the manifold SU(5,1)/SU(5) x U(1). Their Kähler potential (up to a Kähler transformation and field redefinition) is

$$-J(z,z^*) = \ln(y) = -\ln(1-z_i z_i^*), \quad i = 1,\dots,5 \qquad (9.41)$$

When the N=5 gravity multiplet is decomposed into N=2 multiplets, the scalar fields enter a SO(3) triplet of vector multiplets containing three complex scalars which are SO(2) singlets, the two remaining complex scalars being part of a hypermultiplet. First we can show that if we add to the Kähler potential $J(z,z^*)$ a constant superpotential such that the G function reads

$$G(z,z^*) = -\ln(8g'^2) + \ln(1-z_iz_i^*) \tag{9.42}$$

as usual, we recover according to our rules, the N=5 potential. It is very simple, using the SO(5) transformation properties of the scalar fields to establish that the gauge part of the potential is

$$2g'^2 \frac{1}{Y^2} [(z_iz_i^*)^2 - (z_iz_i)(z_j^*z_j^*)], \tag{9.43}$$

while the chiral part of the potential defined by Eqs.(9.5) and (9.42) is

$$-8g'^2 - 16g'^2Y^{-1} \tag{9.44}$$

as is usual in the "minimal" case. The sum of formulae (9.43) and (9.44) is precisely the potential found in Ref.[58]. An obvious way to make clear the relation with N=2 is to split the function G according to

$$J = \ln[\exp(J_1) + \exp(J_2) - 1], \tag{9.45}$$

where

$$J_1 = \ln(1-z_az_a^*), \quad a = 1,2,3, \tag{9.46}$$

gives the Kähler potential for the scalar fields of the three vector multiplets which are coordinates of an SU(3,1)/SU(3) x U(1) manifold, whilst

$$J_2 = \ln(1-z_4z_4^*-z_5z_5^*) \tag{9.47}$$

is the Kähler potential for the scalars of the hypermultiplet, in the manifold SU(2,1)/SU(2) x U(1), which is both Kählerian and quaternionic according to Bagger and Witten [46].

It should be possible to extend our considerations to N=6 and N=8 supergravities. The N=6 theory contains 15 complex scalar fields which are coordinates of the Kählerian manifold $SO^*(12)/U(6)$. When decomposed in terms of N=2 multiplets, the N=6 theory contains seven vector multiplets with scalars living in the manifold $O^*(8) \times SU(1,1)/U(4) \times U(1)$, and four hypermultiplets, the corresponding manifold being $SU(4,2)/SU(4) \times SU(2) \times U(1)$, which is Kählerian and quaternionic. Consequently, our prescription should also apply to the hypermultiplet sector of N=6. The analogue for the N=8 theory is the following: the 70 scalars parametrizing the coset $E_{7,7}/SU(8)$ are split into two pieces. First, 15 complex scalars belonging to 15 N=2 vector multiplets span the Kähler manifold $SO^*(12)/U(6)$. Then, 40 real scalars, which are coordinates of the quaternionic manifold $E_{6,2}/SU(6) \times SU(2)$, are components of 20 N = 2 hypermultiplets. This last coset space is indeed the only non-Kählerian manifold present in the decomposition of all N > 5 extended supergravities. This is related to the fact that $E_{7,7}/SU(8)$ is not a Kähler manifold, and is a consequence of PCT self-conjugation of the N=8 theory and of the 20 hypermultiplets contained in this theory.

10. HIGGS AND SUPERHIGGS EFFECT IN N=2 SUPERGRAVITY-YANG-MILLS SYSTEMS

In the present section we consider the first example of the super-Higgs effect in N=2 supergravity coupled to pure Yang-Mills matter multiplets.

For definiteness we consider the "minimal" case, which corresponds to the particular choice of the G-invariant analytic function f(z) given by Eq.(9.15). The scalar potential is given by Eq.(9.21).

We can summarize the analysis of the present section as follows [55].

If $g' = 0$ there are no extrema which break supersymmetry. We find a set of degenerate vacua which can induce the Brout-Englert-Higgs mechanism and break G in Minkowski space, as was the case of global supersymmetry.

If $g' \neq 0$, the situation is completely different. There is a unique stationary point (a local maximum) which breaks neither supersymmetry nor the gauge group G. It corresponds to massless multiplets in anti de Sitter space. All other stationary points break both n=2 supersymmetry and the gauge group G. They correspond to the combined Higgs and super-Higgs effects of n=2 supergravity. We recall that since the gauge multiplets are not charged with respect to the SO(2) gauge symmetry of n=2 supergravity, this latter symmetry remains unbroken, and consequently both supersymmetries are simultaneously broken.

In the absence of matter multiplets with non-vanishing vacuum expectation values, the cosmological constant is always negative; so the spontaneous breakdown takes place in anti de Sitter space [59]. A situation with vanishing cosmological constant will be considered in the next section.

For G = SU(N), we give all gauge symmetry-breaking patterns. They are characterized by all sets of (non-negative) integers $m_1, m_2, \ldots, m_N$ such that

$$\sum_{n=1}^{N} n m_n = N \qquad (10.1)$$

The residual gauge symmetry which corresponds to such a set is

$$H = SU(m_1) \times SU(m_2) \times \ldots \times SU(m_N) \times U(1)^{p-1}$$

where p is the number of non-zero m_n's. The n=2 supersymmetry is broken, except in the case $m_1 = N$, where H = G. The maximal unbroken symmetry is then $H_{max} = SU(N-2) \times U(1)$.

To check the stability of these non-trivial stationary points, we computed the scalar masses for $H = H_{max}$, from the part of the potential quadratic in the fields. We then used the criterion of stability in anti de Sitter space obtained by Breitenlohner and Freedman [60]. It turns out that these stationary points are stable if the ratio g'^2/g^2 is large enough.

For N > 7, the breaking pattern $SU(N) \to SU(3) \times SU(2) \times U(1)$ is possible[F10]. Interestingly enough, such an unconventional embedding of the usual low-energy gauge group, obtained from the n=2 supergravity potential, still allows us [for instance in the case of SU(7) and SU(8)] to obtain spin 1/2 states with the correct assignments of colour and electroweak quantum numbers, when matter hypermultiplets are added. These fields, which could be identified with quarks and leptons, are unavoidably accompanied by mirror particles and by exotic states, making the fermions not completely realistic at the present stage.

We now proceed to derive and discuss our results.

The n=2 vector multiplet consists of a vector field A_μ, two Majorana spinors $\Omega_\pm$, and a complex scalar field z, all transforming according to the adjoint representation of G. The usual (global) D-term reads

$$D = g[z, z^*] \qquad (10.2)$$

(we adopt a matrix notation z = a + ib with a and b Hermitian). The scalar potential, given by Eq.(9.15), is

$$V(z,z^*) = x_o^4 \, Tr(D^2) - \frac{16g'^2 x_o^2}{k^4} - \frac{8g'^2}{k^4} \qquad (10.3)$$

where

$$x_o = \frac{1}{\sqrt{1-k^2 Tr(zz^*)}} \qquad (10.4)$$

This non-polynomial function of z may be rewritten as a fourth order polynomial by defining

$$x = x_o z \qquad (10.5)$$

Then:

$$V(x,x^*) = g^2 Tr([x,x^*]^2) - 16 \frac{g'^2}{k^2} Tr(xx^*) - 24 \frac{g'^2}{k^4} \qquad (10.6)$$

The first term of this potential corresponds to the gauge potential of n=1 supersymmetry; it is positive definite and vanishes in directions for which $[x,x^*] = 0$, i.e. D is zero [61]. Thus, in the absence of g', supersymmetry remains unbroken since the vacuum expectation value (v.e.v.) of D is zero. The scale of gauge symmetry breaking is, however, arbitrary and the cosmological constant is zero.

When the g' coupling constant is introduced, $V(x,x^*)$ becomes unbounded from below. The v.e.v.'s are given by stationary points of the potential which satisfy[F11)]

$$0 = \frac{dV}{Dx_i} = - 4g^2 f_{ijk} f_{\ell mk} x_j^* x_\ell x_m^* - 16g'^2 x_i^* \qquad (10.7)$$

These stationary points correspond to locally stable solutions in anti de Sitter space if the following stability condition is fulfilled [60]

$$\frac{m^2}{\langle V \rangle} < 3/4 \qquad (10.8)$$

where m^2 is the most negative eigenvalue of the scalar mass matrix given by the quadratic part of the potential, and <V> is the value of the potential for this solution. Equation (10.7) leads immediately to

$$Tr(\langle A \rangle \langle B \rangle) = 0 \qquad (10.9)$$

$$Tr(\langle A \rangle^2) = Tr(\langle B \rangle^2) \qquad (10.10)$$

$$<x_o>^4 Tr(<D>^2) = 8g'^2 Tr(<x><x>^*)$$

$$= 16g'^2 Tr(<A>^2) \qquad (10.11)$$

$$<V> = -24g'^2 (1 + \frac{2}{3} Tr(<A>^2)) \qquad (10.12)$$

for the v.e.v.'s. Two kinds of solutions can occur. First, all v.e.v.'s are zero. Supersymmetry and gauge symmetry remain unbroken and $<V> = = -24g'^2$. The quadratic term in the potential gives a universal squared "mass" $m_u^2 = -16g'^2$ to all scalar fields which verifies $m_u^2 = (2/3)<V>$. This result is precisely what is expected from an actually massless particle in an anti de Sitter space. Moreover, unbroken supersymmetry and gauge symmetry indicate that all scalars have to remain massless. This stable solution is the only stationary point of V in the case where there is no D^2 term, either because g is zero, or if the gauge group is Abelian ($f_{ijk} = 0$).

If, however, some v.e.v.'s are non-zero, supersymmetry and gauge symmetry will be broken. From Eq.(10.7), it is apparent that non-zero v.e.v.'s of scalar fields will be of order g'/g. The supersymmetry-breaking scale $\sqrt{D}$ will then be of order $\sqrt{g'^2/g}$ and the cosmological constant is always negative.

The Goldstone fermions, which after the super-Higgs effect are eaten up by the two gravitinos, are given by

$$\Omega_{G\pm} = Tr(<D>\Omega_\pm)$$

($\Omega_\pm$ are the two multiplets of Majorana fermions belonging to the N=2 Yang-Mills vector multiplet).

Equation (10.7) allows us to work out the general stationary points of $V(x,x^*)$, for a given gauge group G. In the case G = SU(N) we obtain the following solution: $<x> = <A> + i<B>$ can be written in block-diagonal form, with m_n blocks of dimension n, n being an arbitrary integer, and with the obvious constraint [Eq.(10.1)] $N = \Sigma_{n=1}^N nm_n$. Each possible v.e.v. is then characterized by the N integers

$$m_1, m_2, \ldots, m_N$$

Each of these blocks of dimension n, in the basis where <A> is diagonal, reads

$$\langle A\rangle = \frac{g'}{g}\,\mathrm{diag}(n-1,\ n-3, \ldots, -(n-1)) \tag{10.13}$$

$$\langle B\rangle = \begin{pmatrix} 0 & B_1 & 0 & \cdots & 0 & 0 \\ B_1^* & 0 & B_2 & \cdots & 0 & 0 \\ 0 & B_2^* & 0 & \cdots & 0 & 0 \\ \vdots & \vdots & \vdots & & \vdots & \vdots \\ 0 & 0 & 0 & & 0 & B_{n-1} \\ 0 & 0 & 0 & \cdots & B_{n-1}^* & 0 \end{pmatrix} \tag{10.14}$$

with

$$|B_j|^2 = \frac{g'^2}{g^2}\, j(n-j) \tag{10.15}$$

<A> and <B> are obviously traceless in each block and verify:

$$\mathrm{Tr}(\langle A\rangle^2) = \mathrm{Tr}(\langle B\rangle^2) = \frac{g'^2}{3g^2}\,(n+1)n(n-1) \tag{10.16}$$

The invariance of a given solution characterized by $m_1, m_2, \ldots, m_N$ is

$$SU(m_1) \times SU(m_2) \times \ldots \times SU(m_M) \times U(1)^{p-1}$$

where p is the number of non-zero m_n and $M = [N/2]$ is the largest value for which m_M can be larger than 1. Let us give two simple examples.

<u>G = SU(2)</u>: There are two solutions

I : $m_1 = 2 \quad m_2 = 0$: $\langle x\rangle = 0$

SU(2) unbroken.

II: $m_1 = 0 \quad m_2 = 1$:

$$\langle A\rangle = \frac{g'}{g}\begin{pmatrix} 1 & 0 \\ 0 & -1 \end{pmatrix} \qquad \langle B\rangle = \frac{g'}{g}\begin{pmatrix} 0 & e^{i\alpha} \\ e^{-i\alpha} & 0 \end{pmatrix}$$

SU(2) completely broken.

<u>G = SU(3)</u>: There are three solutions:

I : $m_1 = 3 \quad m_2 = 0 \quad m_3 = 0$: $\langle x\rangle = 0$

SU(3) unbroken.

II : $m_1 = 1 \quad m_2 = 1 \quad m_3 = 0$:

$$\langle x\rangle = \frac{g'}{g}\begin{pmatrix} 1 & ie^{i\alpha} & 0 \\ ie^{-i\alpha} & -1 & 0 \\ 0 & 0 & 0 \end{pmatrix} \qquad SU(3) \to U(1)$$

III: $m_1 = 0 \quad m_2 = 0 \quad m_3 = 1$

$$\langle x\rangle = 2\,\frac{g'}{g}\begin{pmatrix} 1 & ie^{i\alpha}/\sqrt{2} & 0 \\ ie^{-i\alpha}/\sqrt{2} & 0 & ie^{i\beta}/\sqrt{2} \\ 0 & ie^{-i\beta}/\sqrt{2} & -1 \end{pmatrix} \qquad SU(3) \text{ completely broken}$$

Notice that the solution $0 = m_1 = m_2 = \ldots = m_{N-1}$, $m_N = 1$ always breaks SU(N) completely. Supersymmetry breaking is thus compatible with a large variety of gauge symmetry-breaking patterns. However, the embedding of the unbroken $SU(m_1) \times SU(m_2) \times \ldots SU(m_M)$ subgroup into SU(N) is restricted to be such that

$$\begin{aligned} \underset{\sim}{N} = \; & (\underset{\sim}{m}_1, \underset{\sim}{1}, \ldots, \underset{\sim}{1}) \\ & + 2(\underset{\sim}{1}, \underset{\sim}{m}_2, \underset{\sim}{1}, \ldots, \underset{\sim}{1}) \\ & + \ldots . \\ & + M(\underset{\sim}{1}, \underset{\sim}{1}, \ldots, \underset{\sim}{1}, \underset{\sim}{m}_M) \end{aligned} \tag{10.17}$$

It is tempting at this point to look for SU(N) breaking into G_o = SU(3) x SU(2) x U(1). It is clear, however, that all embeddings of G_o, as defined by Eqs.(10.13) and (10.14), will lead to exotic quantum numbers when matter hypermultiplets are considered. These multiplets contain a doublet of Weyl fermions and complex scalars transforming under G in conjugate representations. Let us consider the simplest candidate model which breaks into G_o, based on SU(7), with $m_1 = 3$ and $m_2 = 2$. The unbroken SU(3) x SU(2) x U(1) subgroup is defined through the embedding

$$\begin{aligned} \underset{\sim}{7} &= (\underset{\sim}{3}, \underset{\sim}{1}, 2/3) + 2(\underset{\sim}{1}, \underset{\sim}{2}, -1/2) \\ \underset{\sim}{21} &= 2(\underset{\sim}{3}, \underset{\sim}{2}, 1/6) + 3(\underset{\sim}{1}, \underset{\sim}{1}, -1) + (\underset{\sim}{\bar{3}}, \underset{\sim}{1}, 4/3) + (\underset{\sim}{1}, \underset{\sim}{3}, -1) \\ \underset{\sim}{35} &= 3(\underset{\sim}{3}, \underset{\sim}{1}, -1/3) + 2(\underset{\sim}{\bar{3}}, \underset{\sim}{2}, 5/6) + (\underset{\sim}{3}, \underset{\sim}{3}, -1/3) \\ &\quad + 2(\underset{\sim}{1}, \underset{\sim}{2}, -3/2) + (\underset{\sim}{1}, \underset{\sim}{1}, 2) \end{aligned} \tag{10.18}$$

The antisymmetric tensors of SU(7) contain all the fields necessary to classify quarks and leptons (with mixed chirality) and additional SU(2) and U(1) exotic states. Since $\underset{\sim}{7}$ contains only one SU(3)-triplet, there will be no colour exotics. Moreover, the electric charge of all leptonic states [i.e. SU(3)-singlets] is an integer, whilst coloured states have charges $\pm$ 4/3, $\pm$ 1/3, $\mp$ 2/3. It is not a trivial result that symmetry breaking precisely selects a U(1) group which is a candidate for a weak hypercharge. This fact is indeed not a peculiarity of SU(7). The same occurs using SU(8) with, however, a few differences. The embedding of SU(3) x SU(2) x U(1) is the following:

$$\underset{\sim}{8} = 2(\underset{\sim}{3}, \underset{\sim}{1}, 1/6) + (\underset{\sim}{1}, \underset{\sim}{2}, -1/2)$$

$$\underset{\sim}{28} = 3(\underset{\sim}{\bar{3}}, \underset{\sim}{1}, 1/3) + (\underset{\sim}{1}, \underset{\sim}{1}, -1) + (\underset{\sim}{6}, \underset{\sim}{1}, 1/3) + 2(\underset{\sim}{3}, \underset{\sim}{2}, -1/3)$$

$$\underset{\sim}{56} = 3(\underset{\sim}{\bar{3}}, \underset{\sim}{2}, -1/6) + 2(\underset{\sim}{8}, \underset{\sim}{1}, 1/2) + 4(\underset{\sim}{1}, \underset{\sim}{1}, 1/2) + 2(\underset{\sim}{3}, \underset{\sim}{1}, -5/6) + (\underset{\sim}{6}, \underset{\sim}{2}, -1/6)$$

$$\underset{\sim}{70} = 3(\underset{\sim}{3}, \underset{\sim}{1}, 2/3) + 3(\underset{\sim}{\bar{3}}, \underset{\sim}{1}, -2/3) + (\underset{\sim}{6}, \underset{\sim}{1}, -2/3) + (\underset{\sim}{\bar{6}}, \underset{\sim}{1}, 2/3) + 4(\underset{\sim}{1}, \underset{\sim}{2}, 0) + 2(\underset{\sim}{8}, \underset{\sim}{2}, 0) \qquad (10.19)$$

We obtain states with quantum numbers of quarks and leptons, and colour and electric charge exotics. Leptonic and coloured states can have electric charges $\pm$ 1/2, whilst there are some SU(3)-triplets with charges $\pm$ 1/6 and $\pm$ 5/6. The weak hypercharge obtained from SU(8)-breaking seems less attractive than in the SU(7) case. The extension of this analysis to larger SU(N) groups is straightforward.

This discussion of course does not give any indication of the possibility of separating left-handed and right-handed fermions inside an n=2 matter multiplet, whose coupling to n=2 supergravity can certainly modify the analysis of both the symmetry-breaking pattern and the cosmological constant problem.

We would now like to discuss the problem of the stability of the symmetry-breaking stationary points of the potential. To check the stability we need the "mass" spectrum[F12)] of the scalar fields (or at least the most negative squared mass) obtained from the quadratic terms in the Taylor expansion of $V(z,z^*)$, and from the structure of the scalar kinetic terms which read:

$$L_{KIN} = - (M_{KIN})^j_i (\partial_\mu z^i)(\partial^\mu z^*_j) \quad (10.20)$$

with

$$(M_{KIN})^j_i = < 2 \frac{\partial}{\partial z^i} \frac{\partial}{\partial z^*_j} \ln x_o > \quad (10.21)$$

Equivalently,

$$(M_{KIN})^j_i = <x_o>^2 (\delta^j_i + <x>^j <x>^*_i) \quad (10.22)$$

Thus, apart from an over-all $<x_o>^2$ factor, the kinetic terms for directions with non-zero v.e.v.'s are not canonically normalized and the mass matrix has to be correspondingly rescaled.

However, the mass spectrum shows several universal features which do not depend on the choice of the gauge group G. Scalar fields which correspond to generators T^i of the little group of $<z>$, $[T^i,<z>] = 0$, do not receive any mass contribution from the D^2 part of the potential. They only have the "universal squared mass" $-16g'^2$ and always fulfill the stability requirement since

$$\frac{m^2}{<V>} = \frac{2}{3(1 + \frac{2}{3} \mathrm{Tr}(<A>^2)} < \frac{2}{3} \quad (10.23)$$

Notice that the non-zero v.e.v.'s shift the ratio from 2/3, so that the scalars are no longer massless states as the corresponding gauge bosons are. Other general results can be obtained. The "Higgs particle", i.e. the scalar state collinear to the v.e.v. $<z>= <a> + i<b>$ has a squared mass $+32g'^2<x_o>^2$, whilst the orthogonal state spanned by $<b> - i<a>$ [see Eqs.(10.9) and (10.10)] is as usual a Goldstone state. The states collinear to $<b> + i<a>$ and $<a> - i<b>$ have squared "masses" $-32g'^2$, which satisfies the stability criterion if

$$\mathrm{Tr}(<A>^2) > 7/6 \quad (10.24)$$

This inequality leads to a lower bound of order 1 on the ratio g'^2/g^2 since

$$\mathrm{Tr}(<A>^2) = 0(1).g'^2/g^2$$

These last general results solve completely the G = SU(2) [or SO(3)] case for which Eq.(10.7) requires

$$\langle x \rangle_{SU(2)} = \frac{g'}{g} \begin{pmatrix} 1 & i \\ i & -1 \end{pmatrix} \tag{10.25}$$

i.e. $Tr(\langle A \rangle^2) = Tr(\langle B \rangle^2) = 2g'^2/g^2$. This v.e.v. breaks SU(2) completely; one thus obtains three Goldstone states, a Higgs particle with positive squared mass $32g'^2\ (1 + 4g'^2/g^2)$, and two scalar states with squared mass $-32g'^2$. Stability is ensured as long as

$$g'^2/g^2 > 7/12 \tag{10.26}$$

The largest unbroken gauge invariance with broken supersymmetry in the case G = SU(N) is SU(N-2) x U(1). It is obtained by breaking the SU(2) factor in the maximal subgroup SU(N-2) x SU(2) x U(1) of SU(N), by using the $\langle x \rangle_{SU(2)}$ solution [Eq.(10.25)]. The scalar mass spectrum is given in Table 1 [states are classified according to SU(N-2) x SU(2)]. Since the most negative squared mass is the same as in the G = SU(2) case, stability will be ensured when the condition (10.26) holds.

We have shown that all solutions have a negative cosmological constant. The only way to make the cosmological constant possibly vanishing is to add hypermultiplets whose scalars take a non-vanishing vacuum expectation value and (or) to depart from the minimal coupling of supergravity to the Yang-Mills action. The latter situation will be considered in the next section. In this case the breaking pattern of the gauge group could be rather different from the one discussed in the present section.

Table 1

SU(N-2) x SU(2) states	Squared masses
$(\underset{\sim}{1}, \underset{\sim}{3})$	3 Goldstone states 3 states with squared masses $+32g'^2\ (1 + 4g'^2/g^2)$ $-32g'^2$ $-32g'^2$
(Adjoint, $\underset{\sim}{1}$) + $(\underset{\sim}{1}, \underset{\sim}{1})$	$2(N-2)^2$ real states with squared mass $-16g'^2$
$(N-2, \underset{\sim}{2})$ + $(\overline{N-2}, \underset{\sim}{2})$	4(N-2) Goldstone states 4(N-2) states with squared masses $-24g'^2$

11. SUPERHIGGS EFFECT IN N=2 SUPERGRAVITY WITHOUT COSMOLOGICAL CONSTANT: N=2 FLAT POTENTIALS

In the present section examples of N=2 extended supergravity theories with vanishing cosmological constant and spontaneous supersymmetry breaking will be considered.

We will confine ourselves only to situations in which the super-Higgs effect is induced by [44,49] coupling supergravity to vector multiplets. The O(2) group which rotates the two gravitini can in general be gauged by a linear combination of the vector partner of the graviton B_μ and the vector fields A_μ:

$$g'B_\mu + g_i A^i_\mu \tag{11.1}$$

If the vectors A^i_μ gauge a group G, then g_i can be non-zero only for U(1) abelian factors of G. The scalar potential for n vector multiplets is given, in N=1 language, by [44]

$$e^G\, G^i G_j (G^{-1})^j_i - 3 \tag{11.2}$$

with the function G given by

$$G = \ln y - \ln|h(z)|^2 \tag{11.3}$$

with y as given by eq.(9.13) and superpotential h(z)

$$h(z) = 2(g^T.z) \tag{11.4}$$

where $g^T.z = g' + g_i z^i$.

Eq.(10.4) generalises eq.(9.22) for $g_i \neq 0$. In terms of the matrix N_{IJ} defined by eq.(9.15) the overall potential (10.2) reads

$$V = -\frac{1}{2} g^T N^{-1} g - \frac{|g^T.z|^2}{z^T N z^*} \tag{11.5}$$

g is the constant n+1 dimensional vector (g', g_i) and z stands for $(1, z_i)$. The invariance of V under the global transformation [62,63]

$$x \to Ux,\ N \to (U^T)^{-1} N U^{-1},\ g \to (U^T)^{-1} g \tag{11.6}$$

$$(z_i = x_i/x_o) \qquad U \in GL(n+1,R)$$

allows us to write the potential in the $g_i = 0$ gauge by performing a general linear transformation. Under such transformation the analytic function f(z) which defines the vector multiplet coupling to N=2 supergravity transforms as follows

$$f(z) \rightarrow g(z) = (U_o^o + U_o^j z_j)^2 f\left(\frac{U_i^o + U_i^j z_j}{U_o^o + U_o^j z_j}\right) \tag{11.7}$$

In the $g_i = 0$ gauge the scalar potential reads

$$\frac{V}{g'^2} = -\frac{1}{2}(N^{-1})_o^o - \frac{1}{z^T N z^*} \tag{11.8}$$

as given by eq.(9.22).

We consider first the coupling of a single vector multiplet to the N=2 supergravity action. In this case it can be shown that the trace of the scalar mass-matrix satisfied the relation

$$V_{,zz^*} = 2G_{zz^*}V \tag{11.9}$$

at the extremum $V_{,z} = 0$.

Therefore the absence of cosmological constant implies

$$m_A^2 + m_B^2 = 0 \quad \text{i.e.} \quad m_A = m_B = 0 \tag{11.10}$$

In ref.[44] it was shown that the point

$$V_{,z} = V = 0 \tag{11.11}$$

can never be a local minimum of V for any choice of f(z).

Therefore the only locally stable solution of eq.(11.11) with $e^G \neq 0$ is a flat potential i.e. [62]

$$V \equiv 0, \; e^G \neq 0. \tag{11.12}$$

The most general solution for a flat potential of a single complex scalar field z, was given, for N=1 supergravity in section 7 by eq.(7.6)

$$G = \log(\phi(z) + \phi^*(z^*))^3 \tag{11.13}$$

If we demand G as given by eq.(11.13) to be of the form (11.7) we find two solutions corresponding to the following choices of the N=2 invariant function f(z).

$$f_1(z) = -2i(\alpha z+\beta)^3 + iP_2(z)$$

$$f_2(z) = -i(\alpha z+\beta)^{3/2} + iP_2(z) \qquad \alpha,\beta \text{ real} \tag{11.14}$$

and $P_2(z)$ a 2nd degree polynomial with real coefficients. The solutions (11.14) corresponds to the special values for $\phi(z)$ as given by eq.(11.13)

$$\phi_1(z) = -i(\alpha+\beta z) + i\gamma, \; \phi_2(z) = (\alpha z+\beta)^{1/2} + i\gamma \tag{11.15}$$

where γ is an arbitrary real constant. Eq.(11.14) gives the first example of a N=2 supergravity theory with vanishing cosmological constant and spontaneous supersymmetry breaking.

The gravitino mass in these models is undetermined at the tree level since the vacuum expectation value of z is arbitrary. Coupling to other multiplets or radiative corrections may remove this degeneracy as it has been proposed in the literature - [29,63].

Vanishing potentials can be generalized in different ways to the case of many vector multiplets.

One obvious way, which is only possible if the n-vector multiplets gauge abelian group, is just given by choosing the $f(z^i)$ to be

$$f(z) = \sum_i f_{(i)}(z^i) \tag{11.16}$$

in which $f_{(i)}(z^i)$ depends only on z^i and is given by eq.(11.14). A more interesting way is obtained by the generalization to many fields of $f_1(z)$ namely

$$f(z^i) = iC_{ijk}z^iz^jz^k \tag{11.17}$$

in which C_{ijk} is an arbitrary (totally symmetric) constant.

If $g_i \neq 0$ the function $f(z^i)$ is given by

$$f(z^i,g^i) = \frac{1}{g^T.z}\, iC_{ijk}z^iz^jz^k \tag{11.18}$$

and still gives a vanishing potential for arbitrary g_i. It is interesting to observe that for $g_i = 0$ eq.(11.17) corresponds to a Kahler potential

$$\log y \text{ with } y = -\, iC_{ijk}(z^i_i-z^{*i})(z^j-z^{*j})(z^k-z^{*k}) \tag{11.19}$$

This is the most general form of y which can be obtained from the coupling of n-vector multiplets to N=2 supergravity in D=5 dimensions [64]. In that case the most general scalar potential [65] corresponds to a subclass of D=4 potentials defined through eq.(11.3) with h(z) as given by eq.(11.4) but y given by eq.(11.19).

Note that for $g_i \neq 0$ the D=5 vanishing potentials are different from the gauge-transformed vanishing potentials defined by the f-function as given by (11.18). The simplest way to understand this fact is by noticing that the D=5 potentials correspond to the Kahler metric given by eq.(11.19) while the D=4 vanishing potentials correspond to the Kahler metric given by eq.(9.13) in terms of f(z) as defined by eq.(11.18). The two Kahler metrices only coincide when $g_i = 0$. We would like to conclude this final section by a comment on the positivity domain for the kinetic-energy terms defined by the Kahler metric (11.17). If the n-vector multiplets gauge the unitary groups SU(n) (in this case $n = n^2-1$) then C_{ijk} can be identified with the d_{ijk} symbols of SU(N). In this case the positivity-energy domain for the scalar fields is empty [62].

There is a non-empty domain in the case of G = U(N) with

$$f(z) = I(z^3 + zz^{i2} + d_{ijk}z^i z^j z^k) \qquad (11.20)$$

and $\alpha > 0$.

Of course there is a non-empty domain for $G = U(1)^n$ and for all C_{ijk} coeffieients obtained from the D=4 dimensional reductions of D=5 supergravity-Maxwell coupled systems [64,65].

FOOTNOTES

F1) We use the conventions of Ref.2. We also use the following conventions: a = 1,2, two-component spinor indices; m = 1, ..., 4 flat Lorentz indices; $\mu = 1, \ldots, 4$ world indices; $\alpha = 1 \ldots \dim G$ gauge group indices; i,j = 1 ... dim R G-group representation indices, R being a finite unitary representation of G. We also set in most of our formulae the gravitational constant k = 1; k is related to the Planck mass M_P as follows: $k = \sqrt{8\pi}/M_P$.

F2) For a related argument, in the framework of non-linear realization, see Samuel and Wess [24].

F3) An example of the soft-breaking term given by an analytic function h(y) is obtained by integrating out the heavy degrees of freedom of GUTs, which in general modify the soft-breaking terms of light fields [26].

F4) An example which is a particular case of Eq.(6.23) has been considered by Cremmer et al.[27].

F5) To better emphasize the geometric relevance of global flatness, it is worth noticing that Eq.(7.6) is equivalent to the unique solution

$$G = \frac{3}{2} \log(z+z^*)$$

up to field redefinition $z \to f(z)$. The scalar Lagrangian takes the form

$$3\sqrt{-g}\, g^{\mu\nu} \frac{\partial_\mu z \partial_\nu z^*}{(z+z^*)^2}$$

This Lagrangian is the same as the scalar sector of the N=4 supergravity theory [30] and it describes a non-linear σ-model with an SU(1,1) non-compact symmetry

$$z \to \frac{\alpha z + i\beta}{i\gamma z + \delta}$$

with α, β, γ, δ real and $\alpha\delta+\beta\gamma = 1$.

We also note that through the particular redefinition

$$z \to \frac{z+\sqrt{3}}{z-\sqrt{3}}$$

the kinetic term takes the form of the conformal scalar coupling [15,31] with $J = +3 \log(1 - \frac{1}{3} zz^*)$

F6) We use in the present section the following conventions for gauge indices: i,j,k, ... refer to arbitrary representations of G, while a,b,c,... always refer to the adjoint representation. Moreover, $G^i = \partial G/\partial z_i$; $G_i = \partial G/\partial z^{*i}$ and $G^i_j = \partial^2 G/\partial z_i \partial z^{*j}$.

F7) We consider in this section only the case where the SO(2) group is gauged by the vector field B_μ sitting in the graviton multiplet. More general situations have been considered in ref.[49].

F8) $\bar{a}$ denotes derivatives with respect to z^{*a}.

F9) Other gauging of N=4 supergravity can be obtained in N=2 by gauging the SO(2) using a physical vector multiplet [49].

F10) The N=10 case is an exception as will later become clear from the study of the general pattern of SU(N).

F11) We use the following conventions: $x = A + iB$ (with A and B Hermitian) is developed on a set of Hermitian generators T_i, $x = 1/\sqrt{2})\, x_i T^i$, normalized with $Tr(T_iT_j) = 2\delta_{ij}$ and $[T_j,T_k] = 2if_{jkl}T_l$. We then have

$$Tr(xx^*) = x_ix_i^* = A_iA_i + B_iB_i$$

$$[x,x^*] = if_{jkl}x_jx_k^*T_l = 2if_{jkl}A_jB_kT_l$$

From now on, we set k = 1.

F12) Recall that in anti de Sitter space, these "masses" do not correspond to the physical masses of states.

REFERENCES

1. For a review see, for example, P. van Nieuwenhuizen, Phys. Rep. 68, 191 (1981).
2. For recent reviews see, for example, R. Barbieri and S. Ferrara, Surveys in High-Energy Physics 4, 33 (1983); B. Zumino, Berkeley preprint UCB-PTH-83/2 (1983), LBL-15819 (1983), to appear in the Proc. Solvay Conf., Austin, Texas, 1982; H.-P. Nilles, SLAC-PUB-3053, to appear in Proc. Conf. on Problems of Unification and Supergravity, La Jolla, 1983; J. Polchinski, Harvard preprint HUTP-83/A036 (1983); J. Polchinski, Harvard preprint HUTP-83/A036 (1983); S. Ferrara, CERN preprint TH.3570 (1983), to appear in Proc. 4th Silarg Symposium on Gravitation, Caracas, Venezuela, 1982; D.V. Nanopoulos, CERN preprint Th.3699 (1983), to appear in Proc. Europhysics Study Conference on Electroweak Effects at High Energies, Erice, 1983; J. Ellis, CERN preprint TH.3718 (1983), to appear in Proc. Int. Symp. on Lepton and Photon Interactions at High Energies, Cornell, 1983; R. Barbieri, Unconventional Weak interactions, Univ. Pisa preprint (1983), to appear in Proc. Int. Symp. on Lepton and Photon Interactions at High Energies, Cornell, 1983.
3. R. Barbieri, S. Ferrara and D.V. Nanopoulos, Z. Phys. C13, 267 (1982) and Phys. Lett. 116B, 6 (1982); J. Ellis, L. Ibanez and G. Ross, Phys. Lett. 113B, 283 (1982); S. Dimopoulos and S. Raby, Nucl. Phys. B219, 479 (1983); J. Polchinski and L. Susskind, Phys. Rev. D26, 3661 (1982).
4. S. Deser and B. Zumino, Phys. Rev. Lett. 38, 1433 (1977).
5. E. Cremmer, B. Julia, J. Scherk, S. Ferrara, L. Girardello and P. van Nieuwenhuizen, Phys. Lett. 79B, 231 (1978), and Nucl. Phys. B147, 105 (1979).
6. E. Cremmer, S. Ferrara, L. Girardello and A. van Proeyen, Phys. Lett. 116B, 231 (1982), and Nucl. Phys. B212, 413 (1983).
7. E. Gildener, Phys. Rev. D14, 1667 (1976); E. Gildener and S. Weinberg, Phys. Rev. D15, 3333 (1976); L. Maiani, Proc. Supper School on Weak Interactions, Gif-sur-Yvette, 1979 (IN2P3, Paris, 1980), p.3; M. Veltman, Acta Phys. Pol. B12, 437 (1981); E. Witten, Nucl. Phys. B188, 513 (1981); S. Dimopoulos and S. Raby, Nucl. Phys. B199, 353 (1981).
8. S. Ferrara and P. van Nieuwenhuizen, Phys. Lett. 74B, 333 (1978), and Phys. Lett. 76B, 404 (1978); K.S. Stelle and P.C. West, Phys. Lett. 74B, 330 (1978) and Phys. Lett. 77B, 376 (1978).

9. M. Kaku, P.K. Townsend and P. Van Nieuwenhuizen, Phys. Rev. D17, 3179 (1978); B. de Wit, in Supergravity 1982, eds. S. Ferrara, J.G. Taylor and P. van Nieuwenhuizen (World Scientific Publishing, Singapore, 1981), p.85; T. Kugo and S. Uehara, Nucl. Phys. B222, 125 (1983).
10. M.F. Sohnius and P.C. West, Phys. Lett. 105B, 353 (1981).
11. S. Ferrara, L. Girardello, T. Kugo and A. van Proeyen, Nucl. Phys. B223, 191 (1983).
12. See, for example, J. Wess and J. Bagger, Supersymmetry and Supergravity (Princeton University Press, 1983).
13. P. Fayet and J. Iliopoulos, Phys. Lett. 51B, 461 (1974).
14. K.S. Stelle and P.C. West, Nucl. Phys. B145, 175 (1978).
15. R. Barbieri, S. Ferrara, D.V. Nanopoulos and K.S. Stelle, Phys. Lett. 113B, 219 (1982).
16. M.T. Grisaru, M. Rocek and A. Karlhede, Phys. Lett. 120B, 189 (1982).
17. T. Kugo and S. Uehara, Kyoto preprint KUNS 646 (1982), to be published in Nuclear Physics B.
18. J. Bagger, Nucl. Phys. B211, 302 (1983).
19. D.Z. Freedman, Phys. Rev. D15, 1173 (1977); B. de Wit and P. van Nieuwenhuizen, Nucl. Phys. B139, 531 (1979).
20. S. Ferrara, L. Girardello and F. Palumbo, Phys. Rev. D20, 403 (1979).
21. R. Barbieri, S. Ferrara and C.A. Savoy, Phys. Lett. 119B, 343 (1982).
22. L. Girardello and M.T. Grisaru, Nucl. Phys. B194, 65 (1982).
23. S.K. Soni and H.A. Weldon, Phys. Lett. 126B, 215 (1983); R. Barbieri and S. Ferrara, see Ref.2.
24. S. Samuel and J. Wess, Columbia Univ. preprints, CU-TP-258 (1982) and CU-TP-260 (1983).
25. H.-P. Nilles, M. Srednicki and D. Wyler, Phys. Lett. 120B, 346 (1983).
26. L. Hall, J. Likken and S. Weinberg, Phys. Rev. D27, 2359 (1983).
27. E. Cremmer, P. Fayet and L. Girardello, Phys. Lett. 122B, 41 (1983).
28. J. Polony, Budapest preprint KFKI-1977 (unpublished).
29. E. Cremmer, S. Ferrara, C. Kounnas and D.V. Nanopoulos, Phys. Lett. B113, 61 (1983); S. Ferrara and A. van Proeyen, Phys. Lett. 138B, 77 (1984).
30. E. Cremmer, J. Scherk and S. Ferrara, Phys. Lett. 74B, 61 (1978).
31. Ngee-Poug Chang, S. Ouvry and Xizeng Wu, Phys. Rev. Lett. 51, 327 (1983).
32. E. Cremmer, S. Ferrara, L. Girardello, C. Kounnas and A. Masiero, Phys. Lett. 137B, 62 (1984).
33. J.M. Frere, D.R.T. Jones and S. Raby, Nucl. Phys. B222, 11 (1983).
34. P. Fayet, Phys. Lett. 125B, 178 (1983).
35. R. Barbieri and L. Maiani, Rome Univ. preprint No.343 (1983).
36. R. Arnowitt, A.H. Chamseddine and P. Nath, Phys. Rev. Lett. 49, 970 (1982); L. Ibanez, Phys. Lett. 118B, 73 (1972).
37. J. Ellis, J.S. Hagelin, D.V. Nanopoulos and K. Tamvakis, Phys. Lett. 124B, 484 (1983); B.A. Ovrut and S. Raby, Phys. Lett. 125B, 270 (1983); L. Hall and M. Suzuki, Berkeley preprint LBL-16150, UCB-PTH-83/8 (1983).
38. S. Ferrara, D.V. Nanopoulos and C.A. Savoy, Phys. Lett. 123B, 214 (1983).
39. H.-P. Nilles, M. Srednicki and D. Wyler, Phys. Lett. 124B, 337 (1983); A.B. Lahanas, Phys. Lett. 124B, 341 (1983).
40. A. Masiero, D.V. Nanopoulos, K. Tamvakis and T. Yamagida, Phys. Lett. 115B, 380 (1982); B. Grinstein, Nucl. Phys. B206, 387 (1982); S. Dimopoulos and F. Wilczek, Santa Barbara preprint (1982); C. Kounnas, D.V. Nanopoulos, M. Quiros and M. Srednicki, Phys. Lett.

127B, 82 (1983).
41. J. Ellis, D.V. Nanopoulos and K. Tamvakis, Phys. Lett. 121B, 123 (1983); H.-P. Nilles, Nucl. Phys. B214, 366 (1983); A. Alvarez-Gaume, J. Polchinksi and M. Wise, Nucl. Phys. B221, 495 (1983); J. Ellis, J.S. Hagelin, D.V. Nanopoulos and K. Tamvakis, Phys. Lett. 125B, 275 (1983); L.E. Ibanez and C. Lopez, Phys. Lett. 126B, 54 (1983); L.E. Ibanez, Phys. Lett. 126B, 196 (1983); M. Claudson, L.J. Hall and I. Hinchliffe, Berkeley preprint LBL-15948 (1983); C. Kounnas, A.B. Lahanas, D.V. Nanopoulos and M. Quiros, CERN preprint TH.3651 (1983); L. Ibanez and C. Lopez, CERN preprint TH.3650 (1983); J.P. Derendinger and C.A. Savoy, Saclay preprint SPhT/83-68 (1983); C. Kounnas, A.B. Lahanas, D.V. Nanopoulos and M. Quiros, CERN preprint TH.3657 (1983).
42. S. Ferrara and P. van Nieuwenhuizen, Phys. Rev. Lett. 37, 1669 (1976).
43. B. de Wit, P.G. Lauwers, R. Philippe, S.Q. Su and A. van Proeyen, Phys. Lett. 134B, 37 (1984).
44. J.P. Derendinger, S. Ferrara, A. Masiero and A. van Proeyen, CERN preprint TH.3813 (1984), to appear in Phys. Lett. B.
45. P. Breitenlohner and M. Sohnius, Nucl. Phys. B187, 409 (1981).
46. J. Bagger and E. Witten, Nucl. Phys. B222, 1 (1983).
47. P.H. Dondi and M. Sohnius, Nucl. Phys. B81, 317 (1974); R.J. Firth and O.J. Jenkins, Nucl. Phys. B85, 525 (1975); R. Grimm, M. Sohnius and J. Wess, Nucl. Phys. B133, 275 (1978).
48. P. Fayet, Nucl. Phys. B113, 135 (1976) and B149, 137 (1979).
49. B. de Wit and A. van Proeyen, NIKHEF Preprint-H/84-4 (1984).
50. J.F. Luciani, Nucl. Phys. B132, 325 (1978).
51. C.K. Zachos, Phys. Lett. 76B, 329 (1978).
52. M. de Roo, J.W. van Holten, B. de Wit and A. van Proeyen, Nucl. Phys. B173, 175 (1980); P. Breitenlohner and M. Sohnius, Nucl. Phys. B178, 151 (1980); B. de Wit, J.W. van Holten and A. van Proeyen, Phys. Lett. 95B, 51 (1980) and Nucl. Phys. B184, 77 (1981) [E: B122, 516 (1983)]; B. de Wit, R. Philippe and A. van Proeyen, Nucl. Phys. B219, 149 (1983).
53. B. de Wit, P.G. Lauwers, R. Philippe and A. van Proeyen, preprint NIKHEF-H/83-16 (1983), to be published in Physics Letters, B, and in preparation.
54. D.Z. Freedman and A. Das, Nucl. Phys. B120, 221 (1977); E.S. Fradkin and M.A. Vasiliev, Lebedev Institute preprint 197 (1976).
55. J.P. Derendinger, S. Ferrara, A. Masiero and van Proeyen, Phys. Lett. 136B, 354 (1984).
56. B. Zumino, Phys. Lett. 87B, 203 (1979); J. Bagger and E. Witten, Phys. Lett. 118B, 103 (1982).
57. E. Cremmer, S. Ferrara, J. Scherk, unpublished; A. Das, M. Fischler and M. Rocek, Phys. Rev. D16, 3427 (1977).
58. B. de Wit and H. Nicolai, Nucl. Phys. B188, 98 (1981).
59. B. Zumino, Nucl. Phys. B127, 189 (1977).
60. P. Breitenlohner and D.Z. Freedman, Phys. Lett. 115B, 197 (1982); Ann. Phys. 144, 249 (1982).
61. F. Buccella, J.P. Derendinger, S. Ferrara and C.A. Savoy, in Unification of the Fundamental Particle Interactions II (eds. J. Ellis and S. Ferrara) (Plenum Press, New York and London, 1983), p.349; Phys. Lett. 115B, 375 (1982); P.H. Frampton and T. Kephart, Phys. Rev. Lett. 48, 1237 (1982); Nucl. Phys. B211, 239 (1983).
62. E. Cremmer, J.P. Derendinger, B. de Wit, S. Ferrara, L. Girardello, C. Kounnas, A. van Proeyen, in preparation; see also J.P. Derendinger and Ferrara, CERN preprint 3903 (1984), to appear in the Proceedings of the 1984 Spring School in Supergravity and Supersymmetry.
63. J. Ellis, A.B. Lahanas, D.V. Nanopoulos and K. Tamvakis, Phys. Lett. 134B, 429 (1984); J. Ellis, C. Kounnas and D.V. Nanopoulos,

CERN preprint TH-3773 (1983).
64. H. Gunaydin, G. Sierra and P.K. Townsend, Phys. Lett. 133B, 72 (1983); LPTENS 83/32 (1983).
65. H. Gunaydin, G. Sierra and P.K. Townsend, Cambridge Univ. preprint DAMTP (1984).

N=1 SUPERGRAVITY UNIFIED THEORIES AND THEIR EXPERIMENTAL SIGNATURES

Pran Nath, R. Arnowitt and A.H. Chamseddine
Department of Physics
Northeastern University
Boston, MA 02115 USA

ABSTRACT

A review is given of the recent developments of the N=1 Supergravity unified theories. The first part of these lectures contains the underlying formalism while the second part contains a discussion of the phenomenological consequences of the theory. The lectures begin with a brief introduction to the supergravity-matter couplings and a deduction of the Lagrangian which couples an arbitrary number of chiral multiplets and a vector multiplet to supergravity. This is followed by a discussion of the phenomena of gravity induced symmetry breaking via the super Higgs effect. The effective low energy theory that arises after elimination of the super Higgs and heavy fields is exhibited. Generation of the weak scale in the theory at the tree level (T.B. models) or through renormalization group effects (R.G. models) is discussed. Gauge hierarchy of the Supergravity unified theory at the tree level and at the one loop level is discussed. Next a model independent formulation of the low energy theory is given which can accommodate a wide variety of N=1 Supergravity models. Particle content of this model independent theory is exhibited by diagonalization of the mass matrices in the supersymmetric sectors of the gauge bosons and Higgs bosons and in the squark and slepton sectors. The interaction vertices in terms of the physical fields are exhibited. Experimental signatures of N=1 Supergravity unified theory are investigated in a variety of phenomena. These include supersymmetric decays of the $W^{\pm}$ and Z^0 and possible experimental signals for the production of Winos, Zinos, squarks and sleptons at energies currently avaialbe at the $\overline{P}P$ Collider at CERN or in future experiments at SLC, LEP, Tevatron or SSC. Events with special signatures in the W and Z decays are found to be UFO events, lepton-jet events and certain exotic events. Other signatures of the theory are supersymmetric electro-weak effects on g_μ-2 which are generally found to be larger than the electro-weak contributions of the standard Weinberg-Salam theory. Supersymmetric corrections to the ρ parameter, $\sin^2\Theta_W$ and to M_W and M_Z masses are also discussed.

CONTENTS

I. INTRODUCTION

Since the proposal of the Supergravity unified theory two years ago(1,2), there has been a phenomenal further development of the theory(3-19) and an investigation of its low energy implications(20-40). The purpose of these lectures is to present a survey of these developments up to the current research.

We begin in Sec. II with a brief introduction to the coupling of matter multiplets with supergravity, since these couplings are needed for the construction of unified models. The basic rules for such a construction are given by tensor calculus(41,42). Using these rules Cremmer et al.(43) had exhibited the most general coupling of a single chiral multiplet with supergravity. However, for the construction of any realistic model one needs the coupling of an arbitrary number of chiral multiplets and simultaneously a vector multiplet with supergravity. These couplings were given in Refs. (44)-(46) and rederived in Ref. (47). The supergravity-matter coupled Lagrangian is seen to depend on two functions $f_{\alpha\beta}(Z^a)$ and $G(Z^a,Z_a)$ where Z^a are the full set of complex scalar fields arising from the chiral multiplets. The function $f_{\alpha\beta}(Z^a)$ enters in the Yang-Mills sector of the theory (see Eq. (2.3)). Aside from the fact that $f_{\alpha\beta}(Z^a)$ must transform as the symmetric product of two adjoint representations, this function is completely arbitrary and there are no ideas currently available for its theoretical determination. In all our analysis we shall adopt the minimality condition(44)

$$f_{\alpha\beta}(Z^a) = \delta_{\alpha\beta} \tag{1.1}$$

which restores the Yang-Mills sector to its normal form. The analysis of the second function $G(Z^a,Z_a)$ is more complex; it is related to the Kähler potential(49) in the manifold defined by the complex fields Z^a,Z_a and is discussed in Sec. II.

In Sec. III symmetry breaking via supergravity induced effects is discussed and general conditions for the spontaneous breakdown of supersymmetry are obtained. We shall assume here that the super Higgs effect responsible for the breakdown of the Supergravity gauge invariance operates through a gauge singlet super-Higgs field Z(50). The breakdown of Supergravity gauge invariance leads to a mass growth for the gravitino:

$$m_g = \frac{1}{2}\kappa^2 \quad g(Z_0)\exp(\frac{\kappa^2}{4} d) \tag{1.2}$$

where $\kappa = (8\pi G)^{1/2} = 0.41 \times 10^{-18}$ GeV (G is the Newtonian constant), $g(Z_0)$ is the superpotential (see Eq. (A.11)) evaluated at the minimum and d is the potential in the Kähler manifold of Z, Z^+ (see eq. (2.6)). Now $Z_0 \sim O(\kappa^{-1})$ and $g(Z_0)$ (which has the dimensions of $(\text{mass})^3$) is characteristically of the form $g(Z_0) \sim \kappa^{-1}m^2$ where m is an intermediate mass scale. For $m \sim 10^{10}$ GeV, one has

$$m_g \sim O(\kappa m^2) \sim O(10^2 \text{ GeV}) \tag{1.3}$$

The super Higgs effect leads to a mass growth for some of the fields with the masses for these fields characterised by the scale m_g. The mass matrices for the Bose and Fermi sectors of the theory are given in Sec. III.

Supergravity unified theories based on the gauge invariance

$$(\text{N=1 Supergravity}) \times G \tag{1.4}$$

where G is a grand unified gauge group contain also a GUT mass scale M associated with the breakdown of G. The simultaneous appearance of κ and M in the theory leads to a new gauge hierarchy problem(15). This problem is discussed in Sec. IV. Solution to the gauge hierarchy problem at the tree level as well as sufficiency conditions for maintenance of the gauge hierarchy at the one loop level are also discussed in Sec. IV. Once the satisfaction of the gauge hierarchy is guaranteed one may determine the content of the low energy theory. An efficient procedure here is to obtain an effective low energy theory by integrating out both the super-Higgs(2,5) as well as the heavy fields of the GUT sector(14,15). In Sec. IV the effective tree potential resulting from such an integration is exhibited. The residual fields are only the light fields relevant in the low energy domain. The integration of the super-Higgs field generates in the effective action a set of soft breaking terms(2,5). Remarkably one finds that all the scalar fields acquire a universal mass (2,5,23).

The chiral fermions (i.e. the quarks and the leptons) would develop masses through their usual Higgs interactions when SU(2)xU(1) breaks and Higgs develop VEVs. However, under the assumption that Eq. (1.1) holds the gauginos corresponding to SU(3)xSU(2)xU(1) have no tree level masses. Masses, $\tilde{m}_i$, for these gauginos arise at the loop level through their couplings with the heavy fields of the GUT sector[51,8,26]:

$$\tilde{m}_i = m_g \frac{\alpha_i}{4\pi} C \frac{D(R)}{D(A)} \tag{1.5}$$

where $\alpha_i = e_i^2/4\pi, i = 1,2,3$ are the U(1), SU(2) and SU(3) coupling constant, C is the Casimir, D(R) is the dimensionality of the representation exchanged in the loop and D(A) is the dimensionality of the adjoint representation. The radiative masses of Eq. (1.5) are expected to be O(1-10) GeV for the photino and O(5-75) GeV for the gluino.

Sec. V discusses the generation of the weak scale via gravity induced effects. These models fall into the following categories:

(i) Models with singlets[2,9]

(ii) Models with a heavy top quark[6,7,8]

(iii) No scale models[19], and

(iv) Others[17,18,52].

In models of type (i) SU(2)XU(1) breaking is generated at the tree level and radiative corrections do not play a significant role. Satisfaction of the gauge hierarchy at the loop level[53] requires constraints on the GUT sector[9,22,51]. A discussion of the one loop hierarchy is given in Sec. V. Models of type (ii) for the generation fo the weak scale from radiative corrections arising from the effects of a large top quark mass in the renormalization group equations are also discussed in Sec. V. In all of the above methods one relates the weak scale to the gravitino mass i.e. $M_W \sim O(m_g)$. However, the gravitino mass itself remains a theoretically undetermined quantity. No scale models attempt to correct this deficiency and aim to relate both the gravitino mass and the weak scale directly to the Planck mass, i.e., $m_g \sim O(\kappa^{-1}\exp(-1/\alpha))$ etc. In Sec. V we limit ourselves to a discussion of models of types (i) and (ii) and the reader is referred to the literature for models of types (iii) and (iv).

In order to exhaust the content of supergravity unified theory in the low energy domain it is useful to parameterize the theory in a way as to encompass all models. This is the idea behind the model independent formulations[26-29]. Thus the model independent formulations can describe both the tree breaking (T.B.) models (where SU(2)xU(1) breaking occurs at the tree level) and the renormalization group (R.G.) models (where SU(2)XU(1) breaking is induced beyond the tree through radiative corrections). The model independent formulation is presented in Sec. VI. A diagonalization of the mass matrices in this formulation for the supersymmetric partners of the gauge bosons, Higgs bosons, quarks and leptons are carried out. The interaction structure of the low energy theory can then be expressed in terms of the physical fields of the theory. (See Appendix B.)

The physical spectrum of the theory shows some interesting features. Remarkably it can be shown that if the photino mass is not too large, there must exist a charged Dirac state, the Wino $\tilde{W}_{(-)}^{\pm}$ with a mass $\tilde{m}_{-}$ below the mass of the W boson[11,12]

$$\tilde{m}_{-} < M_W \tag{1.6}$$

and a charged Dirac state, the Wino $\tilde{W}_{(+)}^{\pm}$ with a mass $\tilde{m}_{+}$ above the mass of the W boson[12]. Similarly one finds that there exist a neutral Dirac state, the Zino $\tilde{Z}_{(-)}$ with a mass $\tilde{\mu}_{-}$ below the mass of the Z boson and a $\tilde{Z}_{(+)}$ above the mass of the Z boson[12].

The special features of the mass spectra of the superpartners of the W and Z bosons discussed above raise the possibility that these superpartners may appear in the decays of the W and Z bosons for a range of values of the mass spectra. This topic is discussed in Sec. VII. Examples of such decays are

$$W^{\pm} \rightarrow \tilde{W}_{(-)}^{\pm} + \tilde{\gamma} \tag{1.7}$$

$$Z \rightarrow \tilde{W}_{(-)}^{+} + \tilde{W}_{(-)}^{-} \tag{1.8}$$

provided they are energetically allowed. They are followed by the Wino decays

$$\tilde{W}_{(-)}^{+} \rightarrow \ell^{+} + \nu_{\ell} + \tilde{\gamma} \tag{1.9}$$

$$\tilde{W}_{(-)}^{+} \rightarrow u_i + \bar{d}_i + \tilde{\gamma} \tag{1.10}$$

$$\tilde{W}_{(-)}^{+} \rightarrow u_i + \bar{d}_i + \tilde{g} \tag{1.11}$$

The decay of Eq. (1.7) is comparable to that of $W \rightarrow e\nu$ decay while that of Eq. (1.8) is a factor of 4-7 larger than the $Z^0 \rightarrow e^+e^-$ decay. If photino is the lowest mass particle in the decay chain of the supersymmetric particles such as the Winos, the Zinos, the gluinos etc., then supersymmetric decays of the W and Z are characterized by the presence of two photinos in their final states. Further, since the photinos are weakly interacting they should appear as missing energy. Thus the supersymmetric decays of the W and Z all involve missing energy and unbalanced momentum for their final states. Events which appear most accessible for experimental detection in the W and Z decays are

(i) UFO events

(ii) Lepton-jet events

(iii) Exotic events.

Type (i) are hadronic final states where "unidentified fermionic objects" or "UFO"s escape detection. An example of such an event is the decay of Eq. (1.7) followed by the decay of Eq. (1.10) which produces a single broad jet (from the two quark final states) with momentum unbalanced. There exist also two jet UFO events. Type (ii) events consist of decays which have a single broad jet in one hemisphere and a charged lepton in another again with momentum unbalanced. An example of such an event is the decay of Eq. (1.8) followed by one $\tilde{W}_{(-)}$ decaying leptonically via Eq. (1.9) and the other $W_{(-)}$ decaying hadronically via Eq. (1.10). Type (iii) are a variety of exotic events, such as a lepton pair in one hemisphere and a broad single jet in another with missing $p_{\perp}$, which also arise in the supersymmetric W and Z decays. The signatures of these events are discussed in Sec. VII.

In addition to the search for supersymmetry through the direct experimental observation of the supersymmetry particles, there are indirect tests of supersymmetry. An example is the supersymmetric effects on the muon anomalous moment. g_{μ}-2 is experimentally known to one part in 10^8.[54] Kinoshita et al.[55] have recently carried out improved theoretical calculations for this quantity and suggested that a reduction in the experimental error by an order of magnitude should allow one to test the electroweak contributions of the Weinberg-Salam theory to g_{μ}-2[57]. However, supersymmetry contributes importantly to g_{μ}-2[58-60] and its effect are comparable and even significantly larger (depending on the model) than those of the standard theory[50]. Thus supersymmetric electro-weak effects should be easily discernible in experiments designed to test the effects of the standard Weinberg-Salam theory on g_{μ}-2. Another example is provided by the supersymmetric effects on the parameter ρ which is the ratio of the neutral current to the charged current effective couplings in the weak interaction theory. Supersymmetric effects on ρ as well as on $\sin^2\Theta_W$ and on M_W and M_Z are discussed in Sec. VIII. Sec. IX is devoted to conclusions.

II. SUPERGRAVITY AND MATTER COUPLINGS

The basic structure of unified Supergravity models consists of N=1 matter multiplets coupled to N=1 Supergravity. The N=1 matter multiplet can either be a left-handed chiral (F-type) multiplet ($Z^a, X^a{}_L, h^a$) which belongs to a reducible representation of the gauge group G (a is an internal symmetry index, $Z^a = A^a + iB^a$ are complex scalar fields, $X^a{}_L$ are left-handed chiral fields and $h^a = F^a + iG^a$ are auxiliary or constraint fields) or a vector (D-type) multiplet (C, ξ, H, K, V_{μ}, λ, D) which belongs to the adjoint representation of the gauge group G. There is a drastic reduction of the vector multiplet in the Wess-Zumino gauge[48] with only the elements (0, 0, 0, V_{μ}, λ, D) remaining. The simplest choice for the N=1 supergravity multiplet is the multiplet with the minimal set of auxiliary fields ($e_{a\mu}$, ψ_{μ}, S, P, A_{μ}) where $e_{a\mu}$, ψ_{μ} are the spin 2, spin 3/2 fields and S, P, A_{μ} are the auxiliary fields of the minimal set[41,42]. Supergravity couplings to an F-type multiplet are given by [41,42]

$$e^{-1}L_F = \mathrm{Re}[h + uZ + \bar{\psi}_\mu\gamma^\mu X + \bar{\psi}_\mu\sigma^{\mu\nu}\psi_{\nu R}Z] \qquad (2.1)$$

where u = s-iP. Similarly the couplings of D-type multiplet to supergravity are given by(42)

$$e^{-1}L_D = D - \frac{i\kappa}{2}\bar{\psi}_\mu\gamma^5\gamma^\mu\lambda - \frac{2}{3}(SK - PH)$$
$$+ \frac{2}{3}\kappa V_\mu(A^\mu + \frac{3}{8}ie^{-1}\varepsilon^{\mu\rho\sigma\tau}\bar{\psi}_\rho\gamma_\tau\psi_\sigma)$$
$$+ i\frac{\kappa}{3}e^{-1}\bar{\xi}\gamma_5\gamma_\mu R^\mu + \frac{i\kappa^2}{8}\varepsilon^{\mu\nu\rho\sigma}\bar{\psi}_\mu\gamma_\nu\psi_\rho\bar{\xi}\psi_\sigma$$
$$- \frac{2}{3}\kappa^2 C\, e^{-1}L_{S.G.} \qquad (2.2)$$

One may recall that in global supersymmetry only the F and D terms are the allowed terms in the Lagrangian. For the case of Supergravity all elements of the F and D multiplets enter the Lagrangian to guarantee Supergravity gauge invariance.

For the purpose of constructing unified gauge models including N=1 supergravity one needs to couple an arbitrary number of left-handed chiral multiplets and simultaneously a gauge multiplet to N=1 Supergravity. A brief outline of this coupling procedure is given in Appendix A. The basic steps in the construction consist of first forming the most general F and D multiplets from the N=1 matter multiplets and coupling these to Supergravity by rules of eqs. (2.1) and (2.2). The Lagrangian at this stage contains the auxiliary fields from both the matter and the Supergravity sectors. Next one eliminates these auxiliary fields using the field equations followed by Weyl scale transformations to ensure that the Einstein and the Rarita Schwinger terms in the Lagrangian have normalized kinetic energies(44-47).

After elimination of the auxiliary fields and the Weyl scalings, the Lagrangian depends on two arbitrary functions $f_{\alpha\beta}(Z^a)$ and $G(Z^a, Z_a)$. The meaning of $f_{\alpha\beta}(Z^a)$ can be understood by examining the Yang-Mills sector of the theory(45).

$$e^{-1}L(F_{\alpha\beta}) = \frac{1}{2}f_{\alpha\beta}(-\frac{1}{4}F^\alpha_{\ \mu\nu}F^{\mu\nu\beta} - \frac{1}{2}\bar{\lambda}^\alpha \not{D}\lambda^\beta + \frac{1}{2}D^\alpha D^\beta$$
$$+ \frac{1}{4}F^\alpha_{\ \mu\nu}\tilde{F}^{\mu\nu\beta} - \frac{1}{2}D_\mu(\bar{\lambda}^\alpha\gamma^\mu\lambda_R^{\ \beta})) + \mathrm{h.c.} \qquad (2.3a)$$

where

$$F_{\mu\nu}^{\ \ \alpha} = \partial_\mu V_\nu^{\ \alpha} - \partial_\nu V^\alpha_\mu + e_\alpha f^{\alpha\beta\gamma}V^\beta_{\ \mu}V_\nu^{\ \gamma} \qquad (2.3b)$$

$$D_\mu\lambda^\alpha = \partial_\mu\lambda^\alpha + e_\alpha f^{\alpha\beta\gamma}V^\beta_{\ \mu}\lambda^\gamma + \frac{1}{2}\omega_{\mu rs}\sigma^{rs}\lambda^\alpha + i\frac{\kappa}{2}A_\mu\gamma_5\lambda^\alpha;$$
$$\sigma^{rs} \equiv \frac{1}{4}[\gamma^r, \gamma^s] \qquad (2.3c)$$

The Lagrangian of Eq. (2.3) is invariant under the following supersymmetry transformations

$$\delta_s V^\alpha_\mu = \bar{\varepsilon}\gamma_\mu \lambda^\alpha; \qquad \delta_s D^\alpha = -i\bar{\varepsilon}\gamma_5 \hat{\not{D}}\lambda^\alpha$$

$$\delta_s \lambda^\alpha = -\sigma^{\mu\nu}\bar{\varepsilon}\hat{F}^\alpha_{\mu\nu} - i\gamma_5 \varepsilon D^\alpha$$

where

$$\hat{D}_\mu \lambda^\alpha = D_\mu \lambda^\alpha + \frac{\kappa}{2}(\sigma^{\mu\nu}\hat{F}_{\mu\nu}{}^\alpha + i\gamma_5 D^\alpha)\psi_\mu$$

$$\hat{F}_{\mu\nu}{}^\alpha = F_{\mu\nu}{}^\alpha - \frac{\kappa}{2}(\bar{\psi}_\mu \gamma_\nu \lambda^\alpha - \bar{\psi}_\nu \gamma_\mu \lambda^\alpha)$$

In all of our discussions we shall assume the minimality condition(2,44)

$$f_{\alpha\beta}(Z) = \delta_{\alpha\beta} \tag{2.4}$$

The arbitrary function $G(Z^a, Z_a)$ has the definition

$$G(Z^a, Z_a) = -\frac{\kappa^2}{2} d(Z, Z^\dagger) - \ln\left(\frac{\kappa^6}{4}|g(Z^a)|^2\right) \tag{2.5}$$

where d is related to ϕ of Eq. (A.13) by

$$d = -(6/\kappa^2)\ln\left(-\frac{\kappa^2}{3}\phi\right) \tag{2.6}$$

The bose part of the Lagrangian takes the following form on using the minimality condition of Eq. (2.4):

$$\begin{aligned} L_B = &-(e/2\kappa^2)R(e,\omega) + (e/\kappa^2)G,^a{}_b\, \mathcal{D}_\mu Z_a \mathcal{D}^\mu Z^b \\ &+ (e/\kappa^4)\exp(-G)[3 + (G^{-1})^a{}_b G,_a G,^b] \\ &- \frac{1}{4} e\, F^\alpha_{\mu\nu} F^{\mu\nu\alpha} - \frac{e}{8\kappa^4} | e_\alpha G,_a (T^\alpha Z)^a|^2 \end{aligned} \tag{2.7}$$

From Eq. (2.7) we note that $G,^a{}_b$ acts as a metric in a Kähler manifold which is defined by co-ordinates Z^a, Z_b while $d(Z,Z^\dagger)$ has the meaning of a potential in the Kähler manifold(49). A flat Kähler manifold corresponds to

$$G,^a{}_b = \delta^a{}_b \tag{2.8}$$

The choice of Eq. (2.8) may, however, be too restrictive since gravitational loop corrections could modify this condition (or the equivalent condition $d = Z_a Z^a$).

A more reasonable condition on d is to assume it to be a function of $Z_a Z^a$(11,14). Here the potential possesses a U(n) symmetry among the n-chiral fields which is consistent with the fact that to a good approximation, the gravitational loop corrections respect the U(n) symmetry(11,13). The fermionic part of the Lagrangian takes the following form(44)

$$
\begin{aligned}
L_F = & -\tfrac{1}{2}\,\varepsilon^{\mu\nu\rho\sigma}\bar\psi_\mu\gamma_5\gamma_\nu D_\rho(\omega(e,\psi))\psi_\sigma + \frac{e}{\kappa^2}\,G,^a{}_b\,\overline{\chi^a}\,\overleftrightarrow{\mathcal{D}}\chi^b - \frac{e}{2}\,\bar\lambda^\alpha \not{\mathcal{D}}\lambda^\alpha \\
& - e\,\frac{\kappa}{2}\bar\psi_\mu\sigma^{\kappa\lambda}\gamma^\mu\lambda^\alpha F_{\kappa\lambda}{}^\alpha + \frac{1}{8}\,\varepsilon^{\mu\nu\rho\sigma}\bar\psi_\mu\gamma_\nu\psi_\rho\,(G,_a\mathcal{D}_\sigma Z^a - G,^a\mathcal{D}_\sigma Z_a) \\
& - \frac{e}{\kappa}\,G,^a{}_b(\mathcal{D}_\nu Z^b\,\overline{\chi^a}\gamma^\mu\gamma^\nu\psi_\mu + \bar\psi_\mu\gamma^\nu\gamma^\mu\chi^b\mathcal{D}_\nu Z_a) \\
& + \frac{e}{\kappa^2}\,\overline{\chi^a}\gamma^\mu\chi^b((G,^a{}_{bc} - \frac{\kappa^2}{4}G,^a{}_b G,_c)\mathcal{D}_\mu Z^c - (G,^{ac}{}_b - \frac{\kappa^2}{4}G,^a{}_b G,^c)\mathcal{D}_\mu Z_c) \\
& + \frac{e}{8}\,\bar\lambda^\alpha\gamma^\mu\gamma_5\lambda^\alpha(G,_a\mathcal{D}_\mu Z^a - G,^a\mathcal{D}_\mu Z_a) \\
& + \frac{e}{\kappa^3}\exp(-\tfrac{G}{2})[(G,_{ab} - G_a G_b - (G^{-1})^c{}_d G,_c G,^d{}_{ab})\bar\chi_a\chi^b \\
& - \kappa G,_a\bar\psi_\mu\gamma^\mu\chi^a + \kappa^2\bar\psi_\mu\sigma^{\mu\nu}\psi_{\nu R} + h\cdot c] \\
& + ie\,\frac{g_\alpha}{4\kappa}\,(G,_a(T^\alpha Z)^a\bar\psi_\mu\gamma_5\gamma^\mu\lambda^\alpha + ei\,\frac{g_\alpha}{\kappa^2}\,G,^a{}_b((T^\alpha Z)_a\bar\lambda^\alpha\chi^b - \overline{\chi^a}\lambda^\alpha(T^\alpha Z)^b) \\
& - \frac{e}{\kappa^2}\,(G,_{ab}{}^{cd} - (G^{-1})^e{}_f G,_{ab}{}^e G,^{cd}{}_f + \tfrac{1}{2}G,^c{}_a G,^d{}_b)\bar\chi_a\chi^b\overline{\chi^c}\chi_d \\
& + e\,\frac{\kappa^2}{4}\,\bar\lambda^\alpha\gamma^\mu\sigma^{\nu\rho}\psi_\mu\bar\psi_\nu\gamma_\rho\lambda^\alpha + \frac{e}{8}\,G,^a{}_b\overline{\chi^a}\gamma^\mu\chi^b\bar\lambda^\alpha\gamma_\mu\gamma_5\lambda^\alpha \\
& + 3e\,\frac{\kappa^2}{64}\,\bar\lambda^\alpha\gamma^\mu\gamma_5\lambda^\alpha\bar\lambda^\beta\gamma_\mu\gamma_5\lambda^\beta + \frac{1}{4}\,G,^a{}_b\overline{\chi^a}\gamma_\sigma\chi^b(\varepsilon^{\mu\nu\rho\sigma}\bar\psi_\mu\gamma_\nu\psi_\rho - e\bar\psi_\mu\gamma_5\gamma^\sigma\psi^\mu)
\end{aligned}
\tag{2.9}
$$

As for the case for Eq. (2.7) the Lagrangian of Eq. (2.9) also depends only on the arbitrary function G. Since all the auxiliary fields have been completely eliminated, the total Lagrangian $L = L_B + L_F$ is invariant only on the mass shell under supersymmetry transformations which (after elimination of the auxiliary fields) are as follows:

$$\delta_S V_\mu{}^\alpha = -\bar\varepsilon\gamma_\mu\lambda^\alpha$$

$$\delta_S\lambda^\alpha = -\tfrac{1}{2}\,(G,_a\bar\varepsilon\chi^a - G,^a\overline{\chi^a}\varepsilon)\lambda^\alpha - \sigma^{\mu\nu}\,\varepsilon\hat F_{\mu\nu}{}^\alpha - \frac{ie_\alpha}{2\kappa^2}(\gamma_5\varepsilon)(G,_a(T^\alpha Z)^a)$$

$$\delta_S Z^a = 2\bar\varepsilon\chi^a$$

$$\delta_s \chi^a = \gamma^\mu \varepsilon_R \hat{\mathcal{D}}_\mu Z^a - \frac{1}{2}(G,_b \bar{\varepsilon}\chi^b - G,^b \bar{\chi}^b \varepsilon)\chi^a$$

$$+ (G^{-1})^a{}_b G,^b{}_{cd} \overline{\chi_c \chi^d} \varepsilon_L - \kappa^{-1}(G^{-1})^a{}_b G,^b \exp(-\frac{G}{2})\varepsilon_L$$

$$\delta_s e_\mu{}^r = \kappa \bar{\varepsilon}\gamma^r \psi_\mu$$

$$\delta_s \psi_{\mu L} = 2\kappa^{-1} D_\mu(e,\psi_\nu)\varepsilon_L + \kappa^{-2}\exp(-\frac{G}{2})\gamma_\mu \varepsilon_R + \frac{1}{2}(G,_a \bar{\varepsilon}\chi^a - G,^a \overline{\chi^a}\varepsilon)\psi_{\mu L}$$

$$+ \kappa^{-1}(\sigma_{\mu\nu}\varepsilon_L) G,^a{}_b \overline{\chi^a}\gamma^\nu \chi^b - \frac{\kappa}{8}(\delta_\mu{}^\nu + \gamma^\nu \gamma_\mu)\varepsilon_L \bar{\lambda}^\alpha \gamma_\nu \gamma_5 \gamma^\alpha$$

$$- \frac{\kappa^{-1}}{2}(G,_a \mathcal{D}_\mu Z^a - G,^a \mathcal{D}_\mu Z_a)\varepsilon_L \tag{2.10}$$

The potential V of the scalar fields given by Eq. (2.7) is

$$e^{-1}V = -\frac{1}{\kappa^4}\, e^{-G}[3 + (G^{-1})^a{}_b G,_a G,^b] + \frac{e_\alpha{}^2}{8\kappa^4}(G,_a (T^\alpha Z)^a)^2 \tag{2.11}$$

It is sometimes convenient to express the potential in terms of the Kähler potential $d(Z,Z^\dagger)$ so that

$$e^{-1}V = \frac{1}{2}\exp(\frac{\kappa^2}{2} d)[(d^{-1})^a{}_b\, G_a G^b - \frac{3}{2}\kappa^2|g|^2] + \frac{e_\alpha{}^2}{32}\,|(d,_a (T^\alpha Z)^a)|^2 \tag{2.12}$$

where $(d^{-1})^a{}_b$ is the inverse of the matrix $(d),_a{}^b$ and G_a is given by

$$G_a \equiv g_{,a} + \frac{\kappa^2}{2} d,_a g \tag{2.13}$$

III. GRAVITY INDUCED SYMMETRY BREAKING

The extrema equations arising from Eq. (2.12) have the form

$$(\frac{\partial G_b}{\partial Z_a} + \frac{\kappa^2}{2}\; Z^a G_b)\, G^b - \kappa^2 g\; G^a = 0 \tag{3.1}$$

for a flat Kähler manifold. For the real (PC conserving) manifold of VEVs eq. (3.1) takes the form

$$T_{ab}\; G_b = 0 \tag{3.2}$$

where

$$T_{ab} = \frac{\partial^2 g}{\partial Z^a \partial Z^b} + \frac{\kappa^2}{2}\left(Z_a \frac{\partial g}{\partial Z^b} + Z_b \frac{\partial g}{\partial Z^a}\right) + \frac{\kappa^4}{4} Z_a Z_b g - \kappa^2 \delta_{ab} g \qquad (3.3)$$

When solutions to Eq. (3.2) are of the form that G_b vanish in all channels, one has the case of preserved supersymmetry but the possibility of a broken gauge group. When at least for one channel $G_b \neq 0$, there is a spontaneous breakdown of supersymmetry. To exhibit this phenomenon we assume a chiral multiplet (Z, χ_L, h) (which is a singlet under the gauge group G) and a superpotential $g_2(Z)$ of the form.

$$g_2(Z) = m^2 k^{-1} f_2(kZ) \qquad (3.4)$$

where

$$f_2(kZ) = f_2^{\,o} + kZf_2^{\,1} + \ldots \qquad (3.5)$$

$f_2(kZ)$ is a dimensionless function of the dimensionless quantity κZ. m appearing in Eq. (3.4) is a new mass parameter. Assuming that non-trivial solutions to Eq. (3.2) exist with $G_2 \neq 0$, one has

$$\langle Z\rangle \sim O(k^{-1}); \quad \langle g_2\rangle \sim O(k^{-1}m^2) \qquad (3.6)$$

Consequences of Eq. (3.6) are understood by considering the vacuum expectation values of $\delta\chi^a$ and $\delta\psi_{\mu L}$ in Eq. (2.10). For a flat Kähler manifold Eq. (2.10) gives

$$\langle\delta\chi^a\rangle_o = -\exp\left(\frac{\kappa^2}{4} Z_o^{\,a} Z_{oa}\right) G_a(Z_o)\varepsilon_L$$

$$\langle\delta\psi_{\mu L}\rangle_o = \frac{\kappa}{2}\exp\left(\frac{\kappa^2}{4} Z_o^{\,a} Z_{oa}\right) g(Z_o)\gamma_\mu \varepsilon_R + 2\kappa^{-1}\partial_\mu \varepsilon_L \qquad (3.7)$$

From Eq. (3.7) one finds that the non-trivial solutions to Eq. (3.2) with $G_Z \neq 0$ break supersymmetry. This breaking leads to a mass growth for some of the fields in the theory. We analyse next the mass matrices for the fields Z^a, χ^a, $V_\mu^{\ \alpha}$, λ^α and ψ_μ.

For a flat Kähler manifold the mass terms for the fields χ^a and ψ_μ are given by (see Eq. (2.9))

$$L_F^{\,m} = \frac{e}{\kappa^3}\exp\left(-\frac{G}{2}\right)\left[(G,_{ab} - G,_a G,_b)\overline{\chi_a}\chi^b - \kappa G,_a \bar{\psi}_\mu \gamma^\mu \chi^a + \kappa^2 \bar{\psi}_\mu \sigma^{\mu\nu}\psi_{\nu R} + h\cdot c\right]$$

$$- ei\,\frac{e_\alpha}{8}\,\kappa (Z_a (T^\alpha Z)^a)\bar{\psi}_\mu \gamma_5 \gamma^\mu \lambda^\alpha - ei\,\frac{e_\alpha}{2}\left((T^\alpha Z)_a \bar{\lambda}^\alpha \chi^a - \overline{\chi^a}\lambda^\alpha (T^\alpha Z)^a\right) \qquad (3.8)$$

It is convenient to introduce the matrices

$$m_{ab} = \frac{2}{\kappa^3} \exp(-\frac{G}{2})(G,_{ab} - \frac{1}{3} G,_a G,_b)$$

$$m^{\alpha}a = i \frac{g_{\alpha}}{2} Z_b (T^{\alpha})^b{}_a$$

$$m^{\alpha\beta} = - \exp(-\frac{G}{2}) \frac{e_{\alpha}^2 \kappa^3}{24} (z_a(T^{\alpha}Z)^a)(Z_b(T^{\beta}Z)^b) \qquad (3.9)$$

In terms of these the mass terms of Eq. (3.8) take the form

$$L_F{}^m = \frac{e}{\kappa} \exp(-\frac{G}{2})(\overline{\psi'_{\mu}}\sigma^{\mu\nu}\psi'_{\nu L}) - \frac{e}{2} (m_{ab}\overline{\chi_a}\chi^b) - em^{\alpha}a\overline{\lambda}^{\alpha}\chi^a$$

$$- \frac{e}{2} m^{\alpha\beta}\overline{\lambda}_L{}^{\alpha}\lambda_L{}^{\beta} + h \cdot c \qquad (3.10)$$

where ψ'_{μ} is a linear combination of ψ_{μ}, χ^a and λ^{α} and represents the physical gravitino field:

$$\psi'_{\mu R} = \psi_{\mu R} - \frac{\kappa^{-1}}{3} \gamma_{\mu}(G,_a\chi^a + i \frac{e_{\alpha}\kappa^3}{4} \exp(-\frac{G}{2})(Z_a(T^{\alpha}Z)^a)\lambda_L{}^{\alpha}) \qquad (3.11)$$

The gravitino mass is given by Eq. (1.2). For the spin zero modes one may compute the mass matrices for the (real) fields A^a and B^a. These fields will show a mass splitting when supersymmetry is broken. We can expand these mass matrices (in terms of the complex field derivatives of V) as follows:

$$(M^2)^A{}_{ab} = (V,^a{}_b + V,^b{}_a + 2V,_{ab})_o$$

$$(M^2)^B{}_{ab} = (V,^a{}_b + V,^b{}_a - 2V,_{ab})_o \qquad (3.12)$$

$V,^a{}_b$ and $V,_{ab}$ (for a flat Kähler manifold) are given by[44]

$$V,^a{}_b = \frac{2e}{\kappa^4} \exp(-G)[- \frac{1}{2} G,^a G,_b + \frac{1}{\kappa^2} (G,_{bc} - G,_b G,_c)(G,^{ac} - G,^a G,^c)$$

$$- \frac{\kappa^2}{2} \delta^a{}_b + \frac{1}{2} \delta^a{}_b G,_c G,^c] + \frac{e_{\alpha}^2}{32} Z_b \bar{Z}^d (T^{\alpha})^a{}_d \qquad (3.13)$$

and

$$V,_{ab} = \frac{2e}{\kappa^4} \exp(-G)[\frac{1}{2} (G,_{ab} - G,_a G,_b) + \frac{1}{\kappa^2} (G,_{abc} - 3G,_{(a}G,_{bc)}$$

$$+ G,_a G,_b G,_c)G,^c] + \frac{e_{\alpha}^2}{32} Z_c Z_d (T^{\alpha})^c{}_a (T^{\alpha})^d{}_b \qquad (3.14)$$

Using the minimization condition $V,_a = 0$ and the condition $V = 0$ for the vanishing of the cosmological constant one may easily establish the following relations

$$(Z_a(T^\alpha Z)^a) = 0; \; G,_a G,^a = \frac{3}{2}\kappa^2 \text{ and } G,_{ab}G,^b = \frac{\kappa^2}{2} G,_a \tag{3.15}$$

From Eq. (3.15) and the fact that the contributions of the gauge particles are the same as for the global supersymmetry case one can establish the following sum rule[45]: $\sum(-1)^{2J}(2J+1) \;\; M_J{}^2 = 2(N-1)m^2{}_{3/2}$ where N is the number of chiral multiplets.

VI. LOW ENERGY THEORY

Let us represent by Z^A the full set of complex scalar fields appearing in the matter sector. We may classify these fields into two categories: the fields in the super-Higgs sector denoted by Z and the remaining set of fields Z^a. One has then

$$Z^A = (Z^a, Z) \tag{4.1}$$

We shall allow here the possibility that in addition to the Plank mass scale κ^{-1}, the theory contains also a GUT scale M. In GUT theories, heavy fields with the characteristic mass scale M for masses and VEVs are responsible for the breakdown of the gauge group. We denote these fields by Z_i. Another set of heavy fields $Z_i{}'$ may also appear characterized by masses of O(M) but with vanishing VEVs. Finally, there are left a set of light fields which we denote by Z_α. Thus the fields Z^a have the classification

$$Z^a = (\{Z_i\}, \{Z_i{}'\}, \{Z_\alpha\}) \tag{4.2}$$

Next we examine the minimization conditions and obtain the restrictions required to maintain the hierarchy of Eq. (4.2) at the tree level. As noted in the Introduction the proximity of the GUT mass scale M to the Planck mass κ^{-1} introduces in the theory a new possible gauge hierarchy problem. This problem arises as follows. The minimization equations are in general an infinite order expansion in κ, and thus corrections of the type $(\kappa M)M$, $(\kappa M)^2 M$, ..., $(\kappa M)^6 M$ may appear in the VEV equations for the light fields. Since for SUSY theories $M \sim 3\times10^{16}$ GeV, and thus $\kappa M \sim 10^{-2}$ these corrections must be suppressed in order that the classification of Eq. (4.2) make sense. A direct procedure to examine the tree hierarchy problem is to carry out an expansion of the VEVs themselves in a power series expansion in κ and solve the extrema equations also order by order in κ. It can then be shown that for a broad class of models the classification of Eq. (4.2) holds provided the following restrictions are obeyed at the minimum of the effective potential

$$g_{,\alpha i} \sim O(m_s), \quad g_{,\alpha\beta} \sim O(m_s); \quad m_s \equiv \kappa m^2 \tag{4.3}$$

Eqs. (4.3) produce constraints on the construction of models. In particular, they require that certain types of couplings in the superpotential must be eliminated or unnaturally suppressed. As an example, the couplings

$$\lambda Z_i Z_j Z_\alpha, \quad \lambda' Z_i Z_j Z \tag{4.4}$$

must be suppressed as follows

$$\lambda \sim m_s/M, \; \lambda' \sim \kappa m_s \tag{4.5}$$

Once the conditions for the tree gauge hierarchy have been satisfied, one may proceed to examine the content of the low energy theory that describes the effective interactions of the light fields Z_α by eliminating the heavy fields. This may be accomplished by using their extrema equations (14,15)

$$\frac{\partial V}{\partial Z} = 0; \quad \frac{\partial V}{\partial Z_i} = 0 \tag{4.6}$$

where $V(Z_i, Z, Z_\alpha)$ is the effective potential of the full theory. Eqs. (4.6) allow one to determine Z_i and Z in terms of Z_α:

$$Z_i = Z_i[Z_\alpha]; \quad Z = Z[Z_\alpha] \tag{4.7}$$

Elimination of Z_i and Z in terms of Z_α in the $V(Z_i, Z, Z_\alpha)$ gives us the low energy effective potential $U(Z_\alpha)$:

$$U(Z_\alpha) = V(Z_i[Z_\alpha]; Z[Z_\alpha]; Z_\alpha) \tag{4.8}$$

Alternately one may integrate the extrema equations in the light sector after elimination of the heavy fields. This procedure turns out to be simpler than the construction of Eq. (4.8) because the extrema equations in the light contain already the full content of the low energy protection implied by the gauge hierarchy conditions Eqs. (4.3). These extrema equations guarantee that the effective low energy potentials is correctly $O(m_s^4)$ and terms $O(Mm_s^3)$, $O(M^2M_s^2)$ etc. all vanish. A direct integration of the extrema equations gives

$$U(Z_\alpha, Z_\alpha^\dagger) = \frac{1}{2} \exp(\frac{\kappa^2}{2} |Z_0|^2)[\tilde{g}_{1,\alpha} \tilde{g}^\dagger{}_{1,\alpha} + m_1{}^2 Z_\alpha Z^\alpha$$

$$+ (m_2 g'_1 + m_3 Z_\alpha \tilde{g}_{1,\alpha} + \text{h.c.})] + \frac{e_\alpha^2}{32} |(Z_a, (T^\alpha Z)^a|^2 \quad (4.9)$$

where

$$g_1' = \tilde{g}_1(Z_i^{(0)}, Z^\alpha); \; \tilde{g}_1(Z_i, Z_\alpha) = g_1(Z_i, Z_\alpha) - g_1(Z_i, 0) - b \quad (4.10)$$

The mass parameters m_i, m_2 and m_3 appearing in Eq. (4.9) are defined by

$$m_1^2 = \frac{1}{2} m_s^2 [\bar{G}_z^{(0)} \bar{G}_z^{(0)*} - \bar{g}_2^{(0)} \bar{g}_2^{(0)*}] \quad (4.11)$$

$$m_2 = \frac{1}{2} m_s [z^{(0)} \bar{G}_z^{(0)} - 3\bar{g}_2^{(0)}] \quad (4.12)$$

$$m_3 = \frac{1}{2} m_s \bar{g}_2^{(0)}, \; m_1 = |m_3| \quad (4.13)$$

where we have introduced the dimensionless quantities z, $\bar{G}_z$ and $\bar{g}_2$ as follows

$$z = kZ, \; \bar{g}_2 = (k/m^2) g_2, \bar{G}_z = m^{-2} G_z \quad (4.14)$$

Superscript (0) on z, $\bar{g}_2$ and $\bar{G}_z$ denotes the fact that they are the terms $O(\kappa)^0$ in their κ-expansion. The last condition in Eq. (4.13) arises from the additional imposition that the cosmological constant is required to vanish for the potential of Eq. (4.9). The above analysis holds for a flat Kähler manifold obeying a U(n) symmetry. (For a curved Kähler manifold the number of mass parameters in the effective potential is four(14).)

The effective low energy theory is now completely determined. Lagrangian of this theory which possesses a residual gauge group G´ (a subgroup of G) with the corresponding gauge bosons $V_\mu{}^i$ and gauginos λ^i is given by

$$L = -\frac{1}{2} [\partial_\mu Z^a - \frac{i}{2} e_i (T^i)^a{}_b Z^b V_\mu{}^i][\partial_\mu Z_a + \frac{i}{2} e_i Z_b (T^i)^b{}_a V_\mu{}^i]$$

$$+ i \overline{X^a} \gamma^\mu (\partial_\mu \delta_b{}^a - \frac{i}{2} e_i V_\mu{}^i (T^i)^a{}_b) X^b$$

$$+ \frac{i}{2} \overline{\lambda^i} \gamma^\mu (\partial_\mu \delta^{ik} + e_i f^{ijk} V_\mu{}^j) \lambda^k - \frac{1}{2} \overline{\lambda^i} m_i \lambda^i$$

$$- \frac{1}{2} \exp(\frac{k^2}{2} \langle Z \rangle^2)(g_{,ab} \overline{X^{aC}} X^b + \text{h.c.})$$

$$+ (\frac{i}{2} e_i \overline{X^a} \lambda^i (T^i)^a{}_b Z^b + \text{h.c.}) - U + L(F_{\alpha\beta}) \quad (4.15)$$

where U in Eq. (4.15) is given by Eq. (4.9). $\tilde{m}_i$ in Eq. (4.15) are the radiative gaugino masses determined by Eq. (1.5).

The last item we wish to discuss in this section is the question of gauge hierarchy beyond the tree level and specifically the one loop gauge hierarchy. In general the effective potential possesses a loop expansion

$$V(Z^a) = V^{(0)}(Z^a) + V^{(1)}(Z^a) + \ldots \tag{4.16}$$

and a similar loop expansion holds for the extrema equations

$$V^{(0)}{}_{,a} + V^{(1)}{}_{,a} = 0 \tag{4.17}$$

The VEVs $\phi_a = \langle Z^a \rangle$ also possess a loop expansion

$$\phi_a = \phi_a{}^{(0)} + \phi_a{}^{(1)} + \ldots \tag{4.18}$$

To zeroeth loop order Eq. (4.17) gives

$$V^{(0)}{}_{,a}(\phi^{(0)}) = 0 \tag{4.19}$$

and to first loop order Eq. (4.17) gives

$$V^{(0)}{}_{,ab}(\phi^{(0)})\phi_b{}^{(1)} + V^{(1)}{}_{,a}(\phi^{(0)}) = 0 \tag{4.20}$$

The tree heirarchy conditions (4.3) constrain the mass matrix $V^{(0)}{}_{,ab}$ as follows:

$$V^{(0)}{}_{,ij} \sim O(M^2),\ V^{(0)}{}_{,\alpha i} \sim O(m_s{}^2),\ V^{(0)}{}_{,\alpha\beta} \sim O(m_s{}^2) \tag{4.21}$$

Eqs. (4.20) and (4.21) show that the one loop VEV hierarchy is maintained provided

$$V^{(1)}{}_{,\alpha}(\phi^{(0)}) \sim O(m_s{}^3); \quad V^{(1)}{}_{,i}(\phi^{o}) \sim O(M^2 m_s) \tag{4.22}$$

One would like of course to achieve restrictions directly on the superpotential so as to guarantee the one loop stability. Deduction of these restrictions are highly non-trivial. However, it is possible to exhibit a set of sufficient conditions by a detailed analysis of the one loop potential

$$V^{(1)} = \frac{1}{64\pi^2} \sum_J (-1)^{2J}(2J+1)\mathrm{Tr}M_J{}^4 \ln \frac{M_J{}^2}{\mu^2} \tag{4.23}$$

where $M_J{}^2$ represents the mass matrix for particles with spin J. The requirements Eq. (4.21) lead to the following sufficiency conditions

$$\tilde{g}_{,ik}\tilde{g}_{,k\alpha} \sim O(m_s^{\,2}) \tag{4.24}$$

$$\tilde{g}_{,ij}\tilde{g}_{,j\beta\alpha} \sim O(m_s) \tag{4.25}$$

$$\tilde{g}_{,ijk}\tilde{g}_{,k\alpha} \sim O(m_s^{\,2}/M) \tag{4.26}$$

$$\tilde{g}_{,ij\alpha} \sim O(m_s/M) \tag{4.27}$$

An alternative route for achieving the one loop hierarchy is to have $V^{(1)}{}_{,\alpha}$ vanish by choosing the representation content properly as in the missing partner models. An example of such a model is given by a superpotential of the form

$$g_1(Z^a) = \frac{1}{3}\lambda_1\Sigma^3 + \frac{1}{2}M\Sigma^2 + \lambda_2\,\Theta'\,\Sigma\Theta + \lambda_3\Theta\Sigma H$$

$$+ \lambda'_3\Theta'\Sigma H' + M\Theta'\Theta + g_{yukawa} \tag{4.28}$$

where $\Sigma = 75$, $\Theta = 50$, $\Theta' = \overline{50}$, $H = 5$ and $H' = \overline{5}$. Here the absence of a partner for the Higgs doublet in $\Theta\Sigma H$ to grow a mass leads to $V^{(1)}{}_{,\alpha} = 0$ and to a stable one loop gauge heirarchy.

V. GENERATION OF WEAK SCALE

In this section we shall discuss how the weak scale m_W can be produced through Supergravity induced effects. The basic idea here is to exhibit the breaking of SU(2)xU(1) gauge invariance by the soft terms induced by Supergravity in Eq. (4.9). This phenomenon can occur at the tree level or may be induced through renormalization group effects. We next discuss these models.

(i) TREE MODELS FOR SU(2)XU(1) BREAKING

We consider the supergravity GUT theory based on the gauge group SU(5) of Ref. (2). Here the superpotential g is assumed to have the form

$$g = g_1(Z^a) + g_2(Z) \tag{5.1}$$

where $g_1(Z^a)$ sector of the theory consists of a $\underline{24}$ $(\Sigma^x{}_y)$, $5(H^x)$, $\overline{5}(H_x{}')$ and a singlet (U) of Higgs field and $\overline{5}(M'_x)$ and $10(M^{xy})$ of quark-lepton fields. g_1 is chosen to have the form[(2)]

$$g_1 = \lambda_1(\tfrac{1}{3}\,\mathrm{Tr}\Sigma^3 + \tfrac{M}{2}\,\mathrm{Tr}\Sigma^2) + \lambda_2 H'_x(\Sigma^x{}_y + 3M'\delta^x{}_y)H'^y + \lambda_3 U H'_x H^x$$

$$+ \epsilon_{uvwxy}\, H^u M^{vw} f_1 M^{xy} + H'_x M^{xy} f_2 M'_y + B_1 \tag{5.2}$$

where f_1, f_2 are matrices in the generation space. spontaneous breaking produce three inequivalent vacua which possess the symmetries (i) SU(5), (ii) SU(4)xU(1) and (iii) SU(3)xSU(2)xU(1). For the physically interesting vacua of (iii) one can achieve light Higgs doublets by requiring $M' = M$. The Higgs fields entering the low energy theory are then H^α, H'_α (α = 4,5) and the singlet field U. On using Eq. (4.9) one finds a breaking of SU(2)xU(1) for a range of values of the parameters appearing in Eq. (5.2). However, the singlet field couplings $\lambda_3 U H' H$ in Eq. (5.2) upset the one loop gauge heirarchy. A way to circumvent the loop heirarchy problem is to use the missing partner technique discussed in Sec. IV. However, the model of Eq. (4.28) itself is insufficient since it does not break SU(2)xU(1). We have to introduce a singlet field U for this purpose and then one must double the $50(\overline{50})$ and $5(\overline{5})$ multiplets to maintain the one loop stability. The resultant superpotential has the form[9]

$$g_1 = \tfrac{1}{3}\lambda_1\Sigma^3 + \tfrac{1}{2}M\Sigma^2 + [\lambda_2\Theta'\Sigma\Theta + \lambda_3\Theta\Sigma H + \lambda'_3\Theta'\Sigma H' + M'\Theta'\Theta]$$

$$+ [\tilde{\lambda}_2\tilde{\Theta}'\Sigma\tilde{\Theta}' + \tilde{\lambda}_3\tilde{\Theta}\Sigma\tilde{H} + \tilde{\lambda}_3\tilde{\Theta}'\Sigma\tilde{H}' + \tilde{M}'\tilde{\Theta}'\Theta]$$

$$+ [\lambda_4 U H \tilde{H}' + \tilde{\lambda}_4 U H' \tilde{H} + \lambda_5 U^3] + g_{yukawa} \tag{5.3}$$

(ii) R.G. MODELS FOR SU(2)xU(1) BREAKING

In the tree breaking models one needs the singlet field U to break the SU(2)xU(1) invariance. Next we discuss models which do not involve U and can break SU(2)xU(1) through radiative corrections. The simplest model thus involves only two Higgs doublets, plus quarks and leptons and the general form of the superpotential here is

$$\mu H'_\alpha H^\alpha + g_{yukawa} \tag{5.4}$$

In a class of tree models (such as Eq. (5.2)) the $\mu H' H$ arises naturally at the tree level through $O(m_g)$ corrections (shifts) to the heavy field VEVs. In other class of GUT models (such as Eq. (5.3)) a term of this type is absent at the tree level, but radiative corrections generate such a term with a size typically $O(\alpha m_g)$. We shall assume $O(\alpha m_g)$ to be the characteristic size of μ appearing in Eq. (5.1).

There exist at present a variety of ways in which renormalization group (RG) may generate large corrections sufficiently strong to generate an SU(2)xU(1) breaking (when the underlying tree theory does not afford such a breaking). Large RG corrections can arise from the postulate of a heavy top quark, from large gaugino masses or from enhancement from other parameters that arise in the original tree potential. Here we shall discuss mainly the consequences of a heavy top quark on symmetry breaking.

For simplicity we retain in the following analysis only the Yukawa couplings of the top quark with the Higgs. The effective low energy potential then is

$$U = [m_H^2 \bar{H}^\alpha H^\alpha + m_{H'}^2 \bar{H}'_\alpha H'_\alpha + \bar{m}_3 m_g (H'_\alpha H^\alpha + \bar{H}'_\alpha \bar{H}^\alpha)]$$

$$+ \frac{1}{8} e_1^2 (\bar{H}^\alpha H^\alpha - \bar{H}'_\alpha H'_\alpha)^2 + \frac{1}{8} e_2^2 (\bar{H}\tau_i H - H'\tau_i \bar{H}')^2 + U_t \tag{5.5}$$

$$U_t = h_t^2 |\bar{\tilde{t}}_R \tilde{t}_L|^2 + h_t^2 |\bar{\tilde{t}}_R H^2|^2 + m_g A_t h_t (\tilde{t}_R H^2 \tilde{t}_L + h.c.)$$

$$+ m_{\tilde{t}_R}^2 \bar{\tilde{t}}_R \tilde{t}_R + m_{\tilde{t}_L}^2 \bar{\tilde{t}}_L \tilde{t}_L \tag{5.6}$$

In Eq. (5.5) e_i (i = 1,2,3) are the U(1), SU(2) and SU(3) gauge coupling constants, $\tilde{t}_L$ is the scalar partner of t_L etc. and $\bar{H}^\alpha \equiv H^{\alpha\dagger}$ etc. The parameter A_t entering Eq. (5.6) is determined by the choice of the super-Higgs potential (e.g. for the Polony model $A_t = 3 - \sqrt{3}$).

In Supergravity theories the renormalization group analysis possesses the remarkable feature that unlike the global supersymmetry case the boundary conditions on the RG equations are determined in Supergravity theories. For example, the quantities $m_H^2(\mu)$, $m_{H'}^2(\mu)$ etc. take on their tree values (determined by the super-Higgs effect) at the GUT scale $\mu = M$:

$$m_H^2(M) = m_{H'}^2(M) = m_{\tilde{t}_L}^2(M) = m^2_{\tilde{t}_R}(M) = m_g^2 \tag{5.7}$$

The leading order terms in the RG equations for the quantities appearing in Eqs. (5.5) and (5.6) are[8]

$$\frac{d\psi}{dt} = \frac{\alpha_t}{2\pi} M\psi + \frac{A_t^2 \alpha_t}{2\pi} m_g^2 \psi_1 - \frac{8}{3\pi} \alpha_3 \tilde{m}_3^2 \psi_1' \tag{5.8}$$

$$\frac{d\alpha_t}{dt} = \frac{3}{\pi} \alpha_t^2 - \frac{\alpha_t}{\pi} (\frac{8}{3} \alpha_3 + \frac{3}{2} \alpha_2 + \frac{13}{18} \alpha_1) \tag{5.9}$$

$$\frac{d}{dt} A_t = \frac{3}{\pi} \alpha_t A_t - \frac{8}{3\pi} \alpha_3 \tilde{m}_3 / m_g \tag{5.10}$$

where $\alpha_i = e_i^2/4\pi$ $(i = 1,2,3)$, $\alpha_t = h_t^2/4\pi$, $\psi = (m_H^2, m_{\tilde{t}_R}^2, m_{\tilde{t}_L}^2)$, ψ_1, ψ_1' and the matrix M are given by

$$\psi_1 = \begin{pmatrix} 3 \\ 2 \\ 1 \end{pmatrix} , \quad \psi_1' = \begin{pmatrix} 0 \\ 1 \\ 1 \end{pmatrix} , \quad M = \begin{pmatrix} 3 & 3 & 3 \\ 2 & 2 & 2 \\ 1 & 1 & 1 \end{pmatrix} \tag{5.11}$$

Neglecting the gauge couplings and the gaugino masses (we shall take account of these effects later) one finds[44]

$$m_H^2 = \frac{3}{2} \frac{m_g^2}{1-\xi} [1 + \frac{A_o^2}{3} \frac{\xi}{1-\xi}] - \frac{1}{2} m_g^2 \tag{5.12}$$

$$m_{\tilde{t}_R}^2 = \frac{m_g^2}{1-\xi} [1 + \frac{A_o^2}{3} \frac{\xi}{1-\xi}] \tag{5.13}$$

$$m_{\tilde{t}_L}^2 = \frac{1}{2} \frac{m_g^2}{1-\xi} [1 + \frac{A_o^2}{3} \frac{\xi}{1-\xi}] + \frac{1}{2} m_g^2 \tag{5.14}$$

$$\xi = 3\alpha_o t/\pi \; ; \; t \equiv \ln(\mu/M) \tag{5.15}$$

SU(2)xU(1) breaking occurs when at a mass scale μ_o, $m_H^2(\mu_o) < 0$ and the Higgs fields H^2 and H'_2 gain VEVs. From Eq. (5.12) one finds that as μ moves down from the GUT scale $\mu = M$ (or $\xi = 0$) to low mass scales $\mu \to 0$ (or $\xi \to -\infty$), $m_H^2(\mu)$ decreases from its value of m_g^2 at the GUT scale to a value $-\frac{1}{2} m_g^2$ as $\xi \to -\infty$. Thus m_H^2 turns negative at some intermediate ξ. Suppose then that for some $\xi = \xi_o$ the effective potential $V(\xi)$ is minimized such that $m_H^2 = -\frac{1}{2} M_Z^2$ and

$$B\xi_o^2 + \xi_o(3 - 2B - A_o^2) + B - 3 = 0$$

$$B \equiv 1 - \frac{M_Z^2}{m_g^2} \tag{5.16}$$

The desired root of Eq. (5.16) is $\xi_o < 0$. The RG equations for the Yukawa couplings gives

$$h_t^2(\xi_o) = \frac{4\pi^2}{3} \frac{\xi_o}{(1-\xi_o)t_o} \; ; \; t_o = \ln(\mu_o/M) \tag{5.17}$$

From Eqs. (5.16) and (5.17) and the definition $m_t = vh_t(\xi_o)$ (where v = 177 GeV) one finds[44]

$$\frac{m_t^2}{v^2} = \frac{4\pi^2}{3}\frac{1}{(-t_o)}\left[1 - \frac{2B}{\{(A_o^2 + 2B - 3)^2 + 4B(3-B)\}^{\frac{1}{2}} + 3 - A_o^2}\right] \quad (5.18)$$

For $\mu_o = 10^2$ GeV, $M = 10^{16}$ GeV, $A_o = 3 - \sqrt{3}$ and $B = 0$ ($m_g = M_Z$) one has $m_t = 115$ GeV. With all other parameters the same but with $B = 1$ ($m_g = \infty$), one has $m_t = 80$ GeV. Eq. (5.18) shows that m_t^2 is symmetric under reflections in A_o with a maximum at $A_o = 0$. This symmetry is lost when gaugino masses are included. The inclusion of the gaugino masses tends to help the breaking of SU(2)xU(1) while the inclusion of the gauge couplings tends to inhibit the breaking.

VI. LOW ENERGY PARTICLE SPECTRUM

We shall parameterize the low energy theory so as to encompass all models. We follow closely the analysis of Ref. (27) and assume a superpotential of the form

$$g = \mu_o H_\alpha{}' H^\alpha + \lambda' U H'_\alpha H^\alpha - \frac{1}{6}\lambda'' U^3 + g_M$$

$$g_M = (f_1)_{ij}\tilde{u}'_{Ii}\tilde{q}_j^{I\alpha}H^\beta\varepsilon_{\alpha\beta} + (f_2)_{ij}(\tilde{d}'_{Ii}\tilde{q}_j^{I\alpha} + \tilde{e}'_i\tilde{\ell}_j^{\alpha})H_\alpha{}' \quad (6.1)$$

where $\varepsilon_{\alpha\beta} = -\varepsilon_{\beta\alpha}$ with $\varepsilon_{12} = 1$ and

$$\tilde{q}_i^{I\alpha} = \begin{pmatrix} \tilde{u}^I{}_{iL} \\ \tilde{d}^I{}_{iL} \end{pmatrix} \quad , \quad \tilde{\ell}_i^{\alpha} = \begin{pmatrix} \tilde{\nu}_{iL} \\ \tilde{e}_{iL} \end{pmatrix} \quad (6.2)$$

In Eqs. (6.1) and (6.2), I is the SU(3) color index, i is the generation index while the primed quantities are SU(2) singlets: $\tilde{e}'_i = e_{iR}^{\dagger}$ etc. After spontaneous breaking of SU(2)xU(1), the neutral components of the Higgs doublets develop VEVs and it is found useful to introduce the parametrization

$$\sin\alpha = w/(v^2+w^2)^{1/2}, \quad \cos\alpha = v/(v^2+w^2)^{1/2}, \quad -\pi<\alpha\leq\pi \quad (6.3)$$

$$v \equiv \langle H^2\rangle; \quad w \equiv \langle H_2'\rangle \quad (6.4)$$

One may thus parametrize all low energy models by the constants $\mu \equiv \mu_o + \lambda' \langle U\rangle$, λ', λ'' and α. T.B. models are characterized by $\alpha = 45°$ and this may be modified somewhat by loop corrections, i.e., $\alpha = 40° - 50°$. For R.G. models, the singlet is absent so that $\lambda' \equiv 0 \equiv \mu''$ and both α and μ are small ($\alpha \gtrsim 10° - 25°$ and $\mu \sim \alpha m_g$). However, we shall carry out here the general analysis. First we examine the mass spectra that result in this formulation.

(i) Spin 1 States

Here one has the massless color gluons and the electro-weak bosons consisting of the photon and $W^{\pm}$ and Z^0 bosons with masses

$$M_W = \frac{e_2}{2}(v^2 + w^2)^{1/2}, \quad M_Z = M_W/\cos\Theta_W; \quad e_2 = e/\sin\Theta_W \tag{6.5}$$

(ii) Charged Spin 1/2 States

For the charged spin 1/2 states the mass matrix has the form

$$M = \nu_+ + \nu_-\tau_3 + \frac{1}{2}(\mu + \tilde{m}_2)\tau_1 + i\tau_2\gamma_5\frac{1}{2}(\mu - \tilde{m}_2) \tag{6.6}$$

$$\sqrt{2}\,\nu_{\pm} = M_W(\cos\alpha \pm \sin\alpha) \tag{6.7}$$

$$\tilde{m}_2 = 3\tilde{m}_\gamma/(8\sin^2\Theta_W) \simeq 1.7\,\tilde{m}_\gamma \tag{6.8}$$

in the two component space labelled by

$$\psi = (\lambda_R + i\tilde{H}^1\ ,\ \lambda_L + i\tilde{H}_1{}^{\prime C}) \tag{6.9}$$

and τ_a are the Pauli matrices in this space. Diagonalization of Eq. (6.6) gives the eigenfields which for eigenvalues $\lambda_{\pm}$ are

$$\tilde{W}_{(+)} = \lambda_R\cos\gamma_+ - \lambda_L\sin\gamma_- + i\tilde{H}^1\cos\gamma_- + i\tilde{H}_1{}^{\prime C}\sin\gamma_+ \tag{6.10}$$

and

$$\tilde{W}_{(-)} = (-\gamma_5)^{\Theta}[\lambda_R\sin\gamma_+ - \lambda_L\cos\gamma_- - i\tilde{H}^1\sin\gamma_- - i\tilde{H}_1{}^{\prime C}\cos\gamma_+] \tag{6.11}$$

where $\gamma_{\pm} = \beta_+ \pm \beta_-$ and $\beta_{\pm}$ are determined by

$$\sin 2\beta_{\pm} = (\mu \mp \tilde{m}_2)/[4\nu_{\pm}{}^2 + (\mu \mp \tilde{m}_2)^2]^{1/2}$$

$$\cos 2\beta_{\pm} = 2\nu_{\pm}/[4\nu_{\pm}{}^2 + (\mu \mp \tilde{m}_2)^2]^{1/2}; \quad -\pi < 2\beta_{\pm} \leq \pi \tag{6.12}$$

$\Theta(-\lambda_-)$ appearing in Eq. (6.11) is the step function which depends on the sign of the eigenvalue λ_-:

$$\Theta = 0, \ \sin 2\alpha > 2m_1\tilde{m}_2/M_W{}^2$$

$$\Theta = 1, \ \sin 2\alpha < 2m_1\tilde{m}_2/M_W{}^2 \tag{6.13}$$

The mass eigenvalues $\tilde{m}_{\pm} = |\lambda_{\pm}|$ corresponding to the eigenfields Eqs. (6.10) and (6.11) are

$$\tilde{m}_{\pm} = \frac{1}{2} \left| [4\nu_+^2 + (\mu - \tilde{m}_2)^2]^{1/2} \pm [4\nu_-^2 + (\mu + \tilde{m}_2)^2]^{1/2} \right| \tag{6.14}$$

and obey the product rule

$$\tilde{m}_+\tilde{m}_- = \rho^2 M_W^2; \quad \rho \equiv |\sin 2\alpha - \mu\tilde{m}_2/M_W^2|^{1/2} \tag{6.15}$$

Eq. (6.15) implies that unless the photino mass is too large in almost all models there exists a Wino below the W boson[11,12].

(iii) Zino States ($\tilde{Z}_{(k)}$)

The mass terms in the neutral spin 1/2 sector may be expressed in the form

$$L^m = -\frac{1}{2}\bar{\psi}_i M_{ij}\psi_j \tag{6.16}$$

where ψ_i (i = 0,1,2,3,4) are convenient Majorana combinations of the neutral gaugino and Higgsino fields λ^3, λ^0, $\tilde{H}^2$, $\tilde{H}_2'$, $\tilde{U}$ (and their charge conjugates):

$$\psi_0 = \sin\theta_W\lambda^3 + \cos\theta_W\lambda^0$$

$$\psi_1 = \sin\theta_W\lambda^0 - \cos\theta_W\lambda^3$$

$$\psi_2 = i[\cos\alpha(\tilde{H}^2 - \tilde{H}^{2C}) - \sin\alpha(\tilde{H}_2' - \tilde{H}_2'^C)]$$

$$\psi_3 = -i[\sin\alpha(\tilde{H}^2 - \tilde{H}^{2C}) + \cos\alpha(\tilde{H}_2' - \tilde{H}_2'^C)]$$

$$\psi_4 = i[\tilde{U} - \tilde{U}^C] \tag{6.17}$$

In the representation of Eqs. (6.17), the mass matrix M_{ij} has the form[27]

$$M = \begin{pmatrix} \tilde{m}_\gamma & \tilde{m}_{\gamma Z} & 0 & 0 & 0 \\ \tilde{m}_{\gamma Z} & \tilde{m}_Z & M_Z & 0 & 0 \\ 0 & M_Z & \mu\sin 2\alpha & \mu\cos 2\alpha & 0 \\ 0 & 0 & \mu\cos 2\alpha & -\mu\sin 2\alpha & \mu' \\ 0 & 0 & 0 & \mu' & \mu'' \end{pmatrix} \tag{6.18}$$

where

$$\tilde{m}_\gamma = \sin^2\Theta_W \tilde{m}_2 + \cos^2\Theta_W \tilde{m}_1 \quad (6.19)$$

$$\tilde{m}_Z = \cos^2\Theta_W \tilde{m}_2 + \sin^2\Theta_W \tilde{m}_1; \; \tilde{m}_{\gamma Z} = \cos\Theta_W \sin\Theta_W (\tilde{m}_1 - \tilde{m}_2) \quad (6.20)$$

$$\mu' \equiv \lambda'(v^2 + w^2)^{1/2}, \; \mu'' = \lambda'' \langle U \rangle, \quad (6.21)$$

The orthogonal matrix O_{ik} which diagonalizes Eq. (6.18) is given by[27]

$$O_{ok} = M_Z \tilde{\lambda}_k \tilde{m}_{\gamma Z} (\lambda_k - \tilde{m}_\gamma)^{-1} / D_k$$

$$O_{1k} = M_Z \tilde{\lambda}_k / D_k$$

$$O_{2k} = \tilde{\lambda}_k \tilde{\lambda}_k' / D_k$$

$$O_{3k} = (\mu \cos 2\alpha) \tilde{\lambda}'_k / D_k$$

$$O_{4k} = \mu'(\lambda_k - \mu'')^{-1} O_{3k}$$

$$D_k = [M_Z^2 \tilde{\lambda}_k^2 \{1 + \tilde{m}_{\gamma Z}^2 (\lambda_k - \tilde{m}_\gamma)^{-2}\}$$

$$+ \tilde{\lambda}_k'^2 \{\tilde{\lambda}_k^2 + \mu^2 \cos^2 2\alpha (1 + \mu'^2 (\lambda_k - \mu'')^{-2}\}]^{1/2}$$

$$\tilde{\lambda}_k \equiv \lambda_k + \mu \sin 2\alpha - \mu'^2 (\lambda_k - \mu'')^{-1}; \; \tilde{\lambda}_k = \lambda_k - \tilde{m}_Z - \tilde{m}_{\gamma Z}^2 (\lambda_k - \tilde{m}_\gamma)^{-1} \quad (6.22)$$

where λ_k are the eigenvalues of the mass matrix Eq. (6.38). The Zino eigenfields $\tilde{Z}_{(k)}$ ($k = 0,1,2,3,4$) corresponding to Zino masses $\tilde{\mu}_k = |\lambda_k|$ are

$$\tilde{Z}_{(k)} = (i\gamma_5)^{\Theta_k} \sum_i \psi_i O_{ik} \quad (6.23)$$

In Eq. (6.23) $\Theta_k \equiv \Theta(-\lambda_k)$ are the step functions with values 0 or 1 depending on whether λ_k is positive or negative.

In general the eigenvalues of the mass matrix Eq. (6.18) can be determined only numerically. However, analytic solutions may be achieved under certain approximations. If one neglects $\tilde{m}_{\gamma Z}$ (since $(\tilde{m}_{\gamma Z}/\tilde{m}_\gamma)^2$ is small), one of the eigenvalues is easily seen to be $\tilde{m}_\gamma$. The remaining 4x4 matrix then separates into a 2x2 block diagonal form provided

$$(\mu\cos 2\alpha)^2/M_Z^2 << 1 \tag{6.24}$$

Condition (6.24) is obeyed by the T.B. and the R.G. models and thus this condition may actually hold for a wide class of models. Under these restrictions, the eigenvalues for the remaining Zino states are[27]

$$\lambda_{1,2} = \frac{1}{2}(\tilde{m}_Z + \mu\sin 2\alpha) \pm [M_Z^2 + \frac{1}{4}(\tilde{m}_Z - \mu\sin 2\alpha)^2]^{1/2} \tag{6.25}$$

$$\lambda_{3,4} = \frac{1}{2}(\mu'' - \mu\sin 2\alpha) \mp [\mu'^2 + \frac{1}{4}(\mu'' + \mu\sin 2\alpha)^2]^{1/2} \tag{6.26}$$

$\lambda_{1,2}$ are the eigenvalues corresponding to the Zino states $\tilde{Z}_{(-)}$ and $\tilde{Z}_{(+)}$ with $\tilde{Z}_{(-)}$ mass generally lying below the Z boson mass and $\tilde{Z}_{(+)}$ mass lying above. In the T.B. model the states $\tilde{Z}_{3,4}$ are both relatively heavy. In the R.G. model there is only one eigenvalue remaining for Eq. (6.26) due to the absence of the singlet field ($\mu' = \mu'' = 0$) with the mass eigenvalue of this mode (called the Twilight) being $|\mu\sin 2\alpha|$. Since both α and μ are generally small in the R.G. models, the Twilight is generally relatively light.

(iv) Higgs States

The theory contains ten real scalar fields of which three are absorbed giving masses to the $W^{\pm}$ and Z^0 bosons. The remaining fields decompose into one charged and five neutrals. This is the particle spectrum of the theory for the T.B. model. For the R.G. models the number of remaining neutral fields reduces to three (due the absence of the singlet Higgs field here) and the mass spectrum of these neutrals obey the relation[27]

$$m_0^2 = \frac{m_g^2}{\cos^2\alpha}\left[1 + \frac{\mu^2}{m_1^2} - \frac{1}{2}\frac{\cos 2\alpha M_Z^2}{m_g^2}\right] \tag{6.27}$$

$$m_{(1,2)}^2 = \frac{1}{2}[M_Z^2 + m_0^2 \pm \{(M_Z^2 + m_0^2)^2 - 4(\cos 2\alpha)^2 m_0^2 M_Z^2\}^{1/2}] \tag{6.28}$$

The charged Higgs mass is given by $m_{\pm}^2 = M_W^2 + m_0^2$ and thus lies above the W boson mass. The H_0 Higgs can be very light for a range of values of the parameters appearing in Eq. (6.27) while H_2 lies below the Z boson mass[27,90]

(v) S-matter States

S-matter consists of the superpartners of the left and right handed leptons and quarks. The left and the right s-matter fields are generally mixed due to the soft breaking terms such as $Am_q m_g$ in the squark $\tilde{q}$ mass matrix. Thus, the s-matter eigenstates are not pure L and R but mixtures of these. For example for the squarks the mass eigenvalues are

$$\tilde{m}_{q_{2,1}} = \frac{1}{2}(\tilde{m}_L^2 + \tilde{m}_R^2) \mp [\frac{1}{4}(\tilde{m}_L^2 - \tilde{m}_R^2)^2 + A^2 m_q^2 m_g^2]^{1/2} \tag{6.29}$$

$\tilde{m}_L$ and $\tilde{m}_R$ (for the first and second generation) up squarks are given by(8,27)

$$m^2_{\tilde{u}_L} = m_g^2 + (-\frac{1}{2} + \frac{2}{3}\sin^2\Theta_W)M_Z^2\cos 2\alpha \tag{6.30}$$

$$m^2_{\tilde{u}_R} = m_g^2 + (-\frac{2}{3}\sin^2\Theta_W)M_Z^2\cos 2\alpha \tag{6.31}$$

For the (first and second generation) down squarks one has(8,27)

$$m^2_{\tilde{d}_L} = m_g^2 + (\frac{1}{2} - \frac{1}{3}\sin^2\Theta_W)M_Z^2\cos 2\alpha \tag{6.32}$$

$$m^2_{\tilde{d}_R} = m_g^2 + (\frac{1}{3}\sin^2\Theta_W)M_Z^2\cos 2\alpha \tag{6.33}$$

Eqs. (6.30)-(6.32) are significantly modified for the third generation due to the large top quark mass contributions in the R.G. model. One has(8,27)

$$m^2_{\tilde{t}_L} = \frac{2}{3} m_g^2 + (-\frac{2}{3} + \frac{2}{3}\sin^2\Theta_W)M_Z^2\cos 2\alpha \tag{6.34}$$

$$m^2_{\tilde{t}_R} = \frac{1}{3} m_g^2 + (-\frac{1}{3} - \frac{2}{3}\sin^2\Theta_W)M_Z^2\cos 2\alpha \tag{6.35}$$

$$m^2_{\tilde{b}_L} = \frac{2}{3} m_g^2 + (\frac{1}{3} - \frac{1}{3}\sin^2\Theta_W)M_Z^2\cos 2\alpha \tag{6.36}$$

Finally for the sleptons we have(8,27)

$$m^2_{\tilde{e}_L} = m_g^2 + (\frac{1}{2} - \sin^2\Theta_W)M_Z^2\cos 2\alpha \tag{6.37}$$

$$m^2_{\tilde{e}_R} = m_g^2 + (\sin^2\Theta_W)M_Z^2\cos 2\alpha \tag{6.38}$$

$$m^2_{\tilde{\nu}} = m_g^2 - \frac{1}{2} M_Z^2 \cos 2\alpha \tag{6.39}$$

with similar results holding for the other generations.

VII. SUPERSYMMETRIC DECAYS OF W AND Z

Decays of W and Z which can occur in all models are(11,12,26 - 33)

$$W^{\pm} \to \tilde{W}_{(-)}^{\pm} + \tilde{\gamma}, \quad \tilde{m}_{(-)} + \tilde{m}_{\gamma} < M_W \tag{7.1}$$

$$Z \to \tilde{W}_{(-)}^{+} + \tilde{W}_{(-)}^{-}, \quad 2\tilde{m}_{(-)} < M_Z \tag{7.2}$$

The branching ratios for these are given in Appendix C (see Eqs. (C.1) and (C.9)). The dependence of the branching ratios on the parameters α, μ and $\tilde{m}_{\gamma}$ are studied in Tables 1, 2 and 3. One finds that over a wide range of these parameters the $W \to \tilde{W} + \tilde{\gamma}$ decay branching ratio (relative to $W \to e\nu$) is appreciable while the $Z \to \tilde{W} + \tilde{W}$ decay branching ratio (relative to $Z \to e^+e^-$) is remarkably large, i.e. O(5).

(Degrees)	$W^{\pm} \to \tilde{W}^{\pm}_{(-)} + \tilde{\gamma}$	$Z^0 \to \tilde{W}_{(-)}^{+} + \tilde{W}_{(-)}^{-}$
25	0.552	6.81
30	0.519	6.67
35	0.491	6.50
40	0.473	6.36
45	0.466	6.30

Table 1: Dependence of W and Z decay branching ratios on α when $\tilde{m}_{\gamma}$ = 7 GeV and μ = 120.7 GeV. The W decay branching ratio is relative to $W \to e\nu$ and the Z decay branching ratio is relative to $Z \to e^+e^-$. Table is from Ref. (27).

$\tilde{M}_{\gamma}$ (GeV)	$W \to \tilde{W}_{(-)} + \tilde{\gamma}$	$Z^0 \to \tilde{W}_{(+)}^{+} + \tilde{W}_{(-)}^{-}$
2	0.366	4.65
8	0.354	4.70
14	0.330	4.66
20	0.298	4.53
26	0.260	4.35

Table 2: Dependence of W and Z branching ratios on $\tilde{m}_{\gamma}$ when μ = 20 GeV and α = 15°. Table is from Ref. (27).

μ(GeV)	$W^{\pm} \to \tilde{W}^{\pm}_{(-)} + \tilde{\gamma}$	$Z^0 \to \tilde{W}_{(-)}{}^{+} + \tilde{W}_{(-)}{}^{-}$
10	0.346	4.54
20	0.357	4.70
30	0.375	4.89
40	0.398	5.12
50	0.425	5.35
60	0.454	5.60

Table 3: Dependence of W and Z branching ratios on μ when $m_{\tilde{\gamma}} = 7$ GeV and $\alpha = 15°$. Table is from Ref. (27).

In addition to the decays of Eqs. (7.1) and (7.2) one has decays which depend on the specific choice of the model. For example, the T.B. model allows the decay[26,27]

$$W^{\pm} \to \tilde{W}^{\pm}_{(-)} + \tilde{Z}_{(-)}, \quad \tilde{m}_- + \tilde{\mu}_- < M_W \tag{7.3}$$

For the R.G. model the $\tilde{Z}_{(-)}$ mass is close to the Z^0 mass and the above decay is essentially never realized kinematically. In the R.G. model, however, one may have the decays[12,26,29,31]

$$W^{\pm} \to \tilde{W}^{\pm}_{(-)} + \tilde{Z}_{(3)}, \quad \tilde{m}_- + \tilde{\mu}_3 < M_W \tag{7.4}$$

$$Z^0 \to \tilde{Z}_{(3)} + \tilde{Z}_{(3)}, \quad 2\tilde{\mu}_3 < M_Z \tag{7.5}$$

$\tilde{Z}_{(3)}$ is relatively heavy in the T.B. model (see Eqs. (6.26)) so one expects the decays of Eqs. (7.4) and (7.5) to be kinematically disallowed in the T.B. model. The decay branching ratios for the processes of Eqs. (7.3), (7.4) and (7.5) are given by Eqs. (C.3), (C.6) and (C.12). The branching ratios of Eqs. (7.3) and (7.4) relative to $W \to e\nu$ and of Eq. (7.5) relative to $Z^0 \to e^+e^-$ are found to be generally O(1).

The decays of the final state $\tilde{W}_{(-)}, \tilde{Z}_{(-)}$ and $\tilde{Z}_{(3)}$ are governed by diagrams of Figs. (1), (2) and (3) respectively. The Wino decay modes are given by Eqs. (1.9)-(1.11) and Eqs. (C.14)-(C.19). The decay modes of $\tilde{Z}_{(-)}$ Zino are

$$\tilde{Z}_{(-)} \to \ell^+ + \ell^- + \tilde{\gamma} \tag{7.6}$$

$$\tilde{Z}_{(-)} \to u_i(d_i) + \bar{u}_i(\bar{d}_i) + \tilde{\gamma} \tag{7.7}$$

$$\tilde{Z}_{(-)} \to u_i(d_i) + \bar{u}_i(\bar{d}_i) + \tilde{g} \tag{7.8}$$

$$\tilde{Z}_{(-)} \to \bar{u}_i(\ell^-) + d_i(\bar{\nu}_\ell) + \tilde{W}_{(-)}{}^+ \qquad (7.9)$$

$$\tilde{Z}_{(-)} \to u_i(\ell^+) + \bar{d}_i(\nu_\ell) + \tilde{W}_{(-)}{}^- \qquad (7.10)$$

The decay modes Eqs. (7.9) and (7.10) with a Wino in the final state arise from the W intermediate pole of Fig. (2) and can occur only when $\tilde{W}_{(-)}$ is lighter than $\tilde{Z}_{(-)}$ (as is true in most models provided $m_{\tilde{\gamma}}$ is not too large). The Zino decay branching ratios of Eqs. (7.6) - (7.10) are given by Eqs. (C.20) - (C.30) of Appendix C. The decay modes of the Twilight Zino $Z_{(3)}$ arise from diagrams of Fig. (3). The Twilight Zino is expected to be lighter than the Wino as well as the gluino and thus diagrams with Wino or gluino final states are absent in Fig. (3) and so one has only the following decay modes for $Z_{(3)}$:

$$\tilde{Z}_{(3)} \to \ell^+ + \ell^- + \tilde{\gamma} \qquad (7.11)$$

$$\tilde{Z}_{(3)} \to u_i(d_i) + \bar{u}_i(\bar{d}_i) + \tilde{\gamma} \qquad (7.12)$$

The Twilight Zino decay branching ratios are given by Eqs. (C.31) - (C.33) in Appendix C.

Unlike the decays of the W and Z, the $\tilde{W}_{(-)}$,$\tilde{Z}_{(-)}$, and $\tilde{Z}_{(3)}$ decays depend on the gravitino mass and this dependence is exhibited in Figs. (4) and (5) for the various decay modes. Branching ratios for $\tilde{Z}_{(-)}$ decay for the T.B. model are exhibited in Fig. (4). It is seen that the s-matter poles dominate for the small gravitino masses and the W-pole dominates for large gravitino masses with a cross over between the two domains occuring at a relatively low gravitino mass of $\gtrsim$ 120 GeV. The Θ_W enhancement of the Wino mode due to the factor $(\sin\Theta_W)^{-4}$ in Eq. (C.27) accounts in part for this phenomena.

Branching ratios for $\tilde{W}_{(-)}$ and $\tilde{Z}_{(3)}$ in the R.G. model are exhibited in Fig. (5). Again for large m_g, the W-pole dominates in $\tilde{W}_-$ decay producing the ratio $(h+\tilde{\gamma})$: $(\ell + \nu + \tilde{\gamma}) \simeq 3:1$ for the hadronic to the leptonic final states. For small m_g, the hadronic branching ratio is enhanced. This is due to the fact that in the R.G. model $\tilde{W}_{(-)}$ couples mainly to the $I_3 = 1/2$ components of the left-handed doublets (when γ_+ is small) as is seen from Eqs. (C.14) and (C.15) and $\tilde{u}_L$ mass decreases as m_g decreases. The features of the $Z_{(3)}$ branching ratios in Fig. (5) are also governed by the R.G. model. For large m_g, all s-matter poles become approximately equal and the branching ratios approach a constant. (B.R. $(\ell\nu\tilde{\gamma}) = 0.44$.) For small m_g, the effects of the D-term split the s-matter masses with $\tilde{u}_L$, $\tilde{u}_R$ masses becoming smaller and slepton (and other squark) masses becoming larger. The relative smallness of the $\tilde{u}_L$ mass at small m_g ($m_{\tilde{u}_L} \simeq 1/2\, m_{\tilde{e}_L}$ at $m_g = 60$ GeV) produces the enhancement of the hadronic mode at lower m_g in Fig. 5.

We discuss now the branching ratios for the supersymmetric decays of the W and Z into various channels. From Eqs. (7.1) and (7.2) and Eqs. (1.9) - (1.11), one finds that essentially all models possess the following decays for the W and Z bosons:

$$W \to \tilde{\gamma} + (\tilde{W} \to h + \tilde{\gamma}) \tag{7.13}$$

$$W \to \tilde{\gamma} + (\tilde{W} \to \ell + \nu_\ell + \tilde{\gamma}) \tag{7.14}$$

$$Z \to (\tilde{W} \to h + \tilde{\gamma}) + (\tilde{W} \to h + \tilde{\gamma}) \tag{7.15}$$

$$Z \to (\tilde{W} \to h + \tilde{\gamma}) + (\tilde{W} \to \ell + \nu_\ell + \tilde{\gamma}) \tag{7.16}$$

$$Z \to (\tilde{W}^+ \to \bar{\ell}_1 + \nu_1 + \tilde{\gamma}) + (\tilde{W}^- \to \ell_2 + \bar{\nu}_2 + \tilde{\gamma}) \tag{7.17}$$

In addition to the supersymmetric decays of Eqs. (7.13) - (7.17), one has supersymmetric decays that arise exclusively in the T.B. or the R.G. models. Thus Eq. (7.3) along with Eqs. (1.9) - (1.11) and Eqs. (7.6) - (7.10) produce the following supersymmetric decays for the T.B. model:

$$W \to (\tilde{W} \to h + \tilde{\gamma}) + (\tilde{Z} \to h + \tilde{\gamma}) \tag{7.18}$$

$$W \to (\tilde{W} \to \ell + \nu_\ell + \tilde{\gamma}) + (\tilde{Z} \to h + \tilde{\gamma}) \tag{7.19}$$

$$W \to (\tilde{W} \to h + \tilde{\gamma}) + (\tilde{Z} \to \ell + \bar{\ell} + \tilde{\gamma}) \tag{7.20}$$

$$W \to (\tilde{W} \to \ell_1 + \nu_\ell + \tilde{\gamma}) + (\tilde{Z} \to \ell_2 + \bar{\ell}_2 + \tilde{\gamma}) \tag{7.21}$$

Similarly for the R.G. model there are additional supersymmetric decays of the W and Z arising from Eqs. (7.4) and (7.5), Eqs. (1.9) - (1.11) and Eqs. (7.11) and (7.12):

$$W \to (\tilde{W} \to h + \tilde{\gamma}) + (\tilde{Z}_{(3)} \to h + \tilde{\gamma}) \tag{7.22}$$

$$W \to (\tilde{W} \to \ell + \nu_\ell + \tilde{\gamma}) + (\tilde{Z}_{(3)} \to h + \tilde{\gamma}) \tag{7.23}$$

$$W \to (\tilde{W} \to h + \tilde{\gamma}) + (\tilde{Z}_{(3)} \to \ell + \bar{\ell} + \tilde{\gamma}) \tag{7.24}$$

$$W \to (\tilde{W} \to \ell_1 + \nu_\ell + \tilde{\gamma}) + (\tilde{Z}_{(3)} \to \ell_2 + \bar{\ell}_2 + \tilde{\gamma}) \tag{7.25}$$

$$Z \to (\tilde{Z}_{(3)} \to h + \tilde{\gamma}) + (\tilde{Z}_{(3)} \to h + \tilde{\gamma}) \tag{7.26}$$

$$Z \to (\tilde{Z}_{(3)} \to h + \tilde{\gamma}) + (\tilde{Z}_{(3)} \to \ell + \bar{\ell} + \tilde{\gamma}) \tag{7.27}$$

$$Z \to (\tilde{Z}_{(3)} \to \ell_1 + \bar{\ell}_1 + \tilde{\gamma}) + (\tilde{Z}_{(3)} \to \ell_2 + \bar{\ell}_2 + \tilde{\gamma}) \tag{7.28}$$

Suppressed Z^o Decays: There exist other decays which though kinematically possible are highly suppressed because of the nature of interactions in Supergravity models. Examples of these are

$$Z^{(o)} \to \tilde{Z}_{(3)} + \tilde{\gamma} \tag{7.29}$$

$$Z^{(o)} \to \tilde{\gamma} + \tilde{\gamma} \tag{7.30}$$

The origin of the suppression for these decays resides in the fact that Z^o couples only to neutral Higgsinos but not to neutral gauginos, and due to the small Higgsino content of the photino.

The most prominent experimental signals arising from the supersymmetric decay channels of the W and Z discussed above fall into three categories listed in the discussion following Eq. (1.11). We shall discuss each of these signals in some detail in the following.

(a) UFO Events

One jet UFO events shown in Fig (6a) arise from the decay mode of Eq. (7.13) where a W decays into a Wino and a photino followed by the Wino decay into hadrons and a photino[11,12,26,27,29]. The hadrons are formed from the two quarks in the Wino decay and the hadronization of these quarks would in general form a wide angle jet. The two photinos in the final state interact weakly and escape detection (hence the name Unidentified Fermionic Objects or UFOs) and one is left only with a broad single jet with momentum unbalanced. Two jet UFO events[26,27,29] arise from decay modes of Eqs. (7.15), (7.18) and (7.22). A two jet event corresponding to the decay mode of Eq. (7.15) is shown in Fig. (6a) where a Z decays into two Winos followed by the decay of each of these Winos into a jet and a photino. The two final state photinos escape detection leaving two broad jets in opposite hemispheres with a net missing momentum.

Relative branching ratios for the one jet and two jet UFO events for the W and Z in the T.B. model are shown in Fig. (7) as a function of the gravitino mass. The branching ratios are appreciable for the W decay and remarkably large for the Z decay over a broad range of gravitno masses. Similar features persist for the branching ratios in the R.G. model as exhibited in Fig. 8. In the one jet UFO events of Fig. (6a), a rough estimate (assuming that the final particles in Wino decay share the transverse energy equally) shows nearly $\sim 3/4$ of the events with missing $p_\perp^m$ >15 GeV and gives between 20 percent (for R.G. model) and 30 percent (for T.B. model) of the $W \to e\nu$ decays. A similar analysis for the two jet UFO events with total missing energy $p_\perp^m > 15$ GeV gives 50 - 100 percent as many events as the $W \to e\nu$ decays in both the T.B. and the R.G. models. This is what one expects at the $\bar{P}P$ Collider. At LEP or SLC one expects the two jet UFO events to be ~ 3 times as many as the $Z \to e^+e^-$ events. These results are summarized in Table 4.

Signal	$\bar{P}P$ Collider	LEP
	# Events/100 $W \to e\nu$	# Events/100 $Z \to e^+e^-$
UFO (1 jet)	20 - 30	0
UFO (2 jets)	50 - 100	$\sim$ 300
Lepton-jet	5 - 100	$\sim$ 10 - 70

Table 4. Estimates of UFO and Lepton-jet events at the $\bar{P}P$ Collider and at LEP.

(ii) Lepton-Jet Events

Lepton-jet events may arise from the Z decays of Eq. (7.16) in all models, from the W decay of Eq. (7.19) for the T.B. model and from the W decay of Eq. (7.23) for the R.G. model[26,27]. Each of these events is characterized by a charged lepton in one hemisphere and a single broad jet in the other hemisphere with momentum unbalanced. Branching ratios for these events for the T.B. and the R.G. models are exhibited in Fig. 9 as a function of the gravitino mass. From these branching ratios one finds that the lepton-jet events with $p_\perp > 15$ GeV should be 5-15% of the $W \to e\nu$ rate at the $\bar{P}P$ collider. At LEP or SLC one expects these events to be 10-70% of the $Z \to e^+e^-$ rate depending on the model and the value of m_g. Results are summarized in Table 4.

(iii) Exotic Decays of W and Z

Exotic decays are represented by[26,27] Eqs. (7.17), (7.20), (7.21), (7.24), (7.25), (7.27) and (7.28). The decay of Eq. (7.17) can occur in all models, decays of Eqs. (7.20) and (7.21) are limited to the T.B. model while decays of Eqs. (7.24), (7.25), (7.27) and (7.28) arise only in the R.G. model. Branching ratios for some of the exotic decays are shown in Fig. 10 and one may see that the branching ratios are quite sizable over a wide range of gravitino masses. Because of the appreciable size of these branching ratios, a significant number of the exotic decays should be observable particularly at LEP or SLC even after experimental cuts needed for their observation.

Additional decays of the W and Z are possible if slepton, squark and gluino masses are sufficiently low for them to appear in the final states. For example, one may have[33,34,35,71]

$$W \to (\tilde{e} \to e\tilde{\gamma}) + (\tilde{\nu} \to \nu\tilde{\gamma}),\ \tilde{m}_e + \tilde{m}_\nu < M_W \tag{7.31}$$

$$Z \to (\tilde{e} \to e\tilde{\gamma}) + (\tilde{e} \to e\tilde{\gamma}),\ 2\tilde{m}_e < M_Z \tag{7.32}$$

$$Z \to (\tilde{g} \to q\bar{q}\tilde{\gamma}) + (\tilde{g} \to q\bar{q}\tilde{\gamma}),\ 2\tilde{m}_g < M_Z \tag{7.33}$$

with equations similar to the above also holding with leptons, sleptons replaced by quarks and squarks. (The process of Eq. (7.33) proceeds through matter loops[71].)

There are possible backgrounds to many of the signals arising from supersymmetric decays of the W and Z discussed above. One such background can arise from gluino pair production at the $\bar{p}p$ Collider[72-74]

$$p + \bar{p} \to p + \bar{p} + \tilde{g} + \tilde{g} \tag{7.34}$$

We note that in the minimal Supergravity unified model (see Eq. (1.1) and Ref. (2)) the gluino mass is expected to lie in the range 5-75 GeV and there is a distinct overlap in the allowed ranges for the mass parameters for the Winos, Zinos and gluinos. Thus a possible background for the two jet UFO events arises since each of the gluinos can decay into a photino and a quark pair with the quark pair hadronizing and forming a broad jet. This is a signal similar to the one arising from the decay modes of Eqs. (7.15), (7.18) and (7.22). It should be noted that in the non-minimal Supergravity model where $f_{\alpha\beta}(Z)$ is left arbitrary[45], gaugino masses are left arbitrary as well. <u>Thus observation of gluino masses in the mass range of 5-75 GeV constitutes a verification of the minimal Supergravity model.</u>[2]

A possible background to lepton-jet events arises from the heavy flavor production[75] such as $t\bar{t}$ pairs where $t\bar{t} \to (\ell + \nu_\ell + b) + \bar{t}$ with $\bar{t}$ producing a hadron jet. However, such a decay always involves an additional hadronic debris from the b quark decay and thus a full reconstruction of the events is needed to distinguish between heavy flavor production and the true lepton-jet events of Eqs. (7.16), (7.19) and (7.23).

Supersymmetric particles should also be observable in a variety of final states in e^+e^- machines. Examples of some of the processes are[37, 38,76]

$$e^+e^- \to \tilde{W}_{(-)}{}^+ + \tilde{W}_{(-)}{}^- \tag{7.35}$$

$$e^+e^- \to e^\pm\tilde{\nu}\ \tilde{W}_{(-)}{}^\mp \tag{7.36}$$

$$e^+ + e^- \to \tilde{\gamma} + \tilde{Z} \qquad (7.37)$$

$$e^+e^- \to \tilde{e}\,\tilde{\bar{e}}\;,\;\tilde{q}\,\tilde{\bar{q}}$$

$$\tilde{e} \to e + \tilde{\gamma} \quad \text{etc.} \qquad (7.38)$$

VIII. EFFECTS OF SUPERGRAVITY UNIFIED MODELS ON g_μ-2, ρ, $\sin^2\Theta_W$, M_W AND M_Z

(A) Supersymmetric Electro-Weak Effects on g_μ-2.

The current experimental value of the muon anomalous magnetic moment is[54]

$$(g_\mu-2)_{exp} = (2{,}331{,}848.0 \pm 16.3) \times 10^{-9} \qquad (8.1)$$

Kinoshita et al.[55] have recently carried out an improved theoretical estimate of this quantity including the complete α^4 qed contribution, hadronic contributions and the electro-weak contributions of the standard Weinberg-Salam theory[57]. They give a value[55]

$$(g_\mu-2)_{th} = (2{,}331{,}840.4 \pm 4.0) \times 10^{-9} \qquad (8.2)$$

The hadronic contribution to Eq. (8.2) is[55]

$$(g_\mu-2)(\text{had}) = (140.4 \pm 3.8) \times 10^{-9} \qquad (8.3)$$

while the electro-weak contribution to Eq. (8.2) is[55]

$$(g_\mu-2)(\text{W-S}) = (3.90 \pm .02) \times 10^{-9} \qquad (8.4)$$

The possibility of a reduction of the error in the hadronic contributions and consequently a reduction of the error in $(g_\mu-2)_{th}$ (to 0.6 x 10^{-9} or less) has been suggested in Ref. (55) raising the possibility of a test of Eq. (8.4) if experimental error is reduced by an order of magnitude or more[56].

However, it was recently pointed out[60] that the supersymmetric electro-weak contributions in N=1 Supergravity unified theories in some models are considerably larger than Eq. (8.4). A brief summary of the supersymmetry contributions to g_μ-2 in Supergravity unified models is given in Appendix D. Diagrmas which generate the supersymmetric contributions to g_μ-2 are Figs. (11a) and (11b). Fig. (11a) involves Wino and sneutrino intermediate states while Fig. (11b) involves Zino and smuon intermediate states. The possibility of a large contribution to g_μ-2 arises from the fact that several of the supersymmetric particles entering as intermediate states in Figs. (11a) and (11b) are predicted to lie below the W and Z bosons. In Ref. (60)

it was found convenient to define the supersymmetric electro-weak $\Delta(g_\mu-2)$ to be the sum of the standard electro-weak Weinberg-Salam contribution and the supersymmetric contributions arising from Fig. 11.

$\Delta(g_\mu-2)$ in general depends on the parameters α, μ and m_g and on A (the parameter of the hidden sector) and $\tilde{m}_\gamma$. First we shall discuss the dependence on α, μ and m_g. α is essentially defined by our choice of the model R.G. or T.B. (α is small in the R.G. model and 45° in the T.B. model). Thus for each of these models one has a two dimensional plot which can conveniently be defined in terms of mass parameters $\tilde{m}_\nu$ and $\tilde{m}_-$.

R.G. model results for $\Delta(g_\mu-2)$ for an array of sneutrino masses are exhibited in Fig. (12) as a function of the Wino mass $\tilde{m}_-$. In the R.G. model Fig. (11a) dominates over Fig. (11b). This is so because Fig. (11b) always involves a heavy mass, i.e., the smuon mass, in the loop (note that in the R.G. model smuon masses $\tilde{m}_{L,R}$ must be larger than m_g and $m_g \gtrsim 60$ GeV for vacuum stability) while Fig. (11a) involves the sneutrino and the Wino which have no theoretical lower limits on their masses (though experimental limit on $\tilde{m}_-$ is $\gtrsim 18$ GeV[61]). Fig. (12) shows that the supersymmetry contributions to $g_\mu-2$ are generally considerably larger than the standard electro-weak contribution over a range of parameters. (This feature persists also for a broad range of values of $\tilde{m}_\gamma$ and A.)

T.B. model results for $\Delta(g_\mu-2)$ are displayed in Fig. (13) as a function of the Wino mass $\tilde{m}_-$ again for an array of sneutrino masses. In contrast to the R.G. model here the contributions of diagram (11b) is comparable and even greater than that of diagram (11a) due to the fact that both the smuon and the $\tilde{Z}_{(R)}$ state masses can simultaneously become small. Results of Fig. (13) show that the supersymmetric contributions to $\Delta(g_\mu-2)$ are again very significant.

Photino mass dependence of $\Delta(g_\mu-2)$ is exhibited in Fig. (14) for the T.B. model for an array of sneutrino masses. For small photino masses, the loop integral of Fig. (11b) is governed by the smuon mass and is $O(m_\mu^2/m_{\tilde{\mu}}^2)$ which can get large for small smuon masses. This explains the structure of Fig. (14) for small $\tilde{m}_\gamma$. For large $\tilde{m}_\gamma$ one has an enhancement of $\Delta(g_\mu-2)$ involving significant effects also from Fig. (11a) in the domain of small sneutrino or small Wino masses.

The dependence of $\Delta(g_\mu-2)$ on the parameter A of the hidden sector appears through the smuon mass eigenstates $\tilde{\mu}_\gamma=(\tilde{\mu}_1,\tilde{\mu}_2)$ with the mass eigenvalues given by Eq. (6.29) (with q replaced by μ) and $\tilde{m}_{L,R}$ given by Eqs. (6.37) and (6.38). A significant A dependence of $\Delta(g_\mu-2)$ arises from Eq. (D.18). Expanding the difference in Eq. (D.18) and using Eq. (D.14) one finds

$$\bar{g}_2^{\tilde{Z}} = \frac{1}{4\pi^2} A\, m_\mu^2 m_g \frac{B_k^R B_k^L}{(\tilde{\mu}_{(k)})^3} \left(-\frac{\partial G_2}{\partial x}\right)_{x=x_{1k}} \tag{8.5}$$

We note that one has here a linear dependence on A and thus one can get a large contribution to $\Delta(g_\mu-2)$ as A gets large. The dependence of $\Delta(g_\mu-2)$ on A is exhibited in Fig. (15) and one finds that even the present experimental bounds constrain the allowed values of A.

The important observation emerging from the above analysis is that the supersymmetric electro-weak contributions to $g_\mu-2$ are considerably larger than the standard electro-weak contributions in the R.G. model and quite significant in the T.B. model. (Further, the sign of these contributions is opposite to that of the standard electro-weak contribution.) It is thus suggested[60] that improved experiments[56] which can lower the existing experimental errors by a factor of 3 should discern R.G. model effects and improvements by a factor of 10 should discern T.B. model effects.

(B) The ρ Parameter

Current experimental determinations of ρ (the ratio of the effective strength of the neutral to the charged currents in weak interactions) from deep inelastic scattering give[62,63]

$$\rho_{exp} = 1.01 \pm 0.02 \tag{8.6}$$

$\rho = 1$ at the tree level in the standard theory and the $O(\alpha G_F)$ corrections in this theory are known. One has including $O(\alpha G_F)$ corrections[64,65]

$$\rho = 0.991 \tag{8.7}$$

consistent with the experimental value of Eq. (8.6). It has been pointed out by Veltman[66] that a fermion doublet with a large mass difference gives a sizable contribution to ρ. For the t-b quark doublet (with $m_b = 0$) one has $\delta\rho = 3G_F(8\pi^2\sqrt{2})^{-1}m_t^2$ which puts an upper limit on the top quark mass from Eq. (8.6).

Radiative corrections to ρ in supersymmetric theories have recently been investigated by many authors[8,67-69]. Contributions to ρ at the one loop level arise from corrections to the polarization operators for W and Z, vertex corrections and box diagrams (see Figs. (16)). In the T.B. model one finds that corrections from the polarization operator (in the limit of $m_g/m_t \ll 1$) give[69]

$$\delta\rho_{matter} = \frac{3G_F}{8\pi^2\sqrt{2}}\Big[\Big(1 + 2\frac{m_g^2}{m_t^2}\ln\frac{m_g^2}{m_t^2}\Big)m_t^2 + 2\Big(1 + \frac{m_g^2}{m_t^2}\ln\frac{m_g^2}{m_t^2}\Big)m_g^2\Big] \tag{8.8}$$

(Note that in the limit $m_g \to 0$, Eq. (8.8) reduces to the Veltman result for $\delta\rho$[66].) Consistency of Eq. (8.8) with Eq. (8.6) puts an upper bound on the top quark mass[69].

$$m_t \lesssim 250 - 300 \text{ GeV} \tag{8.9}$$

The full predictions of ρ in the T.B. model of the N=1 Supergravity unified theory including vertex corrections (see Table 5) is given by Eliasson[69] and are exhibited in Fig. (17).

μ (GeV)	$\tilde{m}_\lambda$ (GeV)	m_g (GeV)	$\delta\rho_{vertex}$ (x 10^{-3})
150			0.24
30	5		0.10
300		100	0.20
	2		0.20
	10		0.30
		0.5	4.04
		2	3.68
150		5	2.26
	5	20	1.44
		50	0.60
		200	0.08
		500	0.02

Table 5: Contributions to ρ from vertex corrections in the T.B. model for a variety of gravitino and gaugino masses. Table is from Ref. (70).

(C) M_W, M_Z Mass Relations and $\sin^2\Theta_W$

For the standard model, Eq. (8.7) and a one parameter fit to deep inelastic scattering give[64]

$$(\sin^2\Theta)_{exp} = 0.216 \pm 0.014 \tag{8.10}$$

The above $(\sin^2\Theta)_{exp}$ is related to

$$\sin^2\Theta \equiv 1 - M_W^2/M_Z^2 \tag{8.11}$$

by[64]

$$(\sin^2\Theta)_{exp} = K\sin^2\Theta; \quad K = 0.997 \tag{8.12}$$

where κ takes into account the deep inelastic scattering processes from which $\sin^2\theta$ is extracted. From Eqs. (8.10) and (8.12) one may extract the "theoretical" $\sin^2\theta$ [64]

$$\sin^2\theta = 0.217 \pm 0.014 \tag{8.13}$$

Eqs. (8.11) and (8.13) provide one relation for M_W and M_Z. A second relation can be obtained from the μ-decay rate. Taking radiative corrections of the μ-decay into account one gets [65]

$$M_W = \frac{38.64 \text{ GeV}}{\sin\theta} \tag{8.14}$$

Eqs. (8.13) and (8.14) give [65]

$$M_W = 83.0 \pm 3.0 \text{ GeV}$$

$$M_Z = 93.8 \pm 2.5 \text{ GeV} \tag{8.15}$$

The errors in Eq. (8.15) are essentially those from the large errors in $(\sin^2\theta)_{exp}$ of Eq. (8.10). Instead if one eliminates $(\sin^2\theta)$ one has

$$M_W^2\left(1 - \frac{M_W^2}{M_Z^2}\right) = (38.64 \text{ GeV})^2 \tag{8.16}$$

Eq. (8.16) provides a more accurate test of the standard model than any single relation of Eq. (8.15).

There are modifications of the above analysis for the Supergravity unified models. We shall discuss here the modifications within the framework of the T.B. model [70]. First Eq. (8.7) is modified. Assuming a not too large m_t and a not too small m_g (see Ref. (69)) one has

$$\rho = 0.992 \tag{8.17}$$

A one parameter fit to the deep inelastic data now gives [70] (instead of Eq. (8.10))

$$(\sin^2\theta)_{exp} = 0.217 \pm 0.014 \tag{8.18}$$

The value of κ in Eq. (8.12) is also modified (see Fig. 18). One finds that the supersymmetric corrections tend to cancel the corrections to κ in the standard model and thus one has Eq. (8.13) holding as in the standard model. In effect the supersymmetric corrections to ρ and κ have cancelled out in $\sin^2\theta$. However, there are non-vanishing supersymmetric corrections to μ-decay which give[70]

$$M_W = \frac{38.64\ \text{GeV}}{\sin\Theta}\ (1 + \frac{1}{2} r_{SS}) \tag{8.19}$$

where r_{SS} is the supersymmetric correction which depends on m_g, (See Fig. 19.) r_{SS} is negative over a range of m_g masses and hence the effect of supersymmetric corrections is to shift the value of M_W and M_Z downwards. However, these shifts are small $\sim O(1\%)$. Finally, the supersymmetric version of Fig. (8.16) is

$$M_W^2(1 - \frac{M_W^2}{M_Z^2}) = (38.64\ \text{GeV})^2\ (1 + r_{SS}) \tag{8.20}$$

IX. CONCLUSION

We have given here a brief review of the recent developments of the Supergravity unified models and their phenomenological implications in the low energy domain. The essential theme of these developments is that the electro-weak symmetry breaking is triggered by supergravitational interactions via the super-Higgs effect and thus the weak scale M_W and the gravitino mass scale m_g are related[2]. Some possible tests of these ideas were discussed in Secs. VII and VIII.

There are also other ways of testing Supergravity unified models. Supersymmetric decays of the W and Z produce contributions to their widths which are sizable. Thus the W width is increased by 18% (10%) for the T.B. (R.G.) model and the Z-width is increased by 20% (18%) for the T.B. (R.G.) model over the prediction of the Weinberg-Salam theory with three generations (for $\tilde{m}_\gamma = 7$ GeV). An interesting feature of the R.G. model supersymmetric theories is the existence of at least one neutral Higgs below the Z boson (in fact in the R.G. model, two of the neutral Higgs could be below the Z) which should make it a likely candidate for observation at the $\bar{P}P$ Collider. Further, the whole host of the remaining supersymmetric particles $\tilde{W}_{(+)}$, $\tilde{Z}_{(+)}$, H_1 etc. should become accessible experimentally at SLC, LEP, HERA or SSC. Regarding proton decay[77], supersymmetric GUT theories[79-81] offer characteristically different signals than those in the conventional GUTS[78] where $\pi^o e^+$ is generally the dominant proton decay mode. In supersymmetric GUTS it is the $\nu_\mu K^+$ mode which is the dominant one and both the Wino and the gluino dressings of dimension five operators make significant contributions to the decay[81].

A number of topics were not discussed in these lectures. These include CP violation and the consequences of Supergravity unified models in cosmology and the early universe. The reader is referred to the literature regarding these[82-84]. Possible future directions of the theory point to larger N ($N = 2$[85], $N = 4$[86,87]) supersymmetric unified theories and Kaluza-Klein theories[88]. Also an interesting new direction discussed at this meeting is the subject of ultralight Dirac neutrinos and their possible connection with supergrand scenarios.[89]

Note Added: The UAI collaboration has recently observed certain events at the $\overline{P}P$ Collider which appear difficult to explain in the framework of the standard model[91]. Some of these events may well be the supersymmetric signals such as the production of gauginos, squarks or sleptons [see in this context Refs. (26), (27), (92)-(95)]. In particular, gluino production of Eq. (7.34) (with gluino masses $\approx$ (40 GeV)) appears to simulate the signatures of some of the events seen by UAI[94]. We note that the gluino masses of $\approx O$(40 GeV) fall exactly in the range predicted by the minimal Supergravity unified model[2] through radiative corrections[51,8,26].

ACKNOWLEDGEMENTS

One of us (P.N.) acknowledges the NSF support under Grant No. INT-8400934 for travel to India and wishes to thank the Tata Institute of Fundamental Research Bombay, India for the hospitality accorded him during the period of his visit there.

APPENDIX A: TENSOR CALCULUS AND SUPERGRAVITY MATTER COUPLINGS

The couplings of the gauge multiplets and chiral multiplets to Supergravity are generated through Eqs. (2.1) and (2.2) and by use of tensor calculus which defines the multiplication rules for the matter and gauge multiplets. We shall only briefly review the technique here for generating the Supergravity-matter couplings. For details the reader is referred to Ref. (44) and for equivalent analyses to Refs. (45)-(47).

According to the rules of tensor calculus the product of two multiplets $\Sigma_1 = (Z_1, \chi_1, h_1)$ and $\Sigma_2 = (Z_2, \chi_2, h_2)$ of the same chirality is also a multiplet of the same chirality with components given by

$$\Sigma_1\Sigma_2 = (Z_1Z_2,\ Z_1\chi_2 + Z_2\chi_1,\ Z_1h_2 + Z_2h_1 - 2\overline{\chi_1{}^c}\chi_2 \tag{A.1}$$

Further, multiplication of two multiplets of opposite chiralities yields a vector multiplet

$$\begin{aligned}\Sigma_1 x \Sigma_2 = (&\tfrac{1}{2} Z_1{}^*Z_2,\ i(Z_1{}^*\chi_2 - Z_1\chi_2{}^c),\ - h_1Z_2{}^*,\ \tfrac{i}{2}(Z_1{}^*\hat{D}_\mu Z_2 - z_1\hat{D}_\mu Z_2{}^* - 2\overline{\chi}_1\gamma_\mu\chi_2,\\ &- ih_1{}^*\chi_2 + ih_1\chi_2{}^C - i\hat{\not{D}}Z_1{}^*\chi_2 + i\hat{\not{D}}Z_1\chi_2{}^c,\\ &h_1{}^*h_2 - \hat{D}_\mu Z_1{}^*\hat{D}_\mu Z_2 - \overline{\chi}_1\overleftrightarrow{\hat{\not{D}}}\chi_2) + 1 \leftrightarrow 2\end{aligned} \tag{A.2}$$

where

$$\hat{D}_\mu Z^a = \partial_\mu Z^a - \kappa\bar{\psi}_\mu \chi^a \tag{A.3}$$

$$\hat{D}_\mu \chi^a = [D_\mu \omega(e,\psi) - i\,\frac{\kappa}{2} A_\mu]\chi^a - \frac{\kappa}{2}\, \not{\partial} Z^a \psi_{\mu R} - \frac{\kappa}{2}\, h^a \psi_{\mu L} \tag{A.4}$$

The product of two vector multiplets $V_1 = (C_1, \xi_1, v_1, V_{\mu 1}, \lambda_1, D_1)$ and $V_2 = (C_2, \xi_2, v_2, V_{\mu 2}, \lambda_2, D_2)$ is a vector multiplet

$$\begin{aligned} V_1 \cdot V_2 = (&\tfrac{1}{2}\, C_1 C_2,\ C_1\xi_2, C_1 v_2 - \tfrac{1}{2}\,\bar{\xi}_{1R}\xi_{2L}, C_1 V_{\mu 2} - \tfrac{i}{4}\,\bar{\xi}_1\gamma_\mu\gamma_5\xi_2, \\ &C_1\lambda_2 - \tfrac{1}{2}\,\hat{\not{D}} C_1\xi_2 + \tfrac{1}{2}\,\hat{v}_1^{*}\xi_2 + \tfrac{i}{2}\,\gamma_5\gamma^\mu\xi_2 V_{\mu 1}, \\ &C_1 D_2 - \tfrac{1}{2}\,\hat{D}_\mu C_1 \hat{D}^\mu C_2 - \tfrac{1}{2}\, V_{\mu 1} V_2^{\ \mu} + \tfrac{1}{2}\, v_1^{*} v_2 - \bar{\xi}_1\lambda_2 - \tfrac{1}{2}\,\bar{\xi}_2 \hat{\not{D}}\xi_1) + 1 \leftrightarrow 2 \end{aligned} \tag{A.5}$$

In Eq. (A.5) we have defined

$$\hat{D}_\mu C = \partial_\mu C + i\,\frac{\kappa}{2}\,\bar{\psi}_\mu\gamma_5\xi \ ; \ \hat{v} = K + i\gamma_5 H \tag{A.6}$$

$$\hat{D}_\mu \xi = D_\mu(\omega(e,\psi))\xi - \frac{\kappa}{2}\,(\not{V} + i\gamma_5\hat{\not{D}} C - i\gamma_5\hat{v})\psi_\mu - i\,\frac{\kappa}{2}\, A_\mu\gamma_5\xi \tag{A.7}$$

Using the above rule of multiplication one can obtain the components of the multiplet $\exp(g_\alpha V)$. This multiplet in the Wess-Zumino gauge[48] $(c = \xi = v = 0)$ has the simple form

$$\exp(e_\alpha V) = (1, 0, 0, g_\alpha V_\mu, e_\alpha\lambda, g_\alpha D - \frac{e_\alpha^{\ 2}}{2}\, V_\mu V^\mu) \tag{A.8}$$

The total Lagrangian for the couplings of an arbitrary number of chiral multiplets and simultaneously a gauge multiplet with supergravity is then given by

$$L = L(F_{\alpha\beta}) + L_{pot} + L_D \tag{A.9}$$

$L(F_{\alpha\beta})$ is given by Eq. (2.3). L_{pot} gives the most general couplings of the chiral multiplet with supergravity obtained on using Eq. (2.1). It has the form

$$\begin{aligned} e^{-1} L_{pot} = \tfrac{1}{2}\,(&g,_a h^a - 2g,_{ab}\overline{\chi_a}\chi^b + \kappa u g + \kappa g,_a\bar{\psi}_\mu\gamma^\mu\chi^a \\ &+ \kappa^2 g\,\bar{\psi}_\mu\sigma^{\mu\nu}\psi_{\nu R} + \text{h.c.}) \end{aligned} \tag{A.10}$$

In Eq. (A.10) $g(Z^a)$ is the general superpotential.

$$g(Z^a) = \sum B_{a_1 \ldots a_m} Z^{a_1} \ldots Z^{a_m} \tag{A.11}$$

where $B_{a_1 \ldots a_m}$ are arbitrary constants which maintain the gauge group invariance of g. To obtain L_D in Eq. (A.9) one first forms the most general vector multiplet by coupling chiral multiplets with the gauge multiplet. This general vector multiplet has the form

$$\frac{1}{2}\left(\Sigma_{a_1} \cdots \Sigma_{a_m} (A \exp(g_\alpha V))^{a_1 \cdots a_m}{}_{b_1 \cdots b_n} \Sigma^{b_1} \cdots \Sigma^{b_n} + h \cdot c\right) \quad (A.12)$$

The components of Eq. (A.12) are given in a convenient form in terms of the function $\phi(Z^a, Z_a)$ defined by

$$\phi(Z^a, Z_a) = \Sigma A^{a_1 \cdots a_m}{}_{b_1 \cdots b_n} Z_{a_i} \cdots Z_{a_m} Z^{b_1} \cdots Z^{b_n} \quad (A.13)$$

where $A^{a_1 \cdots a_m}{}_{b_1 \cdots b_n}$ are the arbitrary coupling constants which maintain the gauge group invariance of ϕ. On coupling the vector multiplet of Eq. (A.12) according to the rule of Eq. (2.2) one has[44]

$$\begin{aligned}
e^{-1}L_D = {} & \phi,{}^a{}_b h_a h^b - \phi,{}^{ab}{}_c \overline{\chi^a} \chi_b h^c - \phi,{}_{ab}{}^c \overline{\chi_a} \chi^b h_c \\
& + \phi,{}_{ab}{}^{cd} \overline{\chi_a} \chi^b \overline{\chi^c} \chi_d - \phi,{}^a{}_b \hat{\mathcal{D}}_\mu Z_a \hat{\mathcal{D}}^\mu Z^b \\
& - \phi,{}^a{}_b \overline{\chi^a} \overleftrightarrow{\hat{\mathcal{D}}} \chi^b - \overline{\chi^a} \gamma^\mu \chi^b (\phi,{}^a{}_{bc} \hat{\mathcal{D}}_\mu Z^c - \phi,{}^{ac}{}_b \hat{\mathcal{D}}_\mu Z_c) \\
& - i g_\alpha \phi,{}^a{}_b (\overline{\lambda^\alpha} \chi^b (T^\alpha Z)_a - \overline{\chi^a} \lambda^\alpha (T^\alpha Z)^b) \\
& + \frac{e_\alpha}{2} D^\alpha \phi,{}_a (T^\alpha Z)^a + \frac{\kappa}{2} \phi,{}^a{}_b \overline{\psi}_\mu \gamma_\mu (h_a \chi^b + h^b \chi_a) \\
& - \frac{\kappa}{2} (\overline{\psi}_\mu \gamma^\mu \chi^c \overline{\chi^a} \chi_b \phi,{}^{ab}{}_c - \overline{\chi^c} \gamma^\mu \psi_\mu \overline{\chi_a} \chi^b \phi,{}_{ab}{}^c) \\
& - \frac{\kappa}{2} \phi,{}^a{}_b (\overline{\psi}_\mu \gamma^\mu \gamma^\nu \chi^b \hat{\mathcal{D}}_\nu Z_a - \overline{\chi^a} \gamma^\nu \gamma^\mu \psi_\mu \hat{\mathcal{D}}_\nu Z^b) \\
& - i\kappa \frac{e_\alpha}{2} \overline{\psi}_\mu \gamma_5 \gamma^\mu \lambda^\alpha \phi,{}_a (T^\alpha Z)^a \\
& + \frac{\kappa}{3} [u^* (\phi,{}_a h^a - \phi,{}_{ab} \overline{\chi}_a \chi^b) + u(\phi,{}^a h_a - \phi,{}^{ab} \overline{\chi^a} \chi_b)] \\
& + (i \frac{\kappa}{3} A^\mu + e^{-1} \frac{\kappa^2}{8} \varepsilon^{\mu\nu\rho\sigma} \overline{\psi}_\nu \gamma_\rho \psi_\sigma)(\phi,{}_a \hat{D}_\mu Z^a - \phi,{}^a \hat{D}_\mu Z_a - 2\phi,{}^a{}_b \overline{\chi^a} \gamma_\mu \chi^b) \\
& + 4 \frac{\kappa}{3} (D_\mu \overline{\psi}_\nu \sigma^{\mu\nu} \chi^a \phi,{}_a - \phi,{}^a \overline{\chi^a} \sigma^{\mu\nu} D_\mu \psi_\nu)
\end{aligned}$$

$$- \frac{\kappa^2}{3} \phi\left(- \frac{R}{2\kappa^2} - \frac{e^{-1}}{2} \varepsilon^{\mu\nu\rho\sigma} \bar{\psi}_\mu \gamma_5 \gamma_\nu D_\rho \psi_\sigma - \frac{1}{3} |u|^2 + \frac{1}{3} A_\mu A^\mu\right)$$

$$- \frac{\kappa^3}{8} \varepsilon^{\mu\nu\rho\sigma} e^{-1} \bar{\psi}_\mu \gamma_\nu \psi_\rho (\bar{\psi}_\sigma \chi^a \phi,_a - \bar{\chi}^a \psi_\sigma \phi,^a) \tag{A.14}$$

where

$$u = S - iP \tag{A.15}$$

$$\hat{\mathcal{D}}_\mu Z^a = \hat{D}_\mu Z^a - i \frac{e_\alpha}{2} V^\alpha{}_\mu (T^\alpha)^a{}_b Z^b \tag{A.16}$$

$$\phi,_a = \frac{\partial \phi}{\partial Z^a}, \quad \phi,^a = \frac{\partial \phi}{\partial Z} \tag{A.17}$$

$$D_\mu \psi_\nu = (\partial_\mu + (1/2)\omega_{\mu rs} \sigma^{rs}) \psi_\nu \tag{A.18}$$

$$\omega_{\mu rs} = \omega_{\mu rs}(e) + K_{\mu rs}(e, \psi_\mu) \tag{A.19}$$

$$K_{\mu rs}(e, \psi_\mu) = (\kappa^2/4)(\bar{\psi}_\mu \gamma_r \psi_s - \bar{\psi}_\mu \gamma_s \psi_r + \bar{\psi}_r \gamma_\mu \psi_s) \tag{A.20}$$

$$R = e_r{}^\mu e_s{}^\nu R_{\mu\nu}{}^{rs} \tag{A.21}$$

$$R_{\mu\nu}{}^{rs} = \partial_\mu \omega_\nu{}^{rs} + \omega_\mu{}^{rp} \omega_{\mu p}{}^s - \mu \leftrightarrow \nu \tag{A.22}$$

The Lagrangian of Eq. (A.9) contains the auxiliary fields A_μ, h^a and D^α. To obtain the effective interactions of the physical fields, we eliminate the auxiliary fields using their field equations. Further, Weyl scalings are necessary to normalize the kinetic energies of the fields remaining. Details are given in Ref. (44) and the resulting effective Lagrangian is given by Eqs. (2.7) and (2.9)

APPENDIX B. MODEL INDEPENDENT INTERACTIONS

We list below the interactions arising from the SUSY sector in the SU(3)xSU(2)xU(1) invariant theory broken down spontaneously to SU(3)xU(1) in a model independent way.

(a) e.m. interactions: Here the contributions arise from the Winos and from the squarks and sleptons. These interactions are given by

$$eA_\mu\Big[\bar{\tilde{W}}_{(+)} \gamma_\mu \tilde{W}_{(+)} + \bar{\tilde{W}}_{(-)} \gamma_\mu \tilde{W}_{(-)} - \frac{i}{2} \Big[\frac{2}{3} (\tilde{u}^I{}_{iL}{}^\dagger \overleftrightarrow{\partial_\mu} \tilde{u}^I{}_{iL} + \tilde{u}^I{}_{iR}{}^\dagger \overleftrightarrow{\partial_\mu} \tilde{u}^I{}_{iR})$$

$$- \frac{1}{3} (\tilde{d}^I{}_{iL}{}^\dagger \overleftrightarrow{\partial_\mu} \tilde{d}^I{}_{iL} + \tilde{d}^I{}_{iR}{}^\dagger \overleftrightarrow{\partial_\mu} \tilde{d}^I{}_{iR}) - (\tilde{e}_{iL}{}^\dagger \overleftrightarrow{\partial_\mu} \tilde{e}_{iL} + \tilde{e}_{iR}{}^\dagger \overleftrightarrow{\partial_\mu} \tilde{e}_{iR})\Big] \Big] \tag{B.1}$$

(b) The weak neutral current interaction has the form

$$\frac{2e}{\sin 2\Theta_W} Z_\mu (J_\mu^{N1} + J_\mu^{N2}) \tag{B.2}$$

J_μ^{N1} represents the contribution of the squark and slepton fields to the weak neutral current and is given by

$$j_\mu^{N1} = \frac{i}{2} \{ \frac{1}{2} \tilde{u}^I{}_{iL}{}^\dagger \overleftrightarrow{\partial_\mu} \tilde{u}_{iL}{}^I - \frac{1}{2} \tilde{d}^I{}_{iL}{}^\dagger \overleftrightarrow{\partial_\mu} \tilde{d}^I{}_{iL} + \frac{1}{2} \tilde{\nu}_{iL}{}^\dagger \overleftrightarrow{\partial_\mu} \nu_{iL} - \frac{1}{2} \tilde{e}_{iL}{}^\dagger \overleftrightarrow{\partial_\mu} \tilde{e}_{iL} \}$$

$$+ \sin^2\theta_W \quad [\frac{i}{2} [\frac{2}{3} (\tilde{u}^I{}_{iL}{}^\dagger \overleftrightarrow{\partial_\mu} \tilde{u}^I{}_{iL} + \tilde{u}^I{}_{iR}{}^\dagger \overleftrightarrow{\partial_\mu} \tilde{u}^I{}_{iR})$$

$$- \frac{1}{3} (\tilde{d}^I{}_{iL}{}^\dagger \overleftrightarrow{\partial_\mu} \tilde{d}^I{}_{iL} + \tilde{d}^I{}_{iR}{}^\dagger \overleftrightarrow{\partial_\mu} \tilde{d}^I{}_{iR})$$

$$- (\tilde{e}_{iL}{}^\dagger \overleftrightarrow{\partial_\mu} \tilde{e}_{iL} + \tilde{e}_{iR}{}^\dagger \overleftrightarrow{\partial_\mu} \tilde{e}_{iR})]] \tag{B.3}$$

J_μ^{N2} represents the contribution of the Winos and the Zinos to the weak neutral current. It has the form[27]

$$(J_\mu)^{N2} = - [\bar{\tilde{W}}_{(+)}\gamma_\mu \{(\cos^2\Theta_W - \frac{1}{2}\cos^2\gamma_-)P_L + (\cos^2\Theta_W - \frac{1}{2}\sin^2\gamma_+)P_R\}\tilde{W}_{(+)}$$

$$+ \bar{\tilde{W}}_{(-)}\gamma_\mu \{(\cos^2\Theta_W - \frac{1}{2}\sin^2\gamma_-)P_L + (\cos^2\Theta_W - \frac{1}{2}\cos^2\gamma_+)P_R\}\tilde{W}_{(-)}$$

$$+ \frac{1}{4} \bar{\tilde{W}}_{(-)}\gamma_\mu \{(\sin 2\gamma_-)P_L + (E\sin 2\gamma_+)P_R\}\tilde{W}_{(+)}$$

$$+ \frac{1}{4} \bar{\tilde{W}}_{(+)}\gamma_\mu \{(\sin 2\gamma_-)P_L + (E\sin 2\gamma_+)P_R\}\tilde{W}_{(-)}$$

$$- \frac{1}{4} \sum_{k,k'} \{(0_{2k}0_{2k'} - 0_{3k}0_{3k'})\cos 2\alpha - 2\sin 2\alpha\, 0_{2k}0_{3k'}\}.$$

$$x\ \bar{\tilde{Z}}_{(k)}(-i\gamma_5)^{\Theta_k}(-i\gamma_5)^{\Theta_{k'}}\gamma_\mu \gamma_5 \tilde{Z}_{(k')};\ E = (-1)^\Theta \text{ (See Eq. 6.13)} \tag{B.4}$$

(c) Charged Current Interaction: This interaction has the form

$$\frac{e}{\sqrt{2}\sin\Theta_W} (W_\mu J^{\mu C} + W_\mu{}^+ J^{\mu C+}) \tag{B.5}$$

We list below the contributions to $J_\mu{}^C$ from various sectors.

(i) squark-slepton contribution

$$- \frac{i}{2} (\tilde{u}^{I\dagger}{}_{iL} \overleftrightarrow{\partial_\mu} \tilde{d}^{I}{}_{iL} + \tilde{\nu}^{\dagger}{}_{iL} \overleftrightarrow{\partial_\mu} e_{iL}) \tag{B.6}$$

(ii) $\tilde{W}_{(-)}$ contribution[27]

$$\begin{aligned} &\frac{1}{2} \sum_k \bar{\tilde{W}}_{(-)} \gamma_\mu \{[-\cos\alpha(\sin\gamma_- O_{2k} + E\cos\gamma_+ O_{3k}) - \sin\alpha(E\cos\gamma_+ O_{2k} - \sin\gamma_- O_{3k})] \\ &\quad + \gamma_5 [\cos\alpha(\sin\gamma_- O_{2k} - E\cos\gamma_+ O_{3k}) - \sin\alpha(E\cos\gamma_+ O_{2k} + \sin\gamma_- O_{3k})]\} \\ &\qquad \times (-i\gamma_5)^{\Theta_k} \tilde{Z}_{(k)} \\ &+ \frac{1}{\sqrt{2}} \sum_k \bar{\tilde{W}}_{(-)} \gamma_\mu \{(E\sin\gamma_+ - \cos\gamma_-) + \gamma_5 (E\sin\gamma_+ + \cos\gamma_-)(\cos\Theta_W O_{1k} - \sin\Theta_W O_{ok}\} \\ &\qquad \cdot (-i\gamma_5)^{\Theta_k} \tilde{Z}_{(k)} \end{aligned} \tag{B.7}$$

(iii) $\tilde{W}_{(+)}$ contribution[27]

$$\begin{aligned} &\frac{1}{2} \sum_k \bar{\tilde{W}}_{(+)} \gamma_\mu \{[\cos\alpha(\cos\gamma_- O_{2k} + \sin\gamma_+ O_{3k}) + \sin\alpha(\sin\gamma_+ O_{2k} - \cos\gamma_- O_{3k})] \\ &\quad - \gamma_5 [\cos\alpha(\cos\gamma_- O_{2k} - \sin\gamma_+ O_{3k}) - \sin\alpha(\sin\gamma_+ O_{2k} + \cos\gamma_- O_{3k})]\} \\ &\qquad \times (-i\gamma_5)^{\Theta_k} \tilde{Z}_{(k)} \\ &+ \frac{1}{\sqrt{2}} \sum_k \bar{\tilde{W}}_{(+)} \gamma_\mu \{(\cos\gamma_+ - \sin\gamma_-) + \gamma_5 (\cos\gamma_+ + \sin\gamma_-)\}(\cos\Theta_W O_{1k} - \sin\Theta_W O_{ok}) \\ &\qquad \times (-i\gamma_5)^{\Theta_k} \tilde{Z}_{(k)} \end{aligned} \tag{B.8}$$

(d) Gluon Interactions: Gluon interactions with squarks, sleptons and gluinos are

$$\begin{aligned} &\frac{e_3}{2} V_\mu{}^r [\bar{d}_i{}^I \gamma^\mu (t^r)^J{}_I d_i{}^J + \bar{u}_i{}^I \gamma^\mu (t^r)^I{}_J u_i{}^J - i f_{rst} \bar{\lambda}^s \gamma^\mu \lambda^t \\ &\quad - \frac{i}{2} \{(\tilde{d}^{J\dagger}{}_{iL} D^{\mu I}{}_K \tilde{d}^K{}_{iL} + d^{J\dagger}{}_{iR} D^{\mu I}{}_k \tilde{d}^k{}_{iR} \\ &\quad + \tilde{u}^{J\dagger}{}_{iL} D^{\mu I}{}_k \tilde{u}^K{}_{iL} + \tilde{u}^J{}_{iR} D^{\mu I}{}_k \tilde{u}^K{}_{iR})(t^r)^J{}_I \\ &\qquad - h.c.\}]; \quad r = 1 \ldots 8 \end{aligned} \tag{B.9}$$

where D^{μ} is the Yang-Mills gauge covariant derivative.

Interactions (a) - (d) share the common feature that the essential change over the results of the standard SU(3)xSU(2)xU(1) theory is that the sources of A_{μ}, Z_{μ}, W_{μ} and $V_{\mu}{}^{r}$ develop additional contributions which arise from the SUSY particles i.e., squarks, sleptons, Winos, Zinos and gluinos. In addition to the above, there arise in the theory other types of interactions, required by supersymmetry invariance, which have no analogue in the standard theory. We list these interactions below.

(e) Wino-quark-squark and Wino-lepton-slepton Interactions:[27]

$$\frac{ie}{\sqrt{2}\,\sin\Theta_W}\,[\overline{u^I{}_i}P_R(\cos\gamma_+\tilde{W}_{(+)} + E\sin\gamma_+\tilde{W}_{(-)})\tilde{d}^I{}_{iL}$$

$$- \overline{d^I{}_i}P_R(\sin\gamma_-\tilde{W}^C{}_{(+)} + \cos\gamma_-\tilde{W}^C{}_{(-)})\tilde{u}^I{}_{iL}$$

$$+ \overline{\nu_i}P_R(\cos\gamma_+\tilde{W}_{(+)} + E\sin\gamma_+\tilde{W}_{(-)})\tilde{e}_{iL}$$

$$- \overline{e_i}P_R(\sin\gamma_-\tilde{W}^C{}_{(+)} + \cos\gamma_-\tilde{W}^C{}_{(-)})\tilde{\nu}_{iL}] + \text{h.c.} \qquad \text{(B.10)}$$

these interactions govern the decay of the Winos.

(f) Zino-quark-squark Interactions:[27]

$$ie\sum_k [O_{ok}S_\gamma - \frac{2}{\sin 2\Theta_W}\,O_{1k}(S_3 - \sin^2\Theta_W S_\gamma)]$$

$$(-i\gamma_5)^{\Theta_k}\tilde{Z}_{(k)} \qquad \text{(B.11)}$$

$$S_3 \equiv \frac{1}{2}\,(\bar{u}_L\tilde{u}_L - \bar{d}_L\tilde{d}_L) + \frac{1}{2}\,(\bar{\nu}_L\tilde{\nu}_L - \bar{e}_L\tilde{e}_L) \qquad \text{(B.12)}$$

$$S_\gamma \equiv \frac{2}{3}\,\bar{u}(P_L\tilde{u}_R + P_R\tilde{u}_L) - \frac{1}{3}\,\bar{d}(P_L\tilde{d}_R + P_R\tilde{d}_L)$$

$$- \bar{e}(P_L\tilde{e}_R + P_R\tilde{e}_L) \qquad \text{(B.13)}$$

These interactions contribute to the decay of the Zino as well as to the decay of the squark (slepton) into a quark (lepton) and a photino.

(g) Gluino-quark-squark Interactions:

$$\frac{ie_3}{2} [\overline{u^I{}_i(t^r)^I{}_J}(P_R\tilde{u}^J{}_{iL} + P_L\tilde{u}^J{}_{iR})$$

$$+ \overline{d^I{}_i(t^r)^I{}_J}(P_R\tilde{d}^J{}_{iL} + P_L\tilde{d}^J{}_{iR})]\lambda^r + \text{h.c.} \tag{B.14}$$

These interactions govern the decay of the squark into a quark and a gluino.

In addition, there are quartic interactions arising from gauge covariant derivatives of the s-matter bose fields. Thus the quartic interactions involving W_μ are:

$$- \frac{e^2}{2} [\frac{W^\mu W^\dagger_\mu}{2\sin^2\Theta_W} \{\tilde{e}_{iL}{}^\dagger\tilde{e}_{iL} + \tilde{\nu}_{iL}{}^\dagger\tilde{\nu}_{iL} + \tilde{u}^I{}_{iL}{}^\dagger\tilde{u}^I{}_{iL} + \tilde{d}^I{}_{iL}{}^\dagger\tilde{d}^I{}_{iL}\}$$

$$+ \frac{1}{\sqrt{2}} (\frac{A^\mu}{\sin\Theta_W} + \frac{Z^\mu}{\cos\Theta_W}) \{W_\mu(-\tilde{\nu}_{iL}{}^\dagger\tilde{e}_{iL} + \frac{1}{3}\tilde{u}^I{}_{iL}{}^\dagger\tilde{d}^I{}_{iL}) + \text{h.c.}] \tag{B.15}$$

while the remaining quartic interactions involving A_μ and Z_μ are

$$- \frac{e^2}{2} [A^\mu A_\mu\{(\tilde{e}_{iL}{}^\dagger\tilde{e}_{iL} + \tilde{e}_{iR}{}^\dagger\tilde{e}_{iR}) + \frac{4}{9} (\tilde{u}^I{}_{iL}{}^\dagger\tilde{u}^I{}_{iL} + \tilde{u}^I{}_{iR}{}^\dagger\tilde{u}^I{}_{iR})$$

$$+ \frac{1}{9} (\tilde{d}^{I\dagger}{}_{iL}\tilde{d}^I{}_{iL} + \tilde{d}^{I\dagger}{}_{iR}\tilde{d}^I{}_{iR})\}$$

$$+ \frac{2A^\mu Z_\mu}{\sin 2\Theta_W} \{- \cos 2\Theta_W\tilde{e}_{iL}{}^\dagger\tilde{e}_{iL} + 2\sin^2\Theta_W\tilde{e}_{iR}{}^\dagger\tilde{e}_{iR}$$

$$- \frac{2}{9} (4\cos^2\Theta_W - 1)\tilde{u}^I{}_{iL}{}^\dagger\tilde{u}^I{}_{iL} + \frac{8}{9} \sin^2\Theta_W u^I{}_{iR}{}^\dagger\tilde{u}^I{}_{iR}$$

$$- \frac{1}{9} (2\cos^2\Theta_W + 1)\tilde{d}^I{}_{iL}{}^\dagger\tilde{d}^I{}_{iL} + \frac{2}{9} \sin^2\Theta_W\tilde{d}^I{}_{iR}{}^\dagger\tilde{d}^I{}_{iR}\}$$

$$+ \frac{Z^\mu Z_\mu}{(\sin 2\Theta_W)^2} \{\tilde{\nu}_{iL}{}^\dagger\tilde{\nu}_{iL} + \cos^2 2\Theta_W\, \tilde{e}_{iL}{}^\dagger\tilde{e}_{iL} + 4\sin^4\Theta_W\tilde{e}_{iR}{}^\dagger\tilde{e}_{IR}$$

$$+ \frac{1}{9} (4\cos^2\Theta_W - 1)^2\tilde{u}^I{}_{iL}{}^\dagger\tilde{u}^I{}_{iL} + \frac{16}{9} \sin^4\Theta_W\tilde{u}^I{}_{iR}{}^\dagger\tilde{u}^I{}_{iR}$$

$$+ \frac{1}{9} (2\cos^2\Theta_W + 1)^2\tilde{d}^I{}_{iL}{}^\dagger\tilde{d}^I{}_{iL} + \frac{4}{9} \sin^4\Theta_W\tilde{d}^I{}_{iR}{}^\dagger\tilde{d}^I{}_{iR}] \tag{B.16}$$

APPENDIX C: DECAY WIDTHS OF W AND Z

In this Appendix we give analytic expressions for the decay widths of the W and the Z bosons into various channels in the model independent formalism[11,26-32]. We follow here the analysis of Refs. 26, 27 and 44.

(a) $W \to \tilde{W}_- + \tilde{\gamma}$

The above decay occurs through Eq. (B.7). The decay width of W into a (lower) Wino and a photino relative to $W \to e\nu$ is

$$R(W \to \tilde{W}_{(-)} + \tilde{\gamma}) = 4\sin^2\Theta_W[(\xi_+{}^2 + \xi_-{}^2)$$

$$\times\{1 - (2M_W{}^2)^{-1}(\tilde{m}_-{}^2 + \tilde{m}_\gamma{}^2) - (2M_W{}^4)^{-1}(\tilde{m}_-{}^2 - \tilde{m}_\gamma{}^2)^2\}$$

$$+ (-1)^\Theta 3(\xi_+{}^2 - \xi_-{}^2)\tilde{m}_-\tilde{m}_\gamma M_W{}^{-2}] \cdot [1 - 2(\tilde{m}_-{}^2 + \tilde{m}_\gamma{}^2)M_W{}^{-2}$$

$$+ (\tilde{m}_-{}^2 - \tilde{m}_\gamma{}^2)^2 M_W{}^{-4}]^{-1/2} \tag{C.1}$$

$$\xi_\pm = 2^{-1}(\sin\gamma_+ \mp \cos\gamma_-) \tag{C.2}$$

where Θ in Eq. (C.1) and $\gamma_\pm$ in Eq. (C.2) are defined following Eq. (6.11).

(b) $W \to \tilde{W}_- + \tilde{Z}_-$

The above decay is governed by Eq. (B.7). If energetically allowed, the decay width relative to $W \to e\nu$ is given by

$$R(W^\pm \to \tilde{W}^\pm_{(-)} + \tilde{Z}_{(-)}) = 2[1 - 2(\tilde{m}_-{}^2 + \tilde{\mu}_-{}^2)M_W{}^{-2} + (m_-{}^2 - \tilde{\mu}_-{}^2)^2 M_W{}^{-4}]^{1/2}$$

$$\times\ [\tfrac{1}{2}(A_L{}^2 + A_R{}^2)\{1 - \tfrac{1}{2}(\tilde{m}_-{}^2 + \tilde{\mu}_-{}^2)M_W{}^{-2} - \tfrac{1}{2}(\tilde{m}_-{}^2 - \tilde{\mu}_-{}^2)^2 M_W{}^{-4}\}$$

$$+ 3A_L A_R\, \tilde{m}_-\tilde{\mu}_-\, M_W{}^{-2}] \tag{C.3}$$

$$A_L = -\sqrt{2}\cos\Theta_W \cos\gamma_- O_{1-} - (O_{2-}\cos\alpha - O_{3-}\sin\alpha)\sin\gamma_- \tag{C.4}$$

$$A_R = (-1)^{\Theta+\Theta_-}[\sqrt{2}\cos\Theta_W \sin\gamma_+ O_{1-} - (O_{2-}\sin\alpha + O_{3-}\cos\alpha)\cos\gamma_+] \tag{C.5}$$

where Θ is defined in Eq. (6.13) and Θ_- is defined following Eq. (6.23).

(c) $W \to \tilde{W}_- + \tilde{Z}_{(3)}$

The decay width for this process relative to $W \to e\nu$ is

$$R(W^\pm \to W^\pm_{(-)} + \tilde{Z}_{(3)}) = 2[1 - 2(\tilde{m}_-{}^2 + \tilde{\mu}_{(3)}{}^2)M_W{}^{-2} + (\tilde{m}_-{}^2 + \tilde{\mu}_{(3)}{}^2)^2 M_W{}^{-4}]^{1/2}$$

$$\times \left[\frac{1}{2}(B_L^2 + B_R^2)\{1 - \frac{1}{2}(\tilde{m}_-^2 + \tilde{\mu}_{(3)}^2)M_W^{-2}\right.$$

$$\left. - \frac{1}{2}(\tilde{m}_-^2 - \tilde{\mu}_{(3)}^2)^2 M_W^{-4} + 3B_L B_R \tilde{m}_- \tilde{\mu}_{(3)} M_W^{-2}\right] \quad (C.6)$$

$$B_L = -\sqrt{2}\cos\Theta_W \cos\gamma_- O_{13} - (O_{23}\cos\alpha - O_{33}\sin\alpha)\sin\gamma_- \quad (C.7)$$

$$B_R = (-1)^{\Theta+\Theta_3}[\sqrt{2}\cos\Theta_W \sin\gamma_+ O_{13} - (O_{23}\sin\alpha + O_{33}\cos\alpha)\cos\gamma_+] \quad (C.8)$$

Θ_3 is defined following Eq. (6.23).

(c) $Z^0 \to \tilde{W}_{(-)}^+ + \tilde{W}_{(-)}^-$

The decay of the Z^0 into two lower Winos is governed by Eqs. (B.2) and (B.4). The decay width for this process relative to $Z^0 \to e^+e^-$ is given by

$$R[Z^0 \to \tilde{W}_-^+ \tilde{W}_-^-) = 16\sin^2\Theta_W \cos^2\Theta_W [1 + (1 - 4\sin^2\Theta_W)^2]^{-1}$$

$$[(v^2 + a^2)(1 - \tilde{m}_-^2 M_Z^{-2}) + 3(v^2 - a^2)\tilde{m}_-^2 M_Z^{-2}]$$

$$[1 - 4\tilde{m}_-^2 M_Z^{-2}]^{1/2} \quad (C.9)$$

$$v = 2^{-1}[\cot\Theta_W(\sin^2\gamma_+ + \cos^2\gamma_-) + \cot 2\Theta_W(\cos^2\gamma_+ + \sin^2\gamma_-)] \quad (C.10)$$

$$a = 2^{-1}[\cot\Theta_W(\sin^2\gamma_+ - \cos^2\gamma_-) + \cot 2\Theta_W(\cos^2\gamma_+ - \sin^2\gamma_-)] \quad (C.11)$$

(d) $Z^0 \to \tilde{Z}_{(3)} + \tilde{Z}_{(3)}$

The interactions that govern this decay are Eqs. (B.2) and (B.4). The decay width relative to $Z^0 \to e^+e^-$ is

$$R(Z \to \tilde{Z}_{(3)} + \tilde{Z}_{(3)}) = 2\, g_{(3)}^2[1 + (1 - 4\sin^2\Theta_W)^2]^{-1}$$

$$\times [1 - 4\tilde{\mu}_{(3)}^2 M_Z^{-2}]^{3/2} \quad (C.12)$$

$$g_{(3)} = \cos 2\alpha(O_{23}^2 - O_{33}^2) - 2\sin 2\alpha O_{23}O_{33} \quad (C.13)$$

Next we exhibit the decay branching ratios of the Winos and the Zinos.

(e) $\tilde{W}_{(-)}^{+} \to u + \bar{d} + \tilde{\gamma}$

The interactions for this decay are given by Eqs. (B.10) and (B.11). The decay branching ratio is

$$\Gamma_{\tilde{W}}(u\bar{d}\tilde{\gamma}) = \Gamma_0 \frac{3\alpha^2}{\sin^2\Theta_W} a_\gamma^{\,5}[P_1(a_\gamma)\{\sin^2\gamma_+(\tfrac{1}{3} m_{\tilde{d}_L}^{\,-2} + M_W^{\,-2})^2$$

$$+ \cos^2\gamma_-(\tfrac{2}{3} m_{\tilde{u}_L}^{\,-2} + M_W^{\,-2})^2\} + (-1)^\Theta \frac{\tilde{m}_\gamma}{\tilde{m}_-} P_2(a_\gamma)\cos\gamma_-\sin\gamma_+$$

$$\times \{M_W^{\,-4} + \tfrac{2}{9} m_{\tilde{u}_L}^{\,-2} m_{\tilde{d}_L}^{\,-2} + (\tfrac{2}{3} m_{\tilde{u}_L}^{\,-2} + \tfrac{1}{3} m_{\tilde{d}_L}^{\,-2})M_W^{\,-2}\}] \qquad (C.14)$$

$$P_n(a) = \int_0^1 d\xi\xi^2(1 - \xi)^2(1 - a\xi)^{-n} \qquad (C.15)$$

$$\alpha = e^2/4\pi, \quad a_\gamma = 1 - (\tilde{m}_\gamma/\tilde{m}_-)^2 \qquad (C.16)$$

where Γ_0 is a normalization factor and P_n are phase space integrals.

(f) $\tilde{W}_{(-)}^{+} \to u + \bar{d} + \tilde{g}$

This decay is governed by the interactions of Eqs. (B.10) and (B.14). The decay branching ratio is

$$\Gamma_{\tilde{W}}(u\bar{d}\tilde{g}) = \Gamma_0 \frac{4\alpha\alpha_3}{\sin^2\Theta_W} a_g^{\,5}[P_1(a_g)(\sin^2\gamma_+ m_{\tilde{d}_L}^{\,-4} + \cos^2\gamma_- m_{\tilde{u}_L}^{\,-4})$$

$$- (-1)^\Theta \frac{\tilde{m}_g}{\tilde{m}_-} P_2(a_g)\cos\gamma_-\sin\gamma_+ m_{\tilde{u}_L}^{\,-2} m_{\tilde{d}_L}^{\,-2}] \qquad (C.17)$$

$$a_g = 1 - \tilde{m}_g^{\,2}/\tilde{m}_-^{\,2}, \quad \alpha_3 = e_3^{\,2}/4\pi \qquad (C.18)$$

where α_3 is the SU(3) color coupling constant and $\tilde{m}_g$ is the gluino mass.

(g) $\tilde{W}_{(-)}^{+} \to e^+ + \nu_\ell + \tilde{\gamma}$

The interactions that govern this decay are Eqs. (B.10) and (B.11). The branching ratio is

$$\Gamma_{\tilde{W}}(\ell\, \nu_\ell\, \tilde{\gamma}) = \Gamma_0 \frac{\alpha^2}{\sin^2\Theta_W} a_\gamma^{\,5}[P_1(a_\gamma)\{\sin^2\gamma_+(m_{\tilde{e}_L}^{\,-2} + m_W^{\,-2})^2$$

$$+ \cos^2\gamma_- M_W^{\,-4}\} + (-1)^\Theta \frac{\tilde{m}_\gamma}{\tilde{m}_-} P_2(a_\gamma)\cos\gamma_-\sin\gamma_+(M_W^{\,-4}$$

$$+ m_{\tilde{e}}^{-2} M_W^{-2})] \tag{C.19}$$

(h) $\tilde{Z} \to q + \bar{q} + \tilde{\gamma}$

The Zino decays $\tilde{Z} \to u\bar{u}\tilde{\gamma}$ and $\tilde{Z} \to \bar{d}d\tilde{\gamma}$ are governed by Eq. (B.11). We shall normalize the Zino branching ratios relative to $\tilde{Z} \to u\bar{u}\tilde{\gamma}$ for convenience. We thus have

$$R_{\tilde{Z}}(\bar{d}d\tilde{\gamma}) \equiv \Gamma_{\tilde{Z}}(\bar{d}d\tilde{\gamma})/\Gamma_{\tilde{Z}}(\bar{u}u\tilde{\gamma}) = \frac{(a_d^2 + b_d^2) m_{\tilde{d}}^{-4}}{4(a_u^2 + b_u^2) m_{\tilde{u}}^{-4}} \tag{C.20}$$

$$a_u = \frac{1}{4}\left(\frac{5}{3}\tan\Theta_W - \cot\Theta_W\right); \quad b_u = (2\sin\Theta_W)^{-1} \tag{C.21}$$

$$a_d = \frac{1}{4}\left(\frac{1}{3}\tan\Theta_W - \cot\Theta_W\right); \quad b_d = b_u \tag{C.22}$$

(i) $\tilde{Z} \to q + \bar{q} + \tilde{g}$

The branching ratio for this process relative to $\bar{u}u\tilde{\gamma}$ is obtained from Eqs. (B.10) and (B.14). We have

$$R_{\tilde{Z}}(\bar{u}u\tilde{g}) = \frac{3\alpha_s}{\alpha}\frac{a_g^5}{a_\gamma^5}\left[P_1(a_g) + (-1)^{\Theta_-}\frac{\tilde{m}_g}{2\tilde{\mu}_-}P_2(a_g)\right]$$

$$\times \left[P_1(a_\gamma) + (-1)^{\Theta_-}\frac{\tilde{m}_\gamma}{2\tilde{\mu}_-}P_2(a_\gamma)\right]^{-1} \tag{C.23}$$

and

$$R_{\tilde{Z}}(\bar{d}d\tilde{g}) = R_{\tilde{Z}}(\bar{u}u\tilde{g})\,\frac{(a_d^2 + b_d^2)\, m_{\tilde{d}}^{-4}}{(a_u^2 + b_u^2)\, m_{\tilde{u}}^{-4}} \tag{C.24}$$

(j) $\tilde{Z} \to \ell + \bar{\ell} + \tilde{\gamma}$

The decay is governed by Eqs. (B.10) and (B.14). The branching ratio relative to $u\bar{u}\tilde{\gamma}$ is

$$R_{\tilde{Z}}(\bar{e}e\tilde{\gamma}) = (a_e^2 + b_e^2) m_{\tilde{e}}^{-4}\left[\frac{4}{3}(a_u^2 + b_u^2) m_{\tilde{u}}^{-4}\right]^{-1} \tag{C.25}$$

$$a_e = \frac{1}{4}(3\tan\Theta_W - \cot\Theta_W); \quad b_e = b_u \tag{C.26}$$

(k) $\tilde{Z} \to u + \bar{d} + \tilde{W}_{(-)}$

The interactions relevant for this decay are Eqs. (B.10) and (B.11). The branching ratio is

$$R_{\tilde{Z}}(u\bar{d}\tilde{W}_{(-)}) = R_{\tilde{Z}}(u\bar{u}\tilde{g})\,\frac{3\alpha}{4\alpha_s}\,\frac{A_L{}^2 m_{\tilde{u}}{}^4 M_W{}^{-4}}{2\sin^4\Theta_W O_{1-}{}^2(a_u{}^2 + b_u{}^2)}$$

$$\times \frac{a_W{}^5}{a_g{}^5}\,[P_1(a_W) - \frac{\tilde{m}_-}{2\tilde{\mu}_-} P_2(a_W)][P_1(a_g)$$

$$+ (-1)^{\Theta_-}\frac{\tilde{m}_g}{2\tilde{\mu}_-} P_2(a_g)]^{-1} \qquad (C.27)$$

$$a_W \equiv 1 - (\tilde{m}_-/\tilde{\mu}_-)^2 \qquad (C.28)$$

(l) $\tilde{Z} \to e^+\, \nu\tilde{W}_{(-)}{}^-, \quad e^-\, \bar{\nu}\tilde{W}_{(-)}{}^+$

From Eqs. (B.10) and (B.11) we have

$$\Gamma_{\tilde{Z}}(e^+\, \nu\tilde{W}_{(-)}{}^-) = \frac{1}{3}\,\Gamma_{\tilde{Z}}(u\bar{d}\tilde{W}_{(-)}{}^-) \qquad (C.29)$$

$$\Gamma_{\tilde{Z}}(e^-\, \bar{\nu}\tilde{W}_{(-)}{}^+) = \frac{1}{3}\,\Gamma_{\tilde{Z}}(\bar{u}d\tilde{W}_{(-)}{}^+) \qquad (C.30)$$

(m) $\tilde{Z}_{(3)} \to q + \bar{q} + \tilde{\gamma}$

From Eq. (B.11) we have

$$\Gamma_{\tilde{Z}_{(3)}}(u\bar{u}\tilde{\gamma}) = \Gamma_3\,[\frac{1}{3}\,(\cot\Theta_W - \frac{1}{3}\tan\Theta_W)^2 m_{\tilde{u}_L}{}^{-4} + \frac{16}{27}\tan^2\Theta_W m_{\tilde{u}_R}{}^{-4}] \qquad (C.31)$$

$$\Gamma_{\tilde{Z}_{(3)}}(d\bar{d}\tilde{\gamma}) = \Gamma_3\,[\frac{1}{12}\,(\cot\Theta_W + \frac{1}{3}\tan\Theta_W)^2 m_{\tilde{d}_L}{}^{-4} + \frac{1}{27}\tan^2\Theta_W m_{\tilde{d}_R}{}^{-4}] \qquad (C.32)$$

where Γ_3 is a normalization factor.

(n) $\tilde{Z}_{(3)} \to \ell + \bar{\ell} + \tilde{\gamma}$

Finally from Eq. (B.11) we find that the branching ratio of the leptonic-photino mode of $\tilde{Z}_{(3)}$ is given by

$$\Gamma_{\tilde{Z}_{(3)}}(e\bar{e}\tilde{\gamma}) = \Gamma_3\,[\frac{1}{4}\,(\cot\Theta_W - \tan\Theta_W)^2 m_{\tilde{e}_L}{}^{-4} + \tan^2\Theta_W m_{\tilde{e}_R}{}^{-4}] \qquad (C.33)$$

APPENDIX D: SUPERSYMMETRY EFFECTS ON MUON ANOMALOUS MOMENT

The vertices that enter in the supersymmetric contributions to the anomalous magnetic moment are $\mu\tilde{\nu}\tilde{W}$ and $\mu\tilde{\mu}\tilde{Z}$ and are given by Eqs. (B.10) and (B.11). Contributions arising from supersymmetry are as follows (for details see Refs. (60), (59), (58)).

(1) Wino Contributions: These arise from the following part of Eq. (B.10):

$$L_{\mu\tilde{W}\tilde{\nu}} = i\bar{\mu}(\lambda_+\tilde{W}_{(+)}{}^c + \lambda_-\tilde{W}_{(-)}{}^c)\tilde{\nu} + \text{h.c.} \tag{D.1}$$

where

$$\lambda_a = A_R{}^{(a)}P_R + A_L{}^{(a)}P_L, \quad a = \pm \tag{D.2}$$

$$A_R{}^{(+)} = -\frac{e}{\sqrt{2}\sin\Theta_W}\sin\gamma_- \tag{D.3}$$

$$A_L{}^{(+)} = -\frac{em_\mu}{2M_W\sin\Theta_W\sin\alpha}\sin\gamma_+ \tag{D.4}$$

$$A_R{}^{(-)} = -\frac{e}{\sqrt{2}\sin\Theta_W}\cos\gamma_- \tag{D.5}$$

$$A_L{}^{(-)} = (-1)^{\Theta}\frac{em_\mu}{2M_W\sin\Theta_W\sin\alpha}\cos\gamma_+ \tag{D.6}$$

The Wino contributions to the anomalous moment have the form[60]

$$g^{\tilde{W}} = g_1{}^{\tilde{W}} + g_2{}^{\tilde{W}} \tag{D.7}$$

where

$$g_1{}^{\tilde{W}} = \frac{m_\mu{}^2}{24\pi^2}\frac{A_R{}^{(a)}}{\tilde{m}_a{}^2}F_1(x_a); \quad x_a \equiv \frac{\tilde{m}_\nu{}^2}{\tilde{m}_a{}^2}, \quad a = +, - \tag{D.8}$$

$$g_2{}^{\tilde{W}} = \frac{m_\mu}{4\pi^2}\frac{A_R{}^{(a)}A_L{}^{(a)}}{\tilde{m}_a}F_2(x_a); \quad a = +, - \tag{D.9}$$

and

$$F_1(x) \equiv (1 - 5x - 2x^2)(1-x)^{-3} - 6x^2(1-x)^{-4}\ln x \tag{D.10}$$

$$F_2(x) \equiv (1-3x)(1-x)^{-2} - 2x^2(1-x)^{-3}\ln x \tag{D.11}$$

(2) Zino Contributions: Anologous to Eq. (D.7) one may write[60]

$$g^{\tilde{Z}} = g_1^{\tilde{Z}} + g_2^{\tilde{Z}} \tag{D.12}$$

First for $g_1^{\tilde{Z}}$ we have

$$g_1^{\tilde{Z}} = -\frac{m_\mu^2}{12\pi^2}\frac{1}{\tilde{\mu}_{(k)}}[\{\sin^2\delta(B_k^R)^2 + \cos^2\delta(B_k^L)^2\}G_1(x_{1k})$$

$$+ \{\cos^2\delta(B_k^R)^2 + \sin^2\delta(B_k^L)^2\}G_2(x_{2k})]$$

$$x_{rk} \equiv (\tilde{m}_{\mu_r}/\tilde{\mu}_{(k)})^2, \quad r = 1,2 \tag{D.13}$$

where

$$\cos\delta\sin\delta = \frac{1}{2} Am_\mu m_g[\frac{1}{4}(\tilde{m}_L^2 - \tilde{m}_R^2)^2 + A^2 m_\mu^2 m_g^2]^{-1/2} \tag{D.14}$$

and

$$B_k^R = e[-O_{ok} + \cot 2\Theta_W O_{1k}] \tag{D.15}$$

$$B_k^L = e[-O_{ok} - \tan\Theta_W O_{1k}](-1)^{\Theta_k} \tag{D.16}$$

$$G_1(x) \equiv \frac{1}{2}(2 + 5x - x^2)(1-x)^{-3} + 3x(1-x)^{-4}\ln x \tag{D.17}$$

$g_2^{\tilde{Z}}$ is the chiral interference that arises from diagonalization of the smuon mass matrix. For $g_2^{\tilde{Z}}$ one gets

$$g_2^{\tilde{Z}} = \bar{g}_2^{\tilde{Z}} + g_2^{\tilde{Z}Y} \tag{D.18}$$

where

$$\bar{g}_2^{\tilde{Z}} = \frac{m_\mu}{4\pi^2}\cos\delta\sin\delta\frac{1}{\tilde{\mu}_{(k)}} B_k^R B_k^L[G_2(x_{1k}) - G_2(x_{2k})] \tag{D.19}$$

$$g_2^{\tilde{Z}Y} = -\frac{m_\mu}{4\pi^2}\frac{1}{\tilde{\mu}_{(k)}}[\{(\cos^2\delta)B_k^L - (-1)^{\Theta_k}(\sin^2\delta)B_k^R\}C_k G_2(x_{1k})$$

$$+\ \{(\sin^2\delta)B_k{}^L - (-1)^{\Theta_k}(\cos^2\delta)B_k{}^R\}C_k G_2(x_{2k})] \tag{D.20}$$

Eq. (D.20) arises from the Yukawa terms and C_K are

$$C_k = e\,\frac{m_\mu}{2M_W\sin\alpha\sin\Theta_W}\,[\sin\alpha O_{2k} + \cos\alpha O_{3k}] \tag{D.21}$$

$G_2(x)$ in Eqs. (D.19) and (D.20) is given by

$$G_2(x) \equiv (1+x)(1-x)^{-2} + 2x(1-x)^{-3}\ln x \tag{D.22}$$

REFERENCES

(1) For previous reviews of Supergravity unified theory see P. Nath, R. Arnowitt and A.H. Chamseddine, "Applied N=1 Supergravity," NUB #2613, Lectures at the 1983 Summer Workshop on Particle Physics, Trieste. J. Ellis, in International Symposium on Lepton and Photon Interactions, Cornell 1983. TH. 3718-CERN. H.P. Nilles, "Supersymmetry, Supergravity and Particle Physics," UGVA-DPT 1983/12-412. For a review of Supersymmetry and Supergravity see A. Salam and J. Strathdee, Fortschr. der Phys. 26, 57 (1978). P. Fayet and S. Ferrara, Phys. Rep. 32C, 249 (1977). P. van Nieuwenhuizen, Phys. Reports 68(4), 1981. J. Bagger and J. Wess, Supersymmetry and Supergravity (Princeton University Press, Princeton, NJ, 1983). J. Gates, M. Grisaru, M. Rocek and W. Siegel, Superspace (Benjamin, Reading, Mass., 1983).

(2) A.H. Chamseddine, R. Arnowitt and P. Nath, Phys. Rev. Lett. 49 (1982) 970.

(3) P. Nath, R. Arnowitt and A.H. Chamseddine, Phys. Lett. 121B (1983) 33.

(4) R. Arnowitt, A.H. Chamseddine and P. Nath, Phys. Lett. 120B (1983) 145.

(5) R. Barbieri, S. Ferrara and A. Savoy, Phys. Lett. 119B (1982) 343.

(6) L.E. Ibáñez, Phys. Lett. 118B (1982) 73; L.E. Ibáñez, Universidad Autonoma de Madrid preprint FTUAM/82-8 (1982); L.E. Ibáñez and C. Lopez, Phys. Lett. 126B (1983) 54.

(7) J. Ellis, D.V. Nanopoulos and K. Tamvakis, Phys. Lett. 121B (1983) 123.

(8) L. Alvarez-Gaumé, J. Polchinski and M.B. Wise, Nucl. Phys. B221 (1983) 495.

(9) S. Ferrara, D.V. Nanopoulos and C.A. Savoy, Phys. Lett. 123B (1983) 214.

(10) J. Ellis, J. Hagelin, D.V. Nanopoulos and K. Tamvakis, Phys. Lett. 125B (1983) 275.

(11) S. Weinberg, Phys. Rev. Lett. 50 (1983) 387.

(12) R. Arnowitt, A.H. Chamseddine and P. Nath, Phys. Rev. Lett. 50 (1983) 232.

(13) M.K. Gaillard, AIP Conf. Proc. No. 93 (R.S. Panvini, M.S. Alam, and S.E. Csorna, ed. AIP N.Y. 1982).

(14) L. Hall, J. Lykken and S. Weinberg, Phys. Rev. D. 27 (1983) 2359. See also

L. Hall in these proceedings.
(15) P. Nath, R. Arnowitt and A.H. Chamseddine, Nucl. Phys. B227 (1983) 121.
(16) B.A. Ovrut and S. Raby, Phys. Lett. 130B (1983) 277; 125B (1983) 270; 121B (1983) 381. S.K. Soni and H.A. Weldon, Phys. Lett. 126B, 215 (1983).
(17) M. Claudson, L.J. Hall and I. Hinchliffe, Nucl. Phys. B228 (1983) 501.
(18) C. Kounnas, A.B. Lahanas, D.V. Nanopoulos and M. Quiros, Phys. Lett. 132B (1983) 95; CERN TH. 3651 (1983).
(19) J. Ellis, A.B. Lahanas, D.V. Nanopoulos and K. Tamvakis, Phys. Lett. 134 (1984) 429.
(20) R. Arnowitt, P. Nath and A.H. Chamseddine, Proc. of the International Symposium Gauge Theory and Gravitation, Tezukayama University, Nara, Japan August, 1982.
(21) P. Nath, A.H. Chamseddine and R. Arnowitt, Proc. of APS Meeting at the University of Maryland, College Park, Maryland, October 1982.
(22) P. Nath, A.H. Chamseddine and R. Arnowitt, Proc. of Orbis Scientiae, Miami, Florida, 1983.
(23) J. Ellis and D.V. Nanopoulos, Phys. Lett. 116B (1982) 133.
(24) H. Goldberg, Phys. Rev. Lett. 50 (1983) 1419.
(25) R. Barbieri, L. Girardello and A. Marsiero, Phys. Lett. 127B (1983) 429.
(26) A.H. Chamseddine, P. Nath and R. Arnowitt, Phys. Lett. 129B (1983) 445.
(27) P. Nath, R. Arnowitt and A.H. Chamseddine, Harvard/Northeastern preprint HUTP-83/A077-NUB#2588.
(28) M.-M. Frere and G.L. Kane, Nucl. Phys. B223, 331 (1983). G.L. Kane, in Proc. of the Fourth Workshop on Grand Unification (1983).
(29) J. Ellis, J.S. Hagelin, D.V. Nanopoulos and M. Srednicki, Phys. Lett. 127B (1983) 233.
(30) D.A. Dicus, S. Nandi, and X. Tata, Phys. Lett. 129B (1983) 451.
(31) B. Grinstein, J. Polchinski and M.B. Wise, Phys. Lett. 130B (1983) 285.
(32) P. Fayet, Phys. Lett. 133B, 363 (1983); 117B (1983) 460.
(33) R.M. Barnett, L. Lackner and H.E. Haber, Phys. Rev. Lett. 51 (1983) 176; Phys. Lett. 126B (1983) 64.
(34) N. Cabibbo, L. Maiani and S. Petrarca, Phys. Lett. 132B (1983) 195; R. Barbieri, N. Cabibbo, L. Maiani and S. Petrarca, Phys. Lett. 127B (1983) 458.
(35) G. Altarelli and R. Rückl, Ref. TH. 3814-CERN.
(36) M. Glück and E. Reya, Phys. Lett. 130B (1983) 423.
(37) G. Eilam and E. Reya, TECHNION-PH-84-17.
(38) D.A. Dicus, S. Nandi, W. Repko and X. Tata, Phys. Rev. Lett. 51 (1983) 1030.
(39) S. Dawson, E. Eichten and C. Quigg, FERMILAB-Pub-83/82-THY; LBL-16540 (1983).
(40) J.S. Hagelin, G.L. Kane and S. Raby, Preprint LA-UR-83-711; L.E. Ibáñez, FTUAM-83-28 (1983).
(41) K.S. Stelle and P.C. West, Nucl. Phys. B145, 175 (1978).
(42) P. van Nieuwenhuizen and S. Ferrara, Phys. Lett. 76 (1978) 404.

(43) E. Cremmer, B. Julia, J. Scherk, S. Ferrara, L. Girardello and P. van Nieuwenhuizen, Nucl. Phys. B147, 105 (1979).

(44) A.H. Chamseddine, R. Arnowitt and P. Nath, Phys. Rev. Lett. 49, 970 (1982) and P. Nath, R. Arnowitt and A.H. Chamseddine, NUB #2613.

(45) E. Cremmer, S. Ferrara, L. Girardello and A. van Proeyen, Phys. Lett. 116, B231 (1982); Nucl. Phys. B212 (1983) 413.

(46) E. Witten and J. Bagger, Phys. Lett. 115B (1982) 202. J. Bagger, Nucl. Phys. B211 (1983) 302.

(47) T. Kugo and S. Uehara, Nucl. Phys. B222 (1983) 125. S. Ferrara, L. Girardello, T. Kugo and A. Van Proeyen, CERN TH 3523 (1983). C.S. Aulakh, M. Kaku and R.N. Mohapatra, Phys. Lett. 126B, 183 (1983).

(48) J. Wess and B. Zumino, Nucl. Phys. B70 (1974) 39.

(49) B. Zumino, Phys. Lett. 87B, 203 (1979).

(50) The simplest illustration of the effect is due to J. Polony, University of Budapest Report No. KFKI-1977-93, 1977 (unpublished).

(51) R. Arnowitt, A.H. Chamseddine and P. Nath, Proc. of Workshop on Problems in Unification and Supergravity, LaJolla Institute, Jan. 1983.

(52) M. Cvetič and J.C. Pati, Phys. Lett. 135B (1984) 57.

(53) H.P. Nilles, M. Srednicki and D. Wyler, Phys. Lett. 124B (1983) 337 and 120B (1982) 346. A.B. Lahanas, Phys. Lett. 124B (1983) 341.

(54) J. Bailey et al., Phys. Lett. 68B, 191 (1977); F.J.M. Farley and E. Picasso, Ann. Rev. Nucl. Sci. 29, 243 (1979), F. Combley, F.J.M. Farley, and E. Picasso, Phys. 68, 93 (1981); L. Lyons, Prog. Part. and Nucl. Phys. 10, 227 (1983).

(55) T. Kinoshita, B. Nižić and Y. Okamoto, Phys. Rev. Lett. 53, 717 (1984).

(56) Vernon W. Hughes (Private Communication). A serious consideration of this possibility is planned at the "Workshop on a Possible High Precision Muon G-2 Experiment at the AGS" June 11-June 16, 1984 at BNL.

(57) R. Jackiw and S. Weinberg, Phys. Rev. D5, 2473 (1972). G. Altarelli, N. Cabibbo and L. Maiani, Phys. Lett. 40B 415 (1972). I. Bars and M. Yoshimura, Phys. Rev. D6, 374 (1972). K. Fujikawa, B.W. Lee and A.I. Sanda, Phys. Rev. D6, 2923 (1972). W.A. Bardeen, R. Gastmans and B.E. Lautrup, Nucl. Phys. B46, 319 (1972).

(58) P. Fayet, in Unification of the Fundamental Particle Interactions, eds. S. Ferrara, J. Ellis, and P. van Nieuwenhuizen (Plenum, New York, 1980), p. 587. J.A. Grifols and A. Mendez, Phys. Rev. D26, 1809 (1982). J. Ellis, J.S. Hagelin and D.V. Nanopoulos, Phys. Lett. 116B, 283 (1982). R. Barbieri and L. Maiani, Phys. Lett. 117B, 203 (1982).

(59) D.A. Kosower, L.M. Krauss and N. Sakai, Phys. Lett. 133B, 305 (1983).

(60) T.-C. Yuan, R. Arnowitt, A.H. Chamseddine and P. Nath, NUB #2633 (1984).

(61) W. Bartel et al. (JADE collaboration), Phys. Lett. 114B, 211 (1982)
W. Bartel et al. (CELLO collaboration), Phys. Lett. 114B, 287 (1982)

B. Adeva et al. (MARK J collaboration), Phys. Lett. 115B, 345 (1982).
R. Brandelik (TASSO collaboration), Phys. Lett. 117B, 365 (1982).
E. Fernandez et al. (MAC collaboration), Phys. Rev. Lett. 52, 22 (1984).

(62) J. Kim, P. Langacker, M. Levine and H.H. Williams, Rev. Mod. Phys. 53, 211 (1981).

(63) M. Jonker et al. CHARM Collab. Phys. Lett. 99B, 265 (1981).

(64) A. Sirlin, Rev. Mod. Phys. 50, 573 (1978); Phys. Rev. D22, 971 (1980).

(65) W. Marciano and A. Sirlin, Phys. Rev. D22, 2695 (1980); Nucl. Phys. B189 (1981) 442; Reprint BNL-33819 (1984).

(66) M. Veltman, Nucl. Phys. B123, 89 (1977).

(67) R. Barbieri and L. Maiani, Nucl. Phys. B224, 32 (1983).

(68) T. Inami and N. Sakai, University of Tokyo Preprint INS-Rep.-480 (1983).

(69) E. Eliasson, Northeastern University preprint NUB #2621 (1983).

(70) E. Eliasson, Northeastern University preprint NUB #2640 (1984).

(71) B.A. Campbell, J.A. Scott and M.K. Sundaresan, Phys. Lett. 126B, 376 (1983).

(72) G.L. Kane and J.P. Leveille, Phys. Lett. 112B, 227 (1982); R.R. Harrison and C.H. Llewellyn-Smith, Nucl. Phys. B213, 223 (1983).

(73) N.S. Craigie, K. Hidaka and P. Ratcliffe, Phys. Lett. 129B, 310 (1983).

(74) Proc. of 1982 DPF-Summer Study on Elementary Particle Physics and Future Facilities, SNowmass, Colorado, 1982. Ed. by R. Donaldson, R. Dustafson and F. Paige.

(75) R.M. Godbole, S. Pakvasa and D.P. Roy, Phys. Lett. 50 (1983) 1539. V. Barger, A.D. Martin and R.J.N. Phillips, Phys. Lett. 125B, 339 (1983).

(76) M.K. Gaillard, L. Hall and I. Hinchliffe, Phys. Lett. 116B, 279 (1982).

(77) For the current experimental status of proton decay see Proc. Fifth Grand Unification Workshop at Brown (1984).

(78) J.C. Pati and A. Salam, Phys. Rev. D8, 1240 (1973). H. Georgi and S.L. Glashow, Phys. Rev. Letters 32, 438 (1974). For a recent review see J.C. Pati in Proc. of the ICOMON Conference, Frascati (1983).

(79) J. Ellis, D.V. Nanopoulos and S. Rudaz, Nucl. Phys. B202, 43-62 (1982). V.M. Belyaev and M.I. Vysotsky, Phys. Lett. 127B, 215 (1983). J. Milutnovik, P. Pal and G. Senjavonic, NSF-ITP-84-05.

(80) N. Sakai, TIT/HEP-78 (1983).

(81) R. Arnowitt, C.H. Chamseddine and P. Nath, NUB #2641.

(82) J. Polchinski and M. Wise, Phys. Lett. 125B, 393 (1983). R. Mohapatra, S. Ouvry and G. Senjanovic, Phys. Lett. 126B, 329 (1983). F. del Aguila, M.B. Gavela, J.A. Grifols and A. Mendez, Phys. Lett. 126B, 71 (1983). F. del Aguila, J.A. Grifols, A. Mendez, D. Nanopoulos and M. Srednicki, Phys. Lett. 129B, 77 (1983). J.-M. Gerard, W. Grimus, A. Raychaudauri, and G. Zoupanos, Ref. TH. 3809-CERN.

(83) S. Weinberg, Phys. Rev. Lett. 48, 1303 (1982). J. Ellis, A.D. Linde and

D.V. Nanopoulos, Phys. Lett. 118B (1982) 59. L.M. Krauss, HUTP-83/A009. M. Yu Khlopov and A.D. Linde, Phys. Lett. 138B, 265 (1984). See also A.D. Linde, Talk at the Shelter Island Conference II, 1-3 June 1983. K.A. Olive, in Proc. of the 3rd Moriond Astrophysics Meeting, LaPlagne, France, 1983, Eds. J. Audouze and J. Tran Thanh Van.

(84) M. Dine, W. Fischler and D. Nemeschansky, Preprint 1983. G.D. Coughlan, W. Fischler, E. Kolb, S. Raby and G.G. Ross, Los Alamos Preprint LA-UR83-1423.

(85) F. del Aguila, M. Dugan, B. Grinstein and L. Hall, G.G. Ross and P. West, HUTP-84/A001. See also L. Hall in these proceedings.

(86) R. Arnowitt, C.H. Chamseddine and P. Nath, Proc. of the Fourth Grand Unification Workshop at Brown (1984).

(87) G. Thomas and P. West, Kings College Preprint (1984).

(88) E. Witten, Princeton University Preprint (1983).

(89) See P. Roy in these proceedings and P. Roy and O. Shanker, TIFR/TH-84-6 (1984).

(90) For analyses in SUSY models see K. Inoue, A. Kakuto, M. Komatsu and S. Takeshita, Prog. Theor. Phys. 67, 1889 (1972); 68, 927 (1982). R.A. Flores and M. Sher, Ann. Phys. 148, 95 (1983). B.P. Li and M. Sher, Technical Report No. 84-7.

(91) G. Arnison et al., CERN-EP/84-42.

(92) J. Ellis and H. Kowalski, Ref. TH. 1843-CERN (1984).

(93) G. Altarelli, B. Mele and F. Petrarca, Ref. TH. 3822-CERN.

(94) E. Reya and D.P. Roy, DO-TH. 84/11 (1984).

(95) H.E. Haber and G.L. Kane, UCSC-TH-169-84.

FIGURES

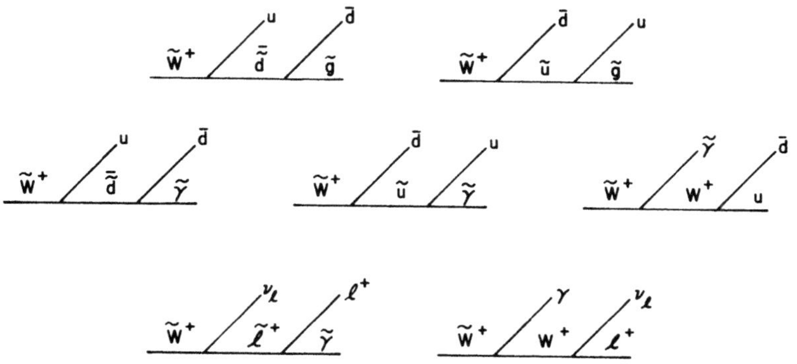

Fig. 1. Decay of the Wino through squark, slepton and W poles.

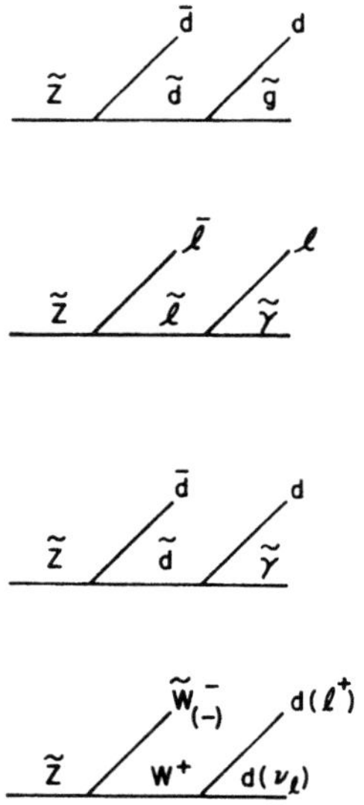

Fig. 2. Decay of the Zino through squark, slepton and W Poles. There are additional charge conjugate diagrams not shown.

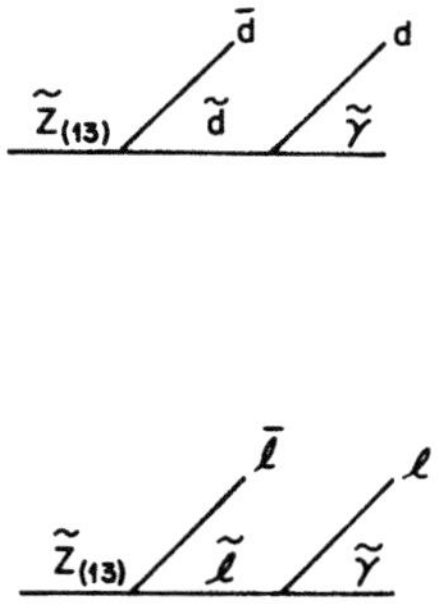

Fig. 3. Decay of the Twilight through squark and slepton poles. The W pole term is absent since a $q_1\bar{q}_2\,\tilde{W}$ final state is kinematically disallowed.

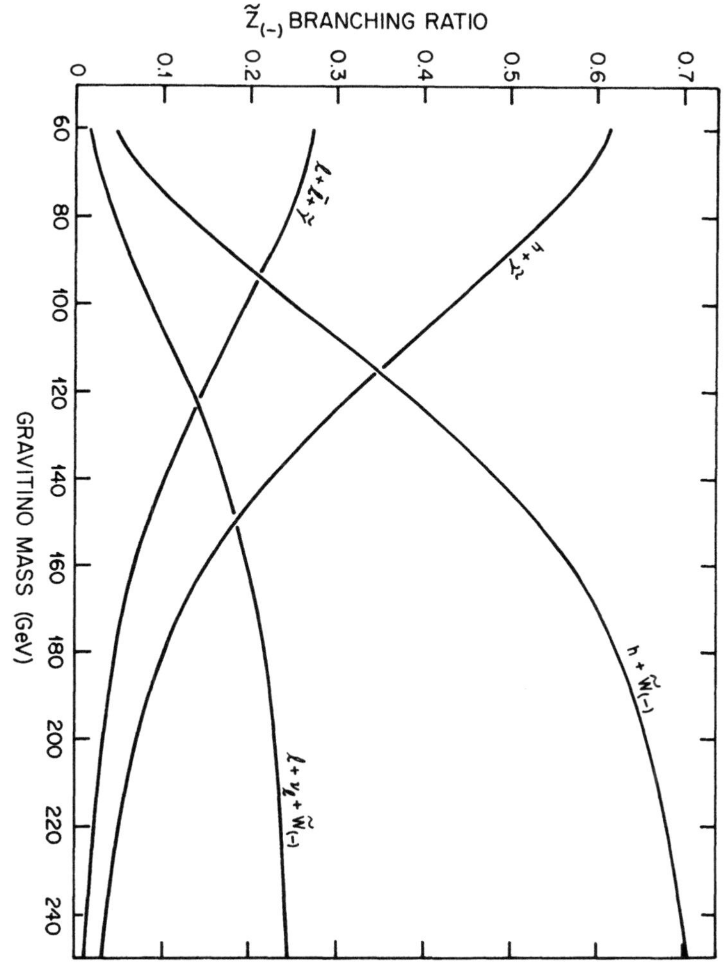

Fig. 4. Branching ratios (from Ref. (27)) for $\tilde{Z}_{(-)}$ decay for the T.B. model ($\alpha = 45°$) for $\tilde{m}_- = 30$ GeV, $\tilde{\mu}_- = 40$ GeV and $\mu = 121$ GeV. The curves are labelled by their final states (ℓ = leptons, h = hadrons) and plotted as a function of the gravitino mass.

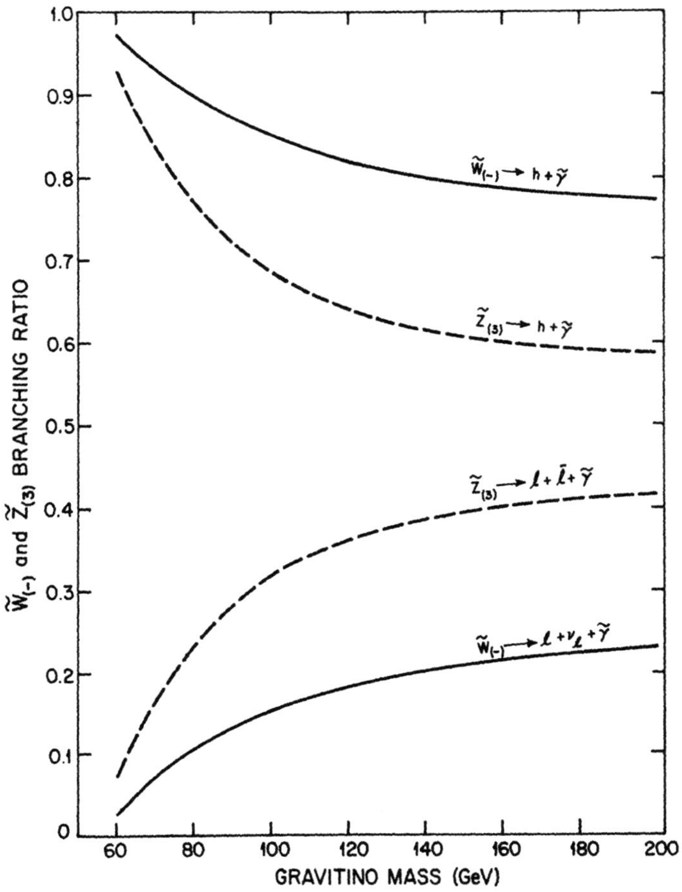

Fig. 5. Branching ratios (from Ref. (27)) for the Wino $\tilde{W}_{(-)}$ and the Twilight Zino $\tilde{Z}_{(3)}$ decay for the R.G. model as a function of the gravitino mass for $\alpha = 17°$, $\tilde{m}_{\gamma} = 7$ GeV and $\mu = 17.9$ GeV.

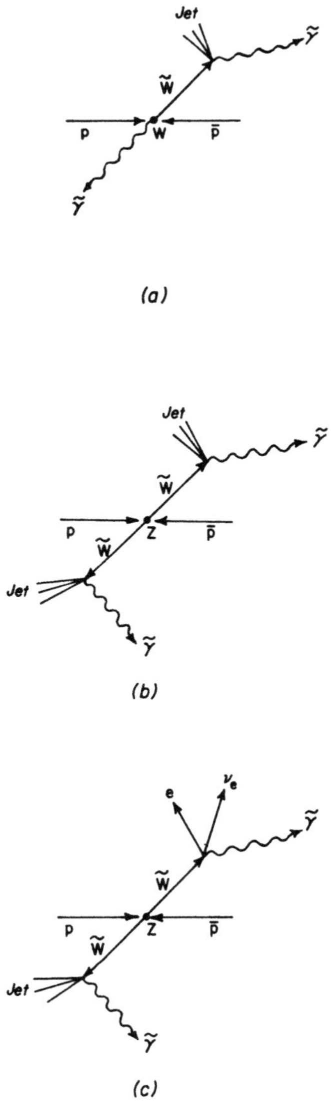

Fig. 6. (a) and (b) are one jet and two jet UFO events in the supersymmetric decays of the W and Z. (c) is a lepton-jet event with the lepton and the jet in opposite hemispheres with unbalanced $p_\perp$.

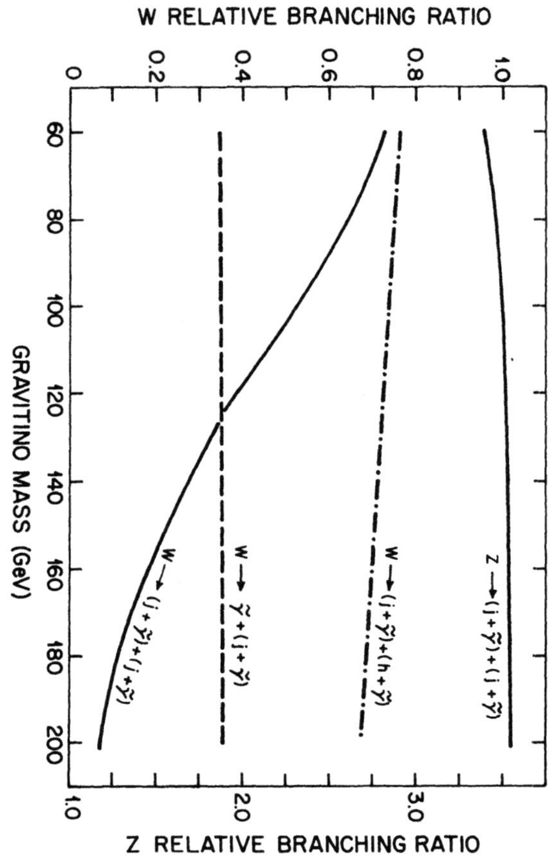

Fig. 7. Relative branching ratios (from Ref. 127)) of W and Z into one jet and two jet UFO events in the T.B. model for $m_{\tilde{\gamma}} = 7$ GeV and $\mu = 121$ GeV. The W branching ratios are relative to $W \to e\nu$ and the Z branching ratios are relative to $Z \to e^+e^-$. The dashed dot curve takes account of additional hadronic events from $W \to \tilde{W}_{(-)} + \tilde{Z}_{(-)}$ where $\tilde{Z}_{(-)} \to q_1 q_2 \tilde{W}_{(-)}$ and $\tilde{W}$ decays hadronically.

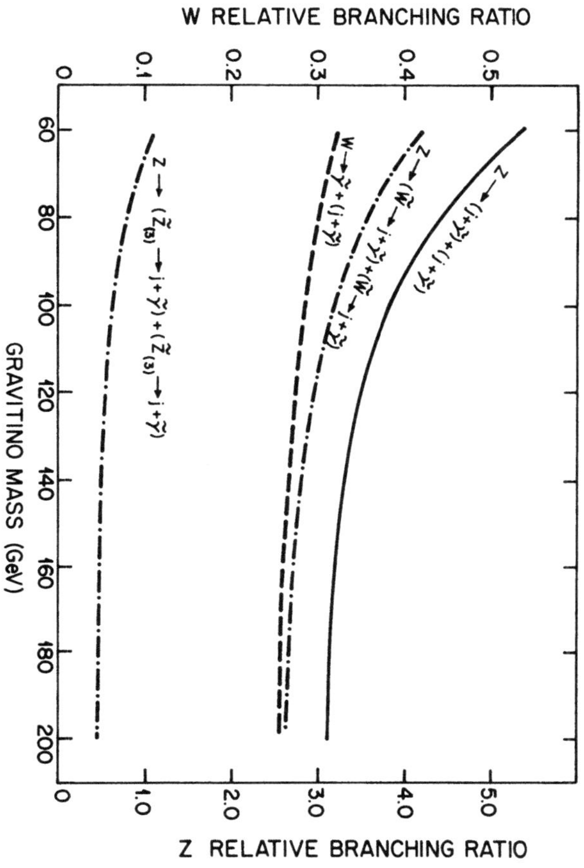

Fig. 8. Relative branching ratios (from Ref. (27)) for the W and Z decays into one jet and two jet UFO events in the R.G. model for $\alpha = 17°$, $\tilde{m}_\gamma = 7$ GeV and $\mu = 17.9$ GeV. The W branching ratios are relative to $W \to e\nu$ and the Z branching ratios are relative to $Z \to e^+e^-$. The solid curve $Z \to (j + \tilde{\gamma}) + (j + \tilde{\gamma})$ is the sum of the dash-dot curves.

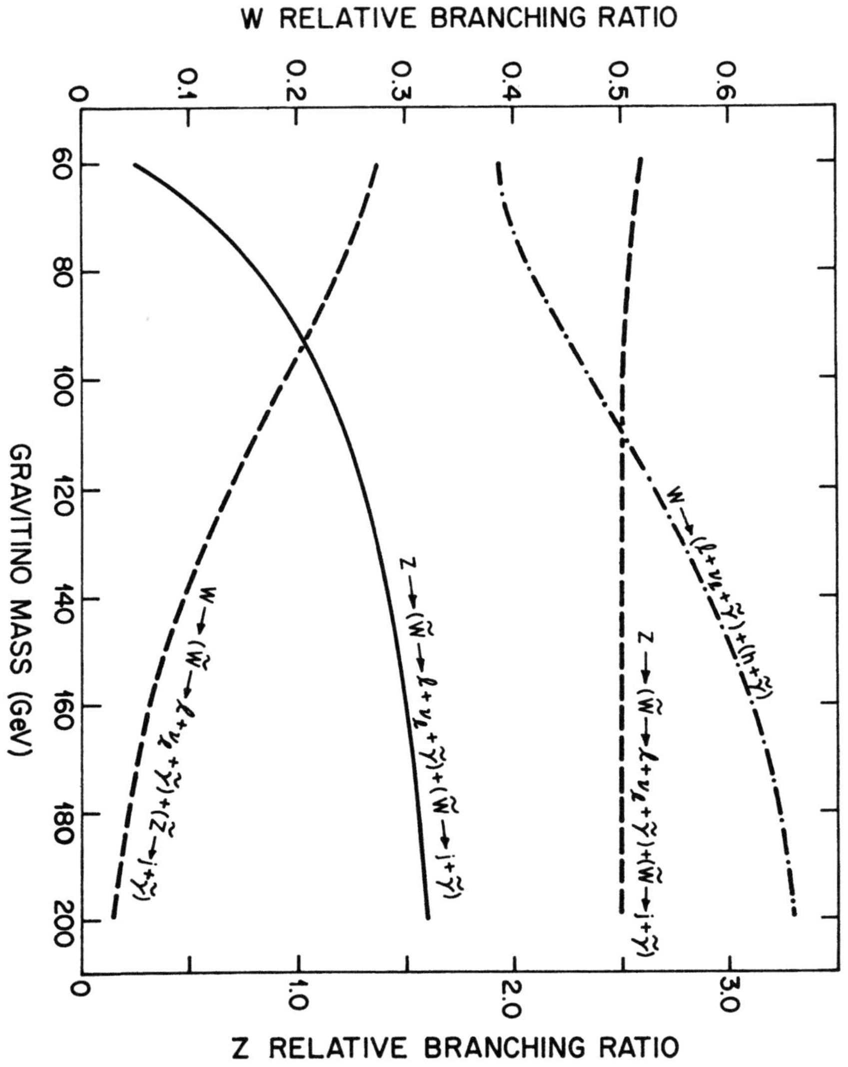

Fig. 9. Relative branching ratios (from Ref. (27)) for the W and Z decays into lepton-jet events. The W branching ratios are relative to $W \rightarrow e\nu$ and the Z branching ratios are relative to $Z \rightarrow e^+e^-$. The Z decay $Z \rightarrow \tilde{W}_{(-)} + \tilde{W}_{(-)}$ generates leptonic-jet events in both the T.B. (the upper dashed curve) and the R.G. (the solid curve) model with one $\tilde{W}$ decaying leptonically and the other hadronically. In the T.B. model $W \rightarrow \tilde{W}_{(-)} + \tilde{Z}_{(-)}$ with the $\tilde{W}_{(-)}$ decaying leptonically and $\tilde{Z}_{(-)}$ decaying hadronically generate lepton-jet events shown by the lower dashed curve. The T.B. model decay $W \rightarrow \tilde{W}_{(-)} + \tilde{Z}_{(-)}$ where $\tilde{Z}_{(-)} \rightarrow q_1\bar{q}_2\tilde{W}_{(-)}$ with $\tilde{W}_{(-)}$ decaying hadronically produces

additional events with hadrons accompanied by a single charged lepton. Inclusion of these events gives the dash-dot curve. Parameters for the T.B. model are $\tilde{m}_\gamma$ = 7 GeV, μ = 121 GeV while for the R.G. model α = 17°, $\tilde{m}_\gamma$ = 7 GeV and μ = 17.9 GeV. In the R.G. model the W decay $W \to \tilde{W}_{(-)} + \tilde{Z}_{(3)}$ where the W decays leptonically and $\tilde{Z}_{(3)}$ decays hadronically generate leptonic-jet events. (This decay mode is not shown here.)

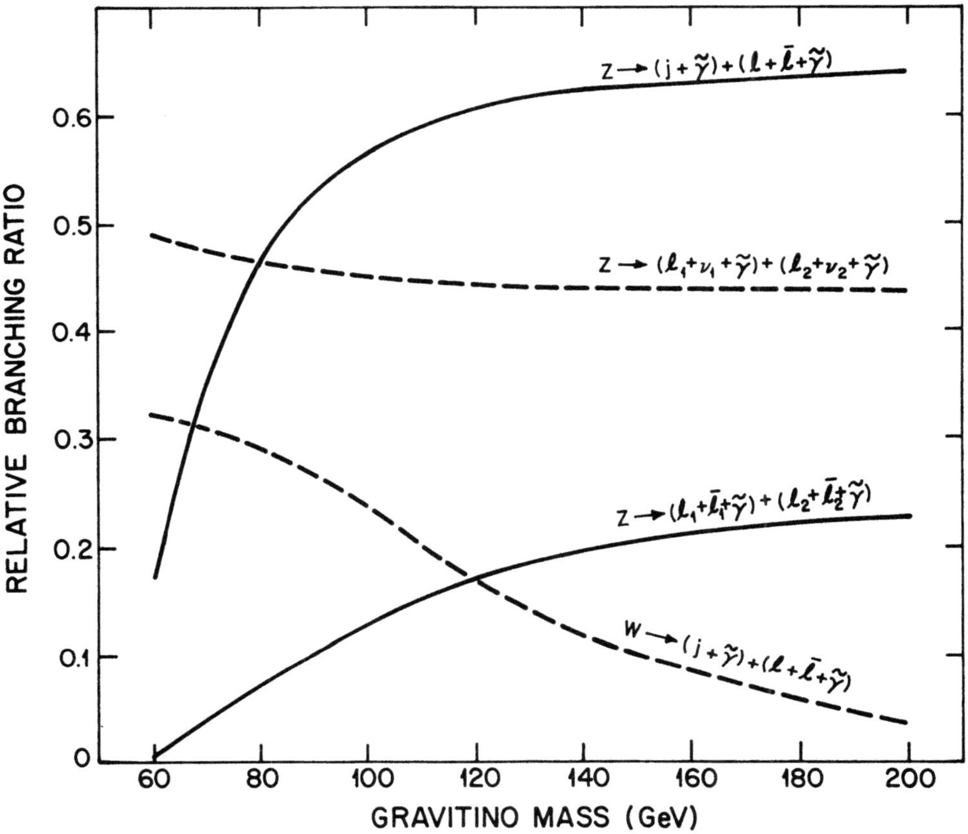

Fig. 10. W and Z decay branching ratios (from Ref. (27)) for "exotic" events. The W branching ratios are relative to $W \to e\nu$ and Z branching ratios are relative to $Z \to e^+e^-$. The solid curves are for the R.G. model with α = 17°, $\tilde{m}_\gamma$ = 7 GeV and μ = 17.9 GeV. The dashed curves are for the T.B. model with $\tilde{m}_\gamma$ = 7 GeV and μ = 121 GeV.

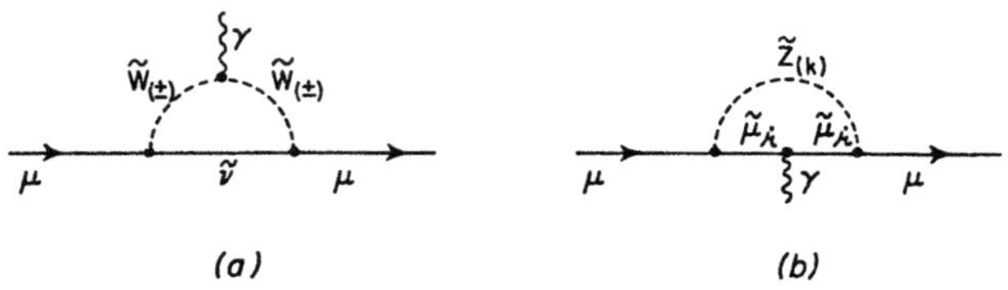

Fig. 11. Supersymmetry contributions to the muon anomalous magnetic moment from one loop diagrams. Contributions arise from the Wino and sneutrino intermediate states (Fig. (11a)) and from the Zino and smuon intermediate states (Fig. (11b)).

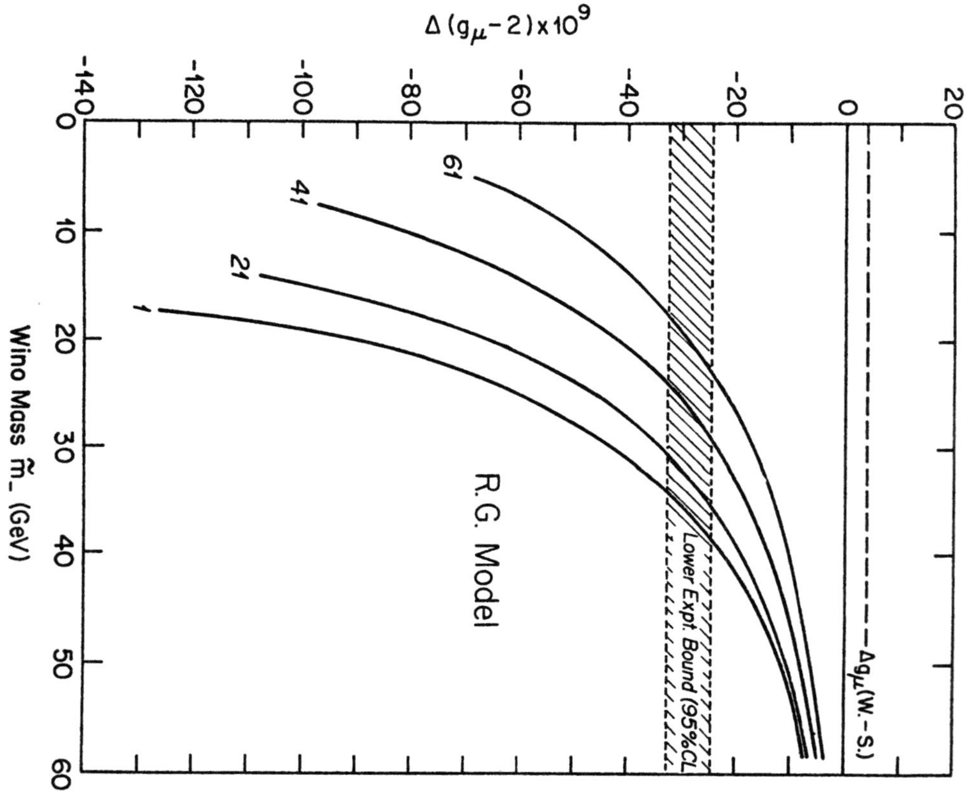

Fig. 12. R.G. model results (from Ref. (60)) for the supersymmetric electro-weak contribution to g_μ-2 as a function of the lower Wino mass $\tilde{m}_-$ with the standard Weinberg-Salam piece included. Analysis is for $A = 3 - \sqrt{3}$, $\tilde{m}_\gamma$ = 7 GeV and μ = 20 GeV. The dashed line shows the standard Weinberg-Salam electro-weak contribution while the cross-hatched area shows the experimental lower bound. Results are shown for sneutrino masses of 1, 21, 41 and 61 GeV.

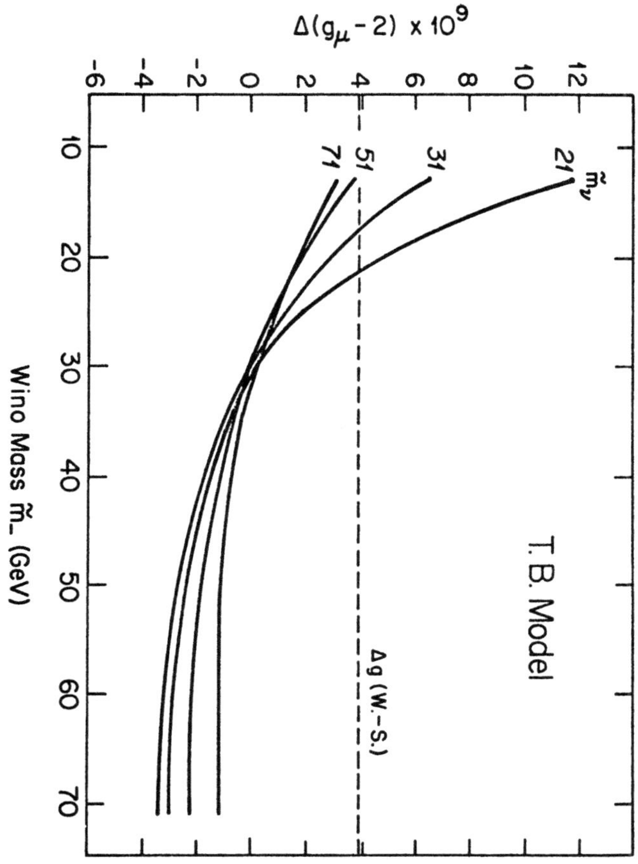

Fig. 13. T.B. model results (from Ref. (60)) for the supersymmetric electro-weak contribution to g_μ-2 as a function of the lower Wino mass $\tilde{m}_-$ with the standard Weinberg-Salam piece included. Analysis is for A = 3 - $\sqrt{3}$ and $\tilde{m}_\gamma$ = 7 GeV. Results are shown for sneutrino masses of 21, 31, 51 and 71 in GeV. The standard Weinberg-Salam electro-weak contribution is shown by the dashed line.

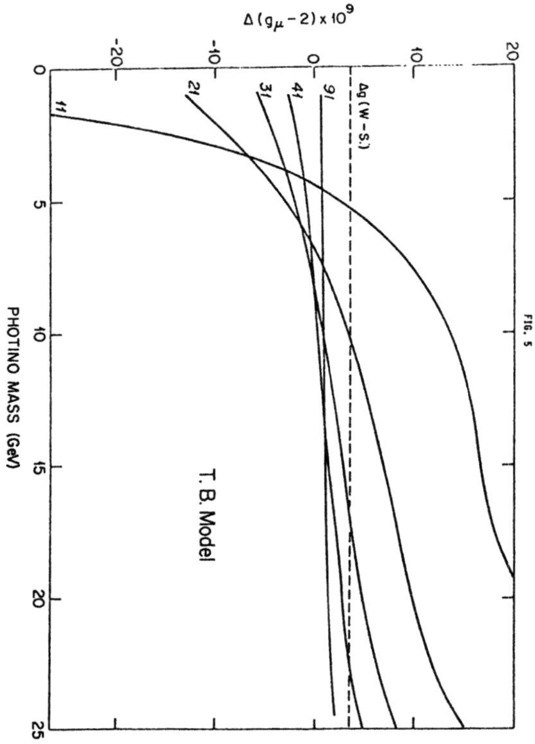

Fig. 14. Photino mass dependence of the supersymmetric electro-weak contribution to g_μ-2 including the standard Weinberg-Salam piece is exhibited for the T.B. model for sneutrino masses of 11, 21, 31, 41 and 91 in GeV and for A = $3 - \sqrt{3}$ and μ = 125 GeV (from Ref. (60)).

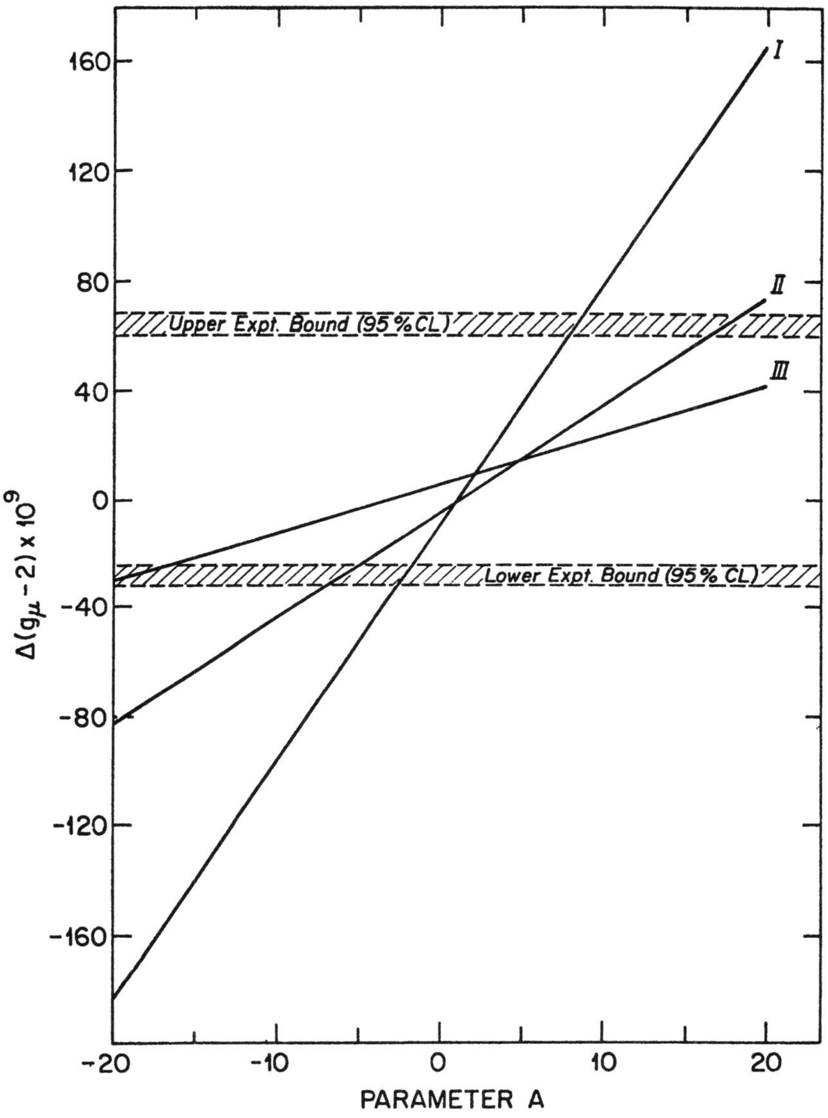

Fig. 15. Dependence of $\Delta(g_{\mu}-2)$ in the T.B. model on the parameter A of the hidden sector (from Ref. (60)). The parameters of the three curves are as follows: curve I: $\tilde{m}_{\gamma}$ = 7 GeV, $\tilde{m}_{\nu}$ = 25 GeV and $\tilde{m}_{-}$ = 25.5 GeV; curve II: $\tilde{m}_{\gamma}$ = 2 GeV, $\tilde{m}_{\nu}$ = 40 GeV and $\tilde{m}_{-}$ = 20.5 GeV; curve III: $\tilde{m}_{\gamma}$ = 2 GeV, $\tilde{m}_{\nu}$ = 40 GeV and $\tilde{m}_{-}$ = 42.6 GeV.

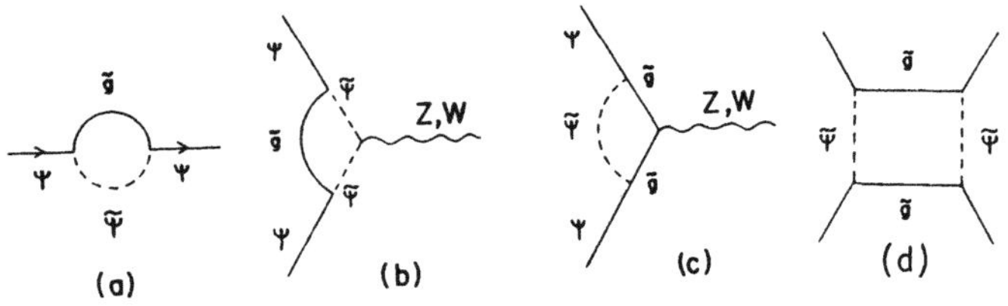

Fig. 16. Supersymmetric one loop contributions to ρ.

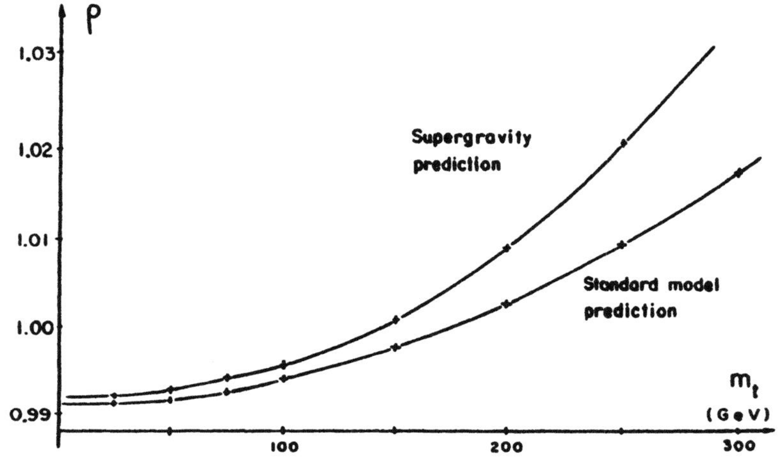

Fig. 17. Prediction for ρ (from Ref. (69)) as a function of the top quark mass for the T.B. model with m_g = 100 GeV. Prediction for ρ for the standard model is also shown.

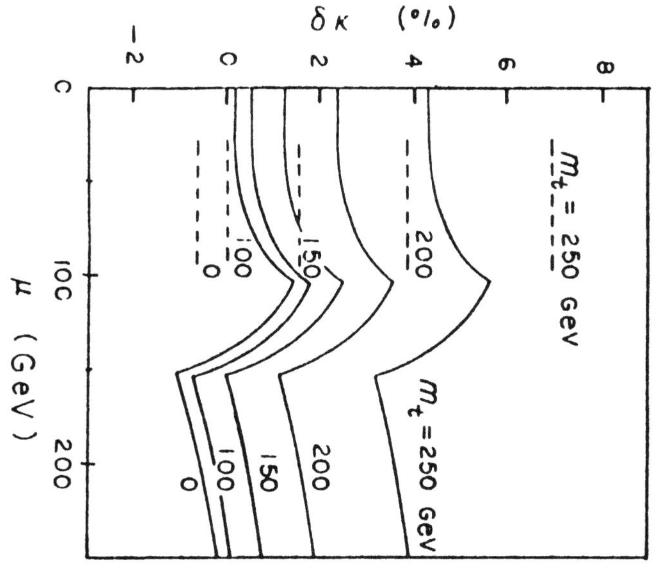

Fig. 18. Corrections to δK (from Ref. (70)) in the T.B. model as a function of μ for m_g = 100 GeV and an array of m_t masses. δK corrections in the standard model are shown by dashed lines.

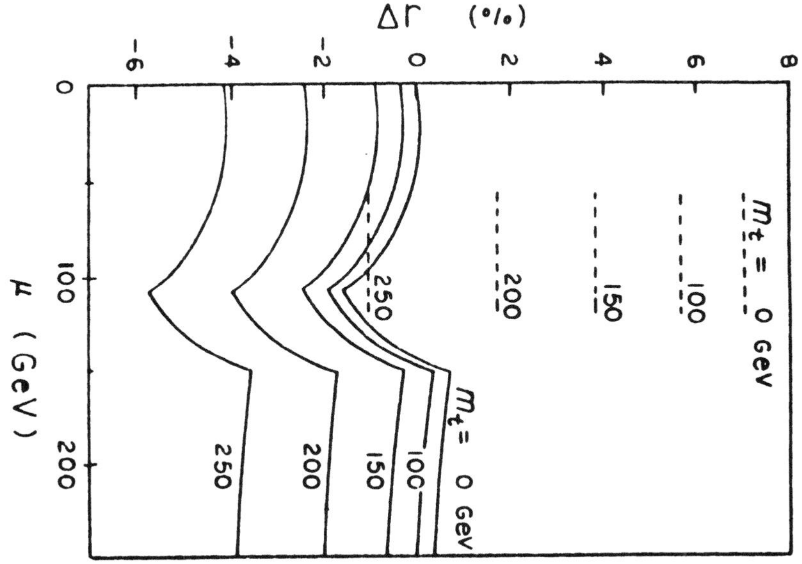

Fig. 19. Corrections to Δr (from Ref. (70)) in the T.B. model as a function of μ for m_g = 100 GeV and an array of m_t masses. Δr corrections in the standard model are shown by dashed lines.

TEV SCALE MODELS WITH TWO SUPERSYMMETRIES

Lawrence J. Hall
Lyman Laboratory of Physics
Harvard University
Cambridge, MA 02138
U.S.A.

Abstract

The construction of models of low energy physics with two supersymmetries is discussed. All such SU(3) x SU(2) x U(1) models appear to be beset by insurmountable problems. A SU(3) x SU(4) x U(1) model is constructed which gives a natural distinction between mirror fermions and ordinary fermions.

Many N=1 supersymmetric models describing particle physics at the TeV scale have been written down [1]. These theories have more particles than the standard model. However they are constrained in new ways, and are less divergent than the standard model. They offer explanations, to varying degrees, of the gauge hierarchy problem, and many would be relieved and excited by the observation of a superpartner particle.

In this talk I wish to go one step further and discuss theoretical prejudices for and against the possibility that experimentalists will soon observe two superpartners for each known particle. Models with two supersymmetries [2] have greatly improved convergence properties [3]. Indeed, theories considered here could be the low energy remnants of a finite grand unified theory. These models have several new constraints. For example, all dimension four interactions are governed by gauge couplings. Constructing fermion mass matrices becomes quite a challenge.

I will show that realistic N=2 supersymmetric models can be built, but only if the gauge group is enlarged beyond SU(3) x SU(2) x U(1). I will only discuss the most interesting aspects of these models; further details can be found in reference [4]. This work was done in collaboration with the authors of reference [4].

The N=2 supersymmetry algebra (without central charges) is

$$\left\{Q^i_\alpha, \bar{Q}^j_{\dot\alpha}\right\} = \sigma^\mu{}_{\alpha\dot\alpha} P_\mu \delta^{ij} \tag{1}$$

where Q^i_α (i = 1,2) are the two supersymmetry generators. The basic multiplets we use in model building are the vector multiplet and the hypermultiplet, shown schematically in Figure 1. It is convenient to

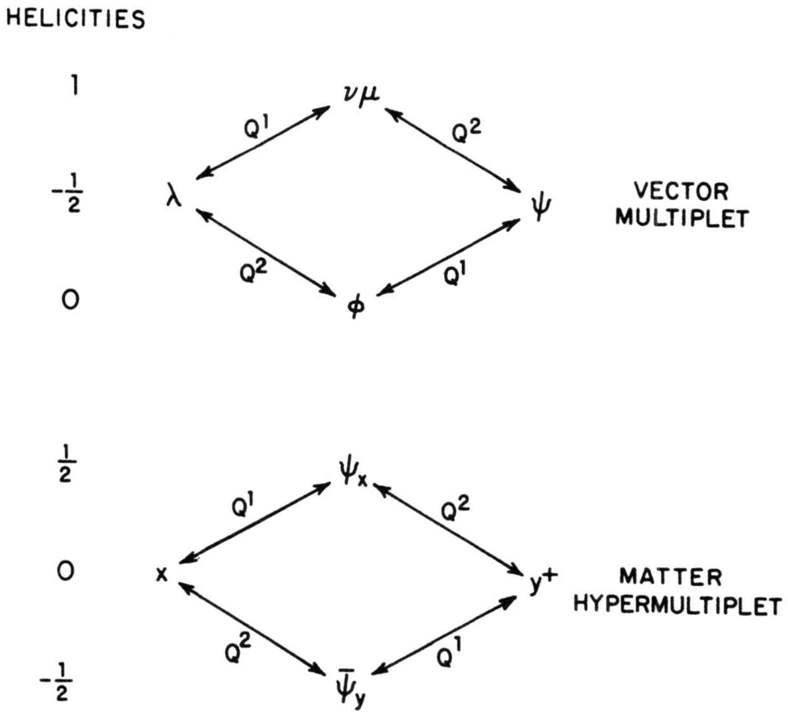

Figure 1

construct the theory in terms of N=1 superfields. The two supersymmetries are treated asymmetrically with one, say Q^1, manifest and the other hidden. The N=2 vector multiplets contains an N=1 vector multiplet $V \supset (V^\mu,\lambda)$ and a chiral multiplet $\Phi \supset (\psi,\phi)$ in the adjoint representation of the gauge group. The N=2 hypermultiplet contains N=1 chiral multiplets $X \supset (\psi_x,x)$ and $Y(\psi_y,y)$ transforming as R and $\bar{R}$ respectively.

In constructing global N=1 supersymmetric models some representation content (V,X) is chosen. When extending the theory to N=2, the additional supersymmetry requires the introduction of (Φ,Y) which we call mirror gauge and mirror matter fields. (For the SU(3) x SU(4) x U(1) models this terminology will be altered slightly.)

The N=2 supersymmetric Lagrangian is given by

$$\mathcal{L} = \frac{1}{8g^2}[WW]_F + [\sqrt{2}\,igY\Phi X]_F + \text{h.c.}$$
$$+ [2\,\mathrm{Tr}\,\phi^+ e^{2gV}\phi e^{-2gV} + X^+ e^{2gV}X + Y^+ e^{-2gV^T}Y]_D \quad (2)$$

where $\Phi = \Phi^\alpha T^\alpha$, $V = V^\alpha T^\alpha$ with generators T^α. Here W is the supersymmetric field strength, and F and D refer to the θ^2 and $\theta^2\bar{\theta}^2$ components of the superfield product. The superpotential has a single trilinear interaction with strength fixed by the gauge coupling constant g; it is the "mirror gauge interaction". It is easy to verify that the superpotential must be fixed in this way by noting that there is an SU(2) symmetry acting on the doublet Q^i of generators. Under this symmetry we have doublets (x_i, y_i^+) and (ψ_ϕ, λ) so that $[\sqrt{2}\,igY\Phi X]_F$ is related to $[X^+e^{2gV}X + Y^+e^{-2gV^T}Y]_D$.

The only coupling constants in these theories are gauge couplings, and their β functions have contributions only at one loop [3]:

$$\beta(g_A) = -\frac{1}{16\pi^2}(2C_2(G_A) - 2T_A(R))g_A^3 \quad (3)$$

for gauge couplings g_A in gauge group $\prod_A G_A$. Here $C_2(G_A)$ is the quadratic Casimir for the adjoint representations of G_A, and $T_A(R)\delta^{\alpha\beta} = \mathrm{Tr}(T_A^\alpha(R)T_A^\beta(R))$. Hence these theories are finite whenever $C_2(G_A) = T_A(R)$. This is always true for N=4 theories which is just the special case of R being the adjoint representation, but many other combinations of hypermultiplets also lead to finite theories. It is clear that any factor in $\prod_A G_A$ is irrelevant in a finite N=2 theory, because all the hypermultiplets must have zero charge for the U(1) β function to vanish.

These theories have three types of global symmetries as well as several discrete symmetries. As the superpotential is purely trilinear, there is an R symmetry. The large number of ways of assigning the R charges corresponds to the presence of θ independent chiral symmetries. Suppose that R decomposes into irreducible representations of G according to $R = \sum_i n_i r_i$, then the chiral symmetry is $\prod_i U(n_i)$ with the $n_i X(r_i)\,(Y(r_i))$ transforming as an $n_i\,(\bar{n}_i)$ under the global $SU(n_i)$.

As we have seen, the algebra of N=2 supersymmetry possesses an SU(2) symmetry which treats the two supersymmetry charges as a doublet and this symmetry is inherited by the interactions. While a formulation in terms of N=2 superfields would make this symmetry manifest, our N=1 superfield formulation has hidden it. Under the global SU(2), the fields ψ_{x_i}, ψ_{y_i}, V and ϕ are singlets, while $D_{iL} = (x_i, y_i^+)$ are doublets and $\bar{D}_2 = (\psi_\phi, \lambda)$ is an antidoublet.

The N=2 theories possess several useful discrete symmetries which can be used to catalogue operators which are never induced at any radiative order, and which serve as a simple method to identify stable particles. It is convenient to divide the hypermultiplets into matter and Higgs hypermultiplets (X_M, Y_M) and (X_H, Y_H). A natural extension of R parity is

$$\theta \to -\theta, \quad X_M \to -X_M, \quad Y_M \to -Y_M \tag{4}$$

with all other fields remaining unchanged. Thus mirror chiral fermions are R even, and mirror gauginos are R odd. R parity does not distinguish ordinary fields from their mirrors. The lightest R odd particle is stable, if R parity is unbroken. A second parity, which does not involve θ, distinguishes between ordinary and mirror particles. We call it mirror parity:

$$Y_M \to -Y_M, \quad Y_H \to -Y_H, \quad \Phi \to -\Phi. \tag{5}$$

In theories with unbroken mirror parity the lightest mirror odd particle is also stable. There is also a mirror interchange symmetry: $X \leftrightarrow Y$, $\Phi \leftrightarrow \Phi^T$, and $V \leftrightarrow -V^T$ which must be broken to obtain a non vector-like low energy theory.

Several problems arise immediately in trying to build N=2 supersymmetric models:

(1) The potentials are flat with infinitely many supersymmetric vacua.

(2) The theories are vectorlike; for every left-handed fermion there is a right-handed fermion.

(3) There are very few parameters. What will be the source of quark and lepton mass matrices and flavor mixing?

(4) Many global and discrete symmetries must be broken.

I will side step the ultimate solutions of these difficulties, and to get going I will simply add soft operators by hand to remedy these problems. Perhaps our procedure can be justified when the structure of N=2 supergravity is better understood. Even allowing this freedom, constructing realistic models is far from trivial, especially if we demand that the soft operators are of a finiteness preserving type.

There are many soft operators which, when added to a finite N=2 theory, preserve the finiteness [3]. These operators fall into five groups [5]:

(1) Any gauge invariant supersymmetric mass term, for example,

$[\Phi^2]_F$, $[X_i\ Y_i]_F$, $[X_i\ X_j]_F\ldots$

(2) Any gauge invariant "A^2-B^2" scalar masses. These are the A components of the above bilinear operators. If the complex scalar fields of the chiral multiplets are written as A+iB, with A and B real, then these terms have the form A^2-B^2.

(3) Restricted "A^2+B^2" scalar mass terms. The operator

$\alpha\ \phi^+\phi + \beta_i x_i^+ x_i + \gamma_i y_i^+ y_i$

is allowed only if $\beta_i + \gamma_i = -\alpha$ for each i.

(4) Restricted trilinear scalar operators and gaugino masses. The operator

$$-\frac{m_0}{2}\lambda\lambda - \sqrt{2}\ i m_0 g \sum_i y_i \phi x_i + \text{h.c.} - |m_0|^2\ \phi^+\phi$$

is allowed for any m_0.

(5) Any gauge invariant scalar trilinear operator, $x_i\ x_j\ x_k$, $x_i\ x_j\ y_k$, $\cdots$ together with an appropriate A^2+B^2 scalar mass $\phi^+\phi$. Note that if any of the scalar fields are conjugated, e.g. $x_i\ x_j\ x_k^+$, the operator does not preserve finiteness.

Operators of class 5 break certain chiral symmetries which would otherwise never be broken. Without them no realistic N=2 SU(5) GUT can be constructed; operators of classes 1 to 4 leave important chiral symmetries unbroken. In particular, rotating the phase of every $10(\overline{10})$ superfield by angle $\theta(-\theta)$ is an exact invariance of any such SU(5) theory. Furthermore, all elements of the 10 are charged so that this U(1) can never be spontaneously broken. Since the down quarks and leptons acquire masses from an operator which is linear in 10 fields, they will be exactly massless. A similar argument can be applied to the corresponding low energy SU(3) x SU(2) x U(1) theory, with U(1) symmetries acting on the right-handed quark and charged lepton fields. This argument makes it clear that any realistic theory with gauge group SU(5) or SU(3) x SU(2) x U(1) must incorporate soft operators of class 5.

However, operators of class 5 yield scalar potentials which are unbounded from below, they cannot be allowed in the theory and consequently N=2 SU(5) and SU(3) x SU(2) x U(1) models cannot be built. Unboundness always results because if a gauge invariant trilinear $x_i x_j x_k$ exists, then each of these scalars may acquire a vev such that the quartic terms of the potential vanish.

Even if we temporarily chose to ignore the vacuum problems of the SU(3) x SU(2) x U(1) theories, we immediately run into further difficulties. If ordinary matter is assigned to N=2 hypermultiplets, it is impossible to obtain tree level SU(2) breaking fermion masses, because the adjoint of the group does not contain any SU(2) doublets. Such theories must produce not only radiative quark and lepton masses, but radiative masses for the mirror quarks and mirror leptons in excess of 20 GeV. Such masses lead to highly unnatural models [4].

One might argue that we are ruling out N=2 SU(3) x SU(2) x U(1) supersymmetric theories rather prematurely. Why not add arbitrary terms of dimension 2 and 3 whether they preserve finiteness or not. Indeed we don't expect the low energy effective theory obtained by integrating out superheavy particles to be finite even if the GUT were. The only constraints of a finite GUT would be the absence of $x_i x_j x_k^+$ soft operators. The presence of these operators would not alter our conclusions. If you prefer, you can forget all about finiteness and just add arbitrary soft operators.

We have argued that the low energy gauge group must be enlarged beyond SU(3) x SU(2) x U(1), and further that the adjoint should contain $SU(2)_L$ doublets so that mirror fermions may acquire tree level masses. The simplest embedding of an N=2 32-plet family, which allows such tree level masses, is with gauge group SU(3) x SU(4) x U(1) and matter representations

$$X_L = \begin{pmatrix} L \\ \bar{L}' \end{pmatrix} \sim (1,4,-1/2) \qquad Y_L = \begin{pmatrix} L' \\ \bar{L} \end{pmatrix} \sim (1,\bar{4},+1/2)$$

$$X_Q = \begin{pmatrix} Q \\ \bar{Q}' \end{pmatrix} \sim (3,4,1/6) \qquad Y_Q = \begin{pmatrix} Q' \\ \bar{Q} \end{pmatrix} \sim (\bar{3},\bar{4},-1/6), \tag{6}$$

where Q,L(Q',L') are quark and lepton (mirror quark and mirror lepton) $SU(2)_L$ doublets and $\bar{Q},\bar{L}(\bar{Q}',\bar{L}')$ are quark and lepton (mirror quark and mirror lepton) $SU(2)_R$ doublets. The interesting feature of this embedding is that conventional fermions are not assigned to some X representation with mirrors filling out the conjugate Y representation. Rather X and Y both contain conventional and mirror particles. Hence mirror parity is no longer $Y \to -Y$, $\Phi \to -\Phi$; although the mirror interchange symmetry is still $X \leftrightarrow Y$, $\Phi \leftrightarrow \Phi^T$.

Introducing Higgs multiplets (X_i,Y_i) with $X_1 \sim (1,4,-1/2)$, $X_2 \sim (1,4,1/2)$, the superpotential is

$$\frac{f}{\sqrt{2}} = ig_3 Y_Q \Phi_3 X_Q + ig_4(Y_Q\Phi_4 X_Q + Y_L\Phi_4 X_L + Y_i\Phi_4 X_i) + ig_1\Big(\frac{1}{6} Y_Q\Phi_1 X_Q - \frac{1}{2} Y_1\Phi_1 X_1 + \frac{1}{2} Y_2\Phi_1 X_2\Big). \tag{7}$$

The electric charge generator is $Q = T_{3L} + T_{3R} + Z$, where $T_{3_R^L}$ are the diagonal generators of $SU(2)_{L \atop R}$, and the U(1) generator Z is 1/2(B-L) on matter fields. The weak mixing angle is given at tree level by $\sin^2\theta = 1/(2 + g_4^2/g_1^2)$, so that $g_1 \sim g_4/1.6$.

It is not surprising that soft operators can be added to the theory to yield $\langle x_i \rangle \neq 0$ which break the gauge group to $SU(3) \times SU(2)_L \times U(1)$. What is remarkable is the ease with which these operators can lead to a global minimum in which ϕ_4 has an off diagonal vev, giving tree level generation independent mirror fermion masses. This is the most interesting feature of the model, and occurs as follows.

We make soft additions to the potential

$$V_{SOFT} = -m_{x_i}^2 |x_i|^2 - m_{y_i}^2 |y_i|^2 + m_{\phi_1}^2 |\phi_1|^2 + m_{\phi_3}^2 |\phi_3|^2 + m_{\phi_4}^2 |\phi_4|^2 + m_{x_{Q_L}}^2 |x_{Q_L}|^2 + m_{y_{Q_L}}^2 |y_{Q_L}|^2 + m_i y_i \phi_4 x_i \quad (8)$$

with each m^2 taken positive, so that these terms stabilize all scalars except x_i and y_i. The $|x_i|^2$ and $|y_i|^2$ terms drive vevs for x_i and y_i whose magnitudes V_i and v_i we can choose by adjusting $m_{x_i}^2$ and $m_{y_i}^2$. The $y_i \phi_4 x_i$ term then drives a vev for ϕ_4, with magnitude $\hat{v}$ which can be chosen by varying m_i. For our purposes the critical question is the relative orientation of the x_i and y_i vevs, and therefore the orientation of the ϕ_4 vev. In the desired limit of $V_i \gg \hat{v} \gg v_i$ these orientations are very simple to determine.

There is only one term in V of order V_i^4 which contributes to an alignment; it is $g_4^2/2 |x_1^+ x_2|^2$, and favors $\langle x_1 \rangle \perp \langle x_2 \rangle$, so that we can choose

$$\langle x_1 \rangle = \begin{pmatrix} 0 \\ 0 \\ V_1 \\ 0 \end{pmatrix}, \quad \langle x_2 \rangle = \begin{pmatrix} 0 \\ 0 \\ 0 \\ V_2 \end{pmatrix}. \quad (9)$$

The next largest terms in V which effect alignment are of order $V_1^2 \hat{v}^2$ and $V_2^2 \hat{v}^2$:

$$g_4^2 (x_1^+ \{\Phi^+, \Phi\} x_1 + x_2^+ \{\Phi^+, \Phi\} x_2). \quad (10)$$

Here $\phi = \phi_4$. If the elements of the y_1 and y_2 vevs which are non-zero are the a and b elements respectively, then the $y_i \Phi x_i$ terms will lead to vevs for Φ_{a3} and Φ_{b4}. The significant feature of (10) is that it is minimized by having $a,b \neq 3,4$; it is this term which ensures that $\langle \Phi \rangle$ is off diagonal as desired. We may choose a=1 by $SU(2)_L$ invariance. The next largest alignment term in the potential, $g_4^2\, Tr([\Phi, \Phi^+]^2)$, gives b=2, so that the alignment terms are minimized by choosing vevs:

$$x_1 = \begin{pmatrix} 0 \\ 0 \\ V_1 \\ 0 \end{pmatrix}, \quad x_2 = \begin{pmatrix} 0 \\ 0 \\ 0 \\ V_2 \end{pmatrix}, \quad y_1 = \begin{pmatrix} v_1 \\ 0 \\ 0 \\ 0 \end{pmatrix}, \quad y_2 = \begin{pmatrix} 0 \\ v_2 \\ 0 \\ 0 \end{pmatrix}, \quad \phi_4 = \begin{pmatrix} 0 & \hat{v}_1 & 0 \\ & 0 & \hat{v}_2 \\ 0 & & 0 \end{pmatrix} \tag{11}$$

Although we have not computed the vevs in terms of the parameters of the potential, this is the correct orientation of the vacuum at the global minimum of the potential. We have checked that parameters can be chosen to make V_i, v_i and $\hat{v}_i$ anything we please (subject to $V_i \gg \hat{v} \gg v_i$).

The scale $g_4 V_i$ is the scale at which SU(4) xU(1) is broken to $SU(2)_L \times U(1)_Y$; it is the scale of m_{x_i} and we call it M_4. Many other soft breaking parameters in the theory will also be taken to have this scale, for example m_{Q_L}, so that we will also call it M_{SUSY}. Phenomenological requirements may force other soft parameters to be less than M_{SUSY}, for example m_{y_i} and m_i, which will not be unnatural unless M_{SUSY} is constrained to be many orders of magnitude above the weak scale.

We have given an explicit model in which the tree level symmetry breaking leads to a fundamental distinction between ordinary and mirror fermions: only mirror fermions acquire tree level masses. The mirror up quarks and mirror neutrinos acquire masses $M_{U'} = \sqrt{2}\, g_4 \hat{v}_1$, while the mirror down quarks and mirror charged leptons acquire masses $M_{D'} = \sqrt{2}\, g_4 \hat{v}_2$. These masses are constrained by the scale of $SU(2)_L$ breaking

$$M_{U'}^2 + M_{D'}^2 = 2M_W^2. \tag{12}$$

To obtain this distinction between ordinary and mirror fermions, it was essential that the mirror interchange symmetry, $X \leftrightarrow Y$, $\Phi \leftrightarrow \Phi^T$, of the dimension four terms be spontaneously broken. This breaking is triggered by the explicit breaking of the interchange symmetry in the dimension two soft terms ($m_{y_i}^2 < m_{x_i}^2$), and we speculate that the ultimate source of such terms could be a hidden sector of the theory coupled to the observable sector via N=2 supergravity.

I would like to briefly mention possible sources for ordinary fermion masses in this theory (for more details see [4]). As it stands, the theory possesses an unbroken U(1) symmetry preventing ordinary fermion masses, and further soft operators must be added to break this symmetry. I'll mention one scheme for ordinary fermion masses in which mirror parity is unbroken, and one where it is broken.

The ordinary fermions may acquire masses via the diagram of Figure 2 if N=1 supersymmetric mass terms for ϕ_1 and ϕ_3 are added to

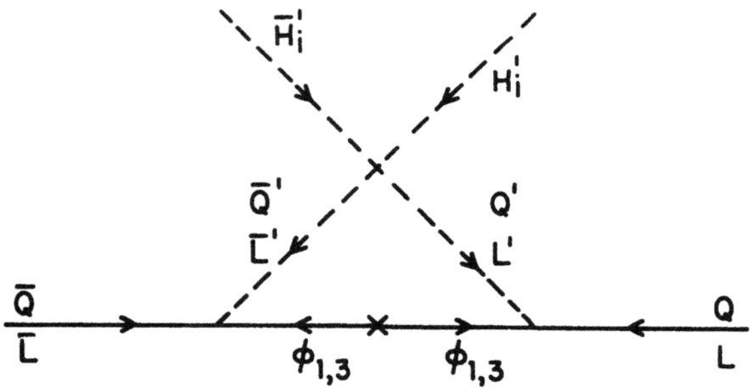

Figure 2

the theory. These mass terms are obviously flavor diagonal, although the dependence on squark masses can lead to different masses for each generation. To obtain flavor mixings, we add the trilinear scalar operators $[Y_{Q_a}\Phi_4 X_{Q_b}]_A$, which allows the diagrams of Figure 3. This

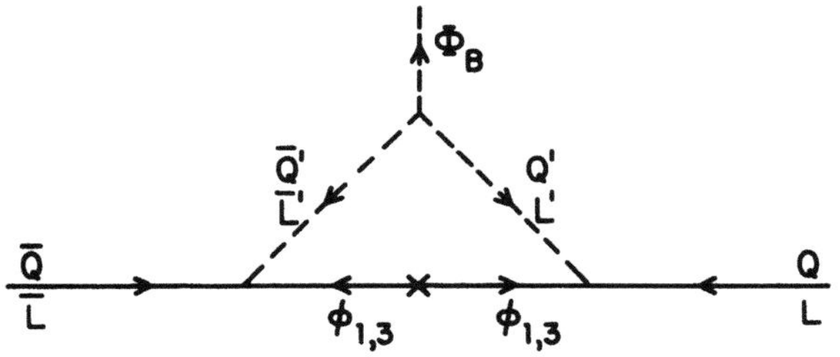

Figure 3

scheme has two setbacks. The coefficient of the scalar trilinear must be rather large for an adequate top quark mass. Although the trilinear is an operator of class 4 and does not make the potential unbounded from below, it may lead to a deeper minimum which breaks color. The second difficulty is very bothersome: Figures 2 and 3 lead to unacceptably large neutrino masses. One way out of this is to break lepton number by adding the soft operator $[Y_{L_a}\phi_1 X_{H_1}]_A$ thus allowing the right-handed neutrino to acquire a Majorana mass from the diagram of Figure 4. For the left-handed neutrino masses to be pushed down to an acceptable level requires the SUSY breaking scale to be made rather large, about 100 TeV.

A second scheme for ordinary fermion masses incorporates the mirror parity breaking operators $[Y_{Q_a}X_{Q_b}]_F$, which mixes ordinary and

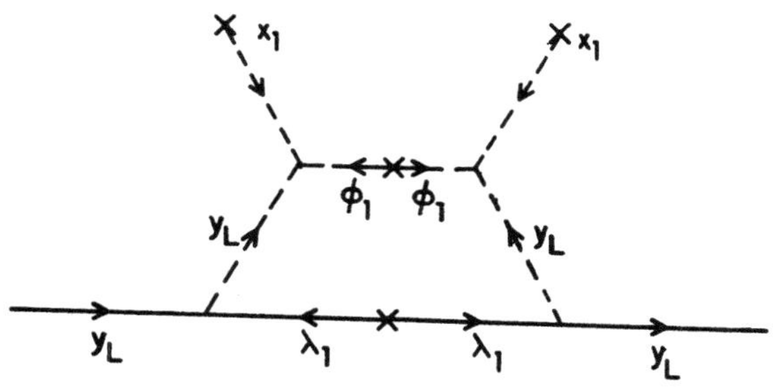

Figure 4

mirror fermions to some degree and greatly facilitates the top quark becoming heavy. The off diagonal ϕ_4 vev can be driven by adding a $[\phi_4\phi_4]_F$ operator instead of the $y_i\phi_4x_i$ scalar trilinear. A large τ mass can then be produced radiatively via the diagram of Figure 5.

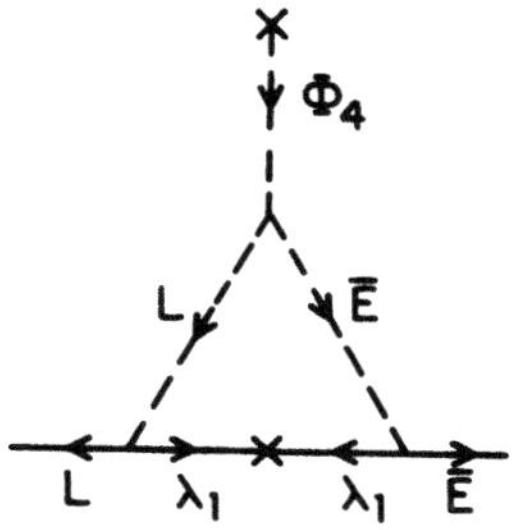

Figure 5

Flavor mixing comes predominantly from Figure 3, and the right-handed neutrino masses from Figure 4. The details are given in reference [4], but it should be clear that the structure of the mass matrices is very rich indeed.

I hope that three main facets of building N=2 supersymmetric models have been made clear in this lecture:

(1) The models have a very rich structure once soft operators are added. This complexity is best dealt with by keeping track of all the global and discrete symmetries in an N=1 superfield formalism.

(2) While SU(3) x SU(2) x U(1) N=2 theories seem impossible to build, an enlargment of the electroweak gauge group allows a very simple distinction between mirror and ordinary fermions.

(3) The ordinary fermion mass matrices cannot be explained in terms of tree level Yukawa couplings. There are a great variety of sources for such masses. The addition of horizontal gauge interac-

tions may lead to relations amongst these radiative fermion masses.

Acknowledgements

This work was done in collaboration with the authors of reference [4]. I thank them for many hours of fruitful and enjoyable conversations. I thank the organizers and participants of this Winter School for producing a magnificent opportunity to exchange ideas.

This research is supported in part by the National Science Foundation under Grant No. PHY-82-15249.

References

[1] For example, see references in J. Ellis, Talk at Cornell Lepton and Photon Interaction Conference, August 1983; P. Nath. R. Arnowitt, and A. Chamseddine, Northeastern University preprint, NUB #2613.

[2] A. Salam and J. Strathdee, Phys. Lett. 51B (1974) 33. P. Fayet, Nucl. Phys. B113 (1976) 135; B149 (1979) 137.

[3] P. Howe, K. Stelle, and P. West, Phys. Lett. 124B (1983) 55. P. West, Talk at La Jolla Workshop, January 1983.

[4] F. del Aguila, M. Dugan, B. Grinstein, L. Hall, G. Ross, and P. West, Harvard University Preprint HUTP-84/A001.

[5] A. Parker and P. West, Phys. Lett. 127B (1983) 353.

ASPECTS OF N=1 SUPERGRAVITY MODELS

Lawrence J. Hall
Lyman Laboratory of Physics
Harvard University
Cambridge, MA 02138

Abstract

The structure of low energy models resulting from N=1 supergravity theories depends on assumptions and choices made for the superpotential and Kähler potential. A review is given of those structures which have been used most frequently.

The minimal model of low energy supergravity is discussed. A review of the various mechanisms for breaking SU(2) x U(1) is given. A few interesting phenomenological features of the model are pointed out.

Building particle physics models which possess global supersymmetry is a rather frustrating business. (For example, see [1-3]). While super Yang Mills theories have a certain elegance and a highly constrained set of interactions, realistic globally supersymmetric models always tend to lose either the elegance or the high degree of constraint.

The many globally supersymmetric models can be categorized according to how supersymmetry is broken. If supersymmetry is broken spontaneously, via the O'Raifeartaigh [2], or Fayet-Iliopoulos mechanism [1], the models contain many extra particles beyond those which are required to supersymmetrize the representation content of the standard model. None of these models, especially the Fayet-Iliopoulos mechanism ones, are very elegant. If supersymmetry is broken softly [3] by adding various operators of dimension 2 and 3 to the Lagrangian, then the theory no longer appears highly constrained. There are a huge number of such operators, and, providing the superpartners of quarks and leptons are not as heavy as many TeV, the soft scalar masses must have a high degree of degeneracy to ensure a super-GIM mechanism to control the super box contributions to the K_L-K_S mass difference.

Over the last year or two many groups have worked on a class of models which arise as the low energy remnants of certain N=1 supergravity theories [4]. These theories require neither multitudes of judiciously chosen soft operators which are added by hand, nor do they require that the low energy theory possess more particles than the minimal set needed in any supersymmetric theory. Indeed, one finds theories in which this minimal set of particles interact via supersymmetric interactions together with a very constrained set of soft supersymmetry breaking interactions. It is rather exciting and non-trivial that these theories can be made realistic.

In these lectures I wish to treat two topics in some detail. Firstly I will discuss some of the assumptions which underlie these supergravity models. Although the soft operators are constrained, exactly how constrained depends sensitively on one's initial starting assumptions. Secondly I will discuss a particular SU(3) x SU(2) x U(1) model which has the fewest number of particles, and which has the most highly constrained set of soft operators which we could hope for.

(I) THE GENERAL STRUCTURE OF MODELS DERIVED FROM N=1 SUPERGRAVITY

Let's consider the generic procedure for passing from N=1 supergravity theories to low energy models of interest:

$$\mathscr{L}(\text{N=1 SUGRAV.}) \xrightarrow[\text{below } M_p]{\text{scales}} \mathscr{L}_{\text{SUSY}} + \mathscr{L}_{\text{SOFT}} \tag{1}$$

The terms of the supergravity Lagrangian [5] which are of most interest to us are the potential

$$V = e^{\kappa^2 d}\left\{F_N (g^{-1})^N{}_M F^{*M} - 3\kappa^2 |f|^2\right\} + \frac{D^2}{2} \tag{2}$$

and gauge kinetic terms $f_{\alpha\beta}W^{\alpha}W^{\beta}$. The notation is: κ is the "rationalized" Planck mass: $\kappa^2 = 8\pi/M_p^2$. $d = d(z,z^*)$ is the Kähler potential of the chiral scalars z_N. $F_N = f_N + \kappa^2 d_N f$ is the Kähler covariant derivative of the superpotential f. Subscripts N on f or d refer to derivatives with respect to z_N. The Kähler metric is $g_N{}^M = d_{NM^*}$. W^{α} is the supersymmetric gauge field strength, with α running over the adjoint of the gauge group. $f_{\alpha\beta}$ is a general function of the chiral fields z_N, and has a piece $\delta_{\alpha\beta}$ which gives the renormalizable kinetic terms.

The passage described by (1) is carried out by making a choice for the "potentials" f(z), $d(z,z^*)$ and $f_{\alpha\beta}(z)$ such that

(1) Supersymmetry is spontaneously broken.

(2) Certain fields associated with this breaking (z_i) decouple.

(3) Certain fields become superheavy (z_A).

(4) Remaining fields are to be observed in low energy theory (z_a).

(In certain cases (2) and (3) may overlap.) After shifting all fields by their vevs, and discarding terms involving decoupled or superheavy fields, the RHS of (1) is obtained at tree level. There are two features of this process which I will concentrate on

(1) $\mathscr{L}_{SOFT}$ depends on the choices made for f, d, and $f_{\alpha\beta}$.

(2) $\mathscr{L}_{SOFT}$ depends on our assumptions about the nature of radiative corrections. I'll leave this more problematic point for later.

Before discussing the wide variety of choices for f, d, and $f_{\alpha\beta}$, I'll give perhaps the simplest model which breaks supersymmetry: the Polonyi model [6]. The model involves just one field $z_i \to z$, together with

$$d = \kappa^2 z^* z$$

$$f = m_{3/2}\kappa^{-2}(\kappa z+\beta) \tag{3}$$

where $m_{3/2}$ and β can be taken real. Choosing $\beta = 2 - \sqrt{3}$ gives a potential with zero cosmological constant, as shown in Figure 1.

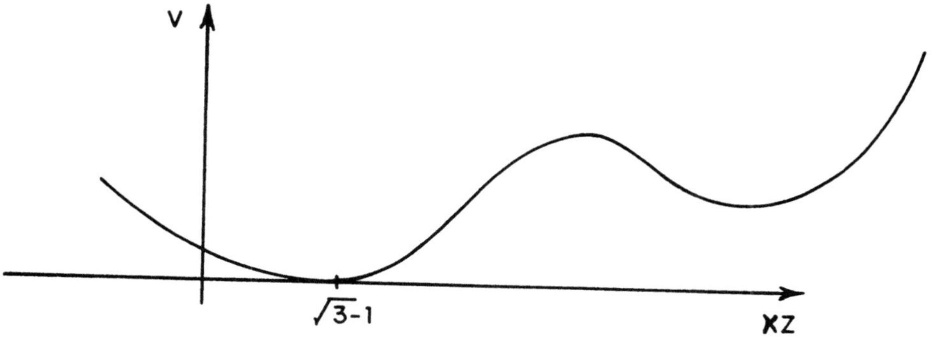

Figure 1. Potential in the Polonyi Model.

The order parameter for supersymmetry breaking is the vev of the Kähler covariant derivative <F> which is non-zero. The gravitino becomes heavy via the superHiggs effect acquiring a mass $\kappa^{-2}\langle f\rangle = m_{3/2}$.

Increasing the number of fields $z \to z_i$ to give a more general supersymmetry breaking sector $f(z_i)$ will not change anything but the details. The scalars will acquire vevs of $O(M_p)$, and if f is written as

$$m_{3/2}\kappa^{-2}\tilde{f}(kz_i) \tag{4}$$

where $\tilde{f}$ is dimensionless, then $m_{3/2}$ will set the scale of supersymmetry breaking as before.

Now we are ready to add superheavy (z_A) and light (z_a) fields to the theory. We will find that V_{SOFT} depends on how we choose to extend $f(z_i)$ to $f(z_i z_A z_a)$ and $d(z_i)$ to $d(z_i, z_A, z_a)$. I will discuss five "reasonable" ways of doing this. The order in which they are discussed has been chosen for pedagogy. The fives schemes are labelled hieroglyphically as A-B-C. Here A will refer to the type of gauge group considered (G for grand unified, LE for low energy), B to the type of Kähler metric (F for flat, etc.) and C to the structure of the superpotential.

(1) THE LE-F-H+0 SCHEME

In the simplest schemes there is no superheavy sector at all, the Kähler potential is flat

$$d = \kappa^2 z_i^* z_i + \kappa^2 z^* z \tag{5}$$

and the superpotential splits into a sum of two terms

$$f = f_H(z_i) + f_0(z_a). \tag{6}$$

H is the "hidden" or supersymmetry breaking sector, and 0 is the "observable" sector, i.e. the part of the theory which corresponds to the world we know. The hidden sector has the structure of (4), so that on substituting $\langle z_i \rangle$ into the potential (2) and keeping only renormalizable terms in z_a the low energy potential is [7]

$$V_{LE} = V_{SUSY} + m_{3/2}^2 z_a^* z_a + m_{3/2}(Af^{(3)} + (A-1)f^{(2)} + (A-2)f^{(1)}) \tag{7}$$

where $m_{3/2}A = \kappa^2 \langle z_i \rangle \langle F_i \rangle$ and $f^{(x)}$ are terms in f_0 of order $(z_a)^x$. Notice that the supersymmetry breaking in the hidden sector is communicated to the low energy observable world by tree level exchange of the supergravity auxiliary fields. This is always the case when f has the structure (H+0) because in a global theory such an f would describe two non-interacting worlds. (Throughout one assumes that if z_i feels any gauge interactions, they are hidden gauge interactions which commute with observable gauge interactions.) There are two remarkable and exciting features of V_{LE} which result from assumptions (5) and (6)

(1) All scalars of the low energy theory have acquired a common tree level supersymmetry breaking mass $m_{3/2}^2$. For supersymmetry to be relevant to the hierarchy problem we must choose $m_{3/2} \lesssim 0(\text{TeV})$ and for supersymmetry to be relevant to physics at allrequires $m_{3/2} \gtrsim 20$ GeV. Perhaps the phenomenological constraints on the scalar mass matrix from the superbox diagram are telling us that $f \sim H+0$ is a sensible starting assumption.

(2) All other soft operators are characterized by a single unknown parameter A, regardless of how complicated the hidden sector is. Furthermore the soft scalar self interactions have a form which strongly reflects the form of the superpotential.

(2) THE G-F-H+0 SCHEME

These are grand unified models (G) with a flat Kähler metric (F) with the superpotential again splitting into a sum of two terms:

$$f = f_H(z_i) + f_0(z_A, z_a). \tag{8}$$

The first GUT model of this type [8] had gauge group SU(5). It may be thought that the addition of superheavy fields (which is all that occurred in passing from scheme 1 to scheme 2) would be irrelevant to V_{LE}, which would again be given by (7). This is not correct [9], as we now show.

It seems most logical to assume that f_0 has one mass scale; that of the GUT, M_G. Then f_0 can contain z_A^3, $z_A^2 z_a$, $z_A z_a^2$, z_a^3 and z_A^2 terms (as well as linear terms for gauge singlet fields). Thus z_a^2 terms must not occur, unless z_a is kept light by fine tuning a $\langle z_A \rangle$ in a $z_A z_a^2$ term. The superheavy vevs can be found in a power series expansion in $(\kappa m_{3/2})$. Thus: $\langle z_A \rangle = z_A^{(0)} + z_A^{(1)} + \cdots$. Defining $f^{(3)}$ as the z_a^3 terms in f, $f^{(2)}$ as the $z_a^2 \; z_A^{(1)}\rangle$ terms, etc. one finds [9]:

$$V_{LE} = V_{SUSY} + m_{3/2}^2 z_a^* z_a + m_{3/2}(Af^{(3)} + 2f^{(2)} + (4-A)f^{(1)}) \tag{9}$$

where A has the same definitions as in (7). Thus in adding superheavy fields z_A, the definitions of $f^{(x)}$ change a little and the dependence of V_{LE} on A is a little different, but the important conclusions are the same. The scalars are degenerate, the soft scalar interactions are very constrained in form and depend on only a few unknown parameters of the hidden sector: $m_{3/2}$ and A, and parameters in $z_A^{(1)}$.

(3) THE G-U-H+0 SCHEME

We now consider a further complexity--that of a non-flat Kähler metric [9]. In order to maintain a relatively constrained form for the tree level soft operators we won't allow d to be arbitrary, rather we take d = d(x)

$$x = \kappa^2 (z_i^* z_i + z_A^* z_A + z_a^* z_a). \tag{10}$$

Thus d is U(N) invariant if there are a total of N chiral superfields. This generalization does not change much in V_{LE}: the common scalar mass becomes slightly different from $m_{3/2}$, and the first order shift in the superheavy vevs, $z_A^{(1)}$, change, but otherwise (9) still applies.

(4) THE LE-F-(H+0+0H) SCHEME

The majority of N=1 supergravity models have used one of the previous three schemes. Soni and Weldon [10] have pointed out that the supersymmetry breaking of the hidden sector can be communicated to the low energy world by non-gravitational mechanisms in addition to those used above. However the situation is rather delicate since one must protect the low energy world from feeling the full force of the supersymmetry breaking. This can be done by allowing a superpotential of the form (neglect superheavy fields):

$$f = f_H(z_i) + f_0(z_a) + f_{0H}(z_i, z_a) \tag{11}$$

with

$$f_{0H} \supset (\kappa m_{3/2})^{n} z_i^{n} z_a^{m}$$

scaled by the appropriate power of the mass scale of the observable sector to make f_{0H} of dimension 3. The interesting feature of these terms in the superpotential which couple hidden and observable sectors non-gravitationally is that they yield terms in V_{LE} in addition to those of (7):

$$V = V_{LE}(\text{eqtn. }(7)) + m_{3/2} M_p B_i \left. \frac{\partial f_{0H}}{\partial z_i} \right|_{<z_i>} . \tag{12}$$

If we know nothing about the hidden sector, this extra term is sufficient to make the soft cubic interactions (z_a^3, z_a^2, z_a) completely arbitrary so that

$$V = V_{SUSY} + m_{3/2}^2 z_a^* z_a + m_{3/2} h(z_a). \tag{13}$$

We have lost much, but not all, of the constrained structure: the $z_a^* z_a$ coefficient is still a independent, and terms of the form $z_a^* z_a^2$ are not allowed.

One can't help feeling that this scheme is contrived: the coefficients of a $z_i z_a^2$ term in f_{0H} are of size $\kappa m_{3/2}$, surely it would be more natural for them to be absent completely. For example, in a N=2 extended supersymmetric model one could use the second supersymmetry to force the f = 0+H split by imposing a gauge group $G_0 \times G_H$. However, the terms in f_{0H} do allow an interesting possibility: that we do away with the hidden sector idea altogether and use the superheavy fields z_A to break supersymmetry. This will lead to superpotentials of the form:

$$f \supset z_A^3, z_A^2, z_a^3, (\kappa m_{3/2}) z_A z_a^2, \ldots \tag{14}$$

which is precisely the form which Stuart Raby explores [11].

(5) <u>THE LE-G-(H+0+0H) SCHEME</u>

We can continue our exploration of how V_{LE} depends on our initial assumptions about the form of f and d. Actually there is only one more stage--that of obtaining an a dependent coefficient for $z_a^* z_a$. This will occur for a general Kähler potential, in particular for one which is not U(N) invariant. For

$$d = \kappa^2 z^*_\alpha z_\beta d_{\alpha\beta}(z_i) + \ldots \qquad (15)$$

the inverse of the Kähler metric in (2) causes a dependence of the $z^*_a z_a$ term [10].

In the following table I summarize the terms which appear in V_{LE} according to the assumptions of the complexity of d and f.

TABLE I: V_{LE} FOR VARIOUS KÄHLER POTENTIALS AND SUPERPOTENTIALS

	GRAND UNIFIED?	KAHLER POTENTIAL	SUPER-POTENTIAL	$z^*_a z_a$	$f^{(3)}$	$f^{(2)}$	$f^{(1)}$
1	x	FLAT	f_H+f_0	$m^2_{3/2}$	A	A-1	A-2
2	✓	FLAT	f_H+f_0	$m^2_{3/2}$	A	2	4-A
3	✓	d(x) U(N) INVARIANT	f_H+f_0	m'^2	A	2	4-A
4	x	FLAT	$f_H+f_0+f_{0H}$	$m^2_{3/2}$	$m_{3/2}h(z_a)$		
5	x	GENERAL	$f_H+f_0+f_{0H}$	m^2_a	$m_{3/2}h(z_a)$		

I would now like to briefly consider whether we can believe that these tree level results give a good approximation to the low energy theory. The most obvious fact is that N=1 supergravity is a non-renormalizable theory; we do not know how to make the theory meaningful at the loop level. I will side step this problem by a device which seems to be the only sensible way of making progress, and which most people in the field would subscribe to. Assume that the N=1 supergravity theory describes the world only up to a scale Λ, and that above Λ a more complete theory emerges which has the effect of cutting off the previously divergent integrals at scale Λ. This means that we should compute all radiative corrections in the N=1 supergravity theory, and just evaluate them numerically by cutting all divergent integrals off at scale Λ.

To start with consider purely gravitational radiative effects. For example, the tree level soft mass $m^2_{3/2} z^*_a z_a$ will have radiative corrections at nth order and ℓ loops of order

$$\frac{(\kappa\Lambda)^n}{(4\pi^2)^\ell} \int d^4\theta\, \kappa^2 \phi_i^+ \phi_i \phi_a^+ \phi_a \qquad (16)$$

where ϕ_i and ϕ_a are superfields whose scalar components are z_i and z_a. We can take two viewpoints to these purely gravitational radiative corrections. If $\kappa\Lambda \gtrsim 4\pi^2$, then radiative effects cause 0(1) changes to tree level terms, and we expect $V_{LE} = V_{SUSY}$ + (ARBITRARY V_{SOFT}). However if $\kappa\Lambda$ and $1/4\pi^2$ are small enough, the tree level results will be a good approximation. The latter case is probably just wishful thinking, but, surprisingly, the case of large $\kappa\Lambda$ does not ruin our analysis. The gravitational interactions respect the U(N) symmetry, so that the radiatively corrected effective action for pure N=1 supergravity up to two derivatives is characterized by $d = d(x)$ and $f=1$. When matter and gauge superfields are added, providing all non-gravitational (i.e. U(N) breaking) interactions are weak ($\sim\varepsilon$) all further radiative corrections are $\propto \varepsilon^n(\kappa\Lambda)^m$ and may be weak. Thus it is possible that strong gravitational radiative corrections may be taken care of by introducing the most general U(N) invariant Kähler potential. Thus we will work in the 3rd scheme of our previous discussion G-U-H+0.

There is one class of non-gravitational radiative corrections which may be large even if ε is small; the ones proportional to $\ell n\Lambda$. For example, the tree level scalar mass of Figure 2(a), which comes from supergravity auxiliary exchange (denoted by dots), is renormalized by the diagram of Figure 2(b) which contributes

$$\delta m_a^2 \simeq \frac{\lambda^2}{4\pi}\, \ell n\, \frac{\Lambda}{\mu}\, m_b^2. \qquad (17)$$

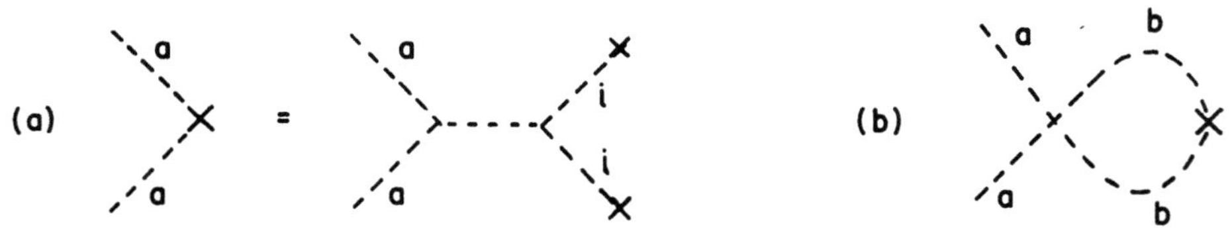

Figure 2

λ is a supersymmetric Yukawa coupling while μ is of order the low energy scale $\sim m_{3/2}$. The leading logarithm terms of order $(\lambda^2/4\pi\ \ell n\,\Lambda/\mu)^n$ must be summed using the renormalization group. This suggests that we use an effective field theory picture. Below M_p we view the theory as an effective globally supersymmetric theory with soft operators which is renormalizable. In this effective theory we can use the renormalization group to compute running parameters: $m_a^2(\mu)$. This effective theory must agree with the full N=1 supergravity theory at scales where both are valid--i.e. just below Λ. Thus the "matching" condition on the effective theory is that $m_a^2(\mu=\Lambda) = m'^2$, the tree level scalar mass in the full theory. I have belabored this a little. It can be a little confusing because, unlike boundaries between effective theories which we are used to, there are no particles of mass of order Λ which we are integrating out. From now on I will work only in the low energy renormalizable effective theory, and simply use the tree level results of the full theory as the boundary condition at $\mu \sim \Lambda \sim M_p$.

II. THE MINIMAL LOW ENERGY MODEL DERIVED FROM N=1 SUPERGRAVITY

Having developed the framework, and stressed the assumptions, of N=1 supergravity models, it is natural to seek the simplest and most predictive such model [12-21]. It is important to notice that the low energy theories derived earlier were very insensitive to the details of the superheavy fields and their interactions. For this reason we will seek the simplest low energy remnants of a N=1 supergravity model independent of any grand unified scheme. In fact many GUTs lead to this low energy theory, and there is no compelling reason to adopt one version of the GUT sector over the other. The nice thing about the low energy model to be presented here is that it involves the smallest gauge group, SU(3) x SU(2) x U(1), and the fewest particles (Table II) possible for such a theory--it is, in this sense, the minimal model of low energy supergravity.

TABLE II: CHIRAL SUPERFIELDS OF THE MINIMAL MODEL

Q(3,2,1/6)	L(1,2,-1/2)	H(1,2,1/2)
$\bar{U}(\bar{3},1,-2/3)$	$\bar{E}(1,1,1)$	H'(1,2,-1/2)
$\bar{D}(\bar{3},1,1/3)$		

SU(3) x SU(2) x U(1) representations are given in brackets.

We will work in the framework which requires the fewest assumptions to obtain a constrained theory; described previously as G-U-0+H. The low energy theory then has the tree level form:

$$\begin{aligned}\mathcal{L}_{LE} = \mathcal{L}(\text{KINETIC + GAUGE}) &+ \frac{M_n}{2}\lambda_n\lambda_n \\ &+ [f(\phi_a)]_F + \text{h.c.} \\ &+ A\, f^{(3)}(z_a) + \text{h.c.} \\ &+ B\, f^{(2)}(z_a) + \text{h.c.} + m_{3/2}^2 z_a^* z_a\end{aligned} \tag{18}$$

where the superpotential $f = f^{(3)} + f^{(2)}$ contains the most general gauge invariant terms which conserve matter parity

$$f = \lambda_U Q\bar{U}H + \lambda_D Q\bar{D}H' + \lambda_E L\bar{E}H' + \mu HH' \qquad (19)$$

where the λ are 3x3 Yukawa coupling matrices in a theory of three generations. The parameters A and B are unknown dimensionful complex constants which depend on the hidden and GUT sectors. The scalars have a common mass $m_{3/2}$ which is of order the gravitino mass. The gaugino masses M_n for group factor n = 1,2,3 for U(1) x SU(2) x SU(3) can have several sources. I will assume two things about the dominant contributions to M_n: firstly that they occur at tree level from the vev $\mathrm{Re}\langle f_{\alpha\beta,i}\rangle$. Secondly I will assume that there is a unification of the low energy gauge group into a non-Abelian group so that the tree level contribution to M_n is n independent. It should be remembered that contributions which depend on n occur at 1 loop at the super-heavy threshold, and can be quite large.

Since I am assuming that a GUT sector exists the mass scale associated with $f_0(z_A, z_a)$ is that of the GUT scale. It may be thought that μ should be zero since otherwise f_0 would have two fundamental scales. In fact μ can arise at $O(m_{3/2})$ quite naturally as follows. Suppose H and H' are parented by GUT representations G and G' and that $f_0 \supset XGG'$ where X is some other GUT representation. When the GUT potential is minimized in the approximation $m_{3/2} \to 0$ some unbroken symmetry (for example R symmetry) may give $\langle X^{(0)}\rangle = 0$. With $m_{3/2} \neq 0$ this symmetry may be broken by the hidden sector which induces shifts in the superheavy vevs of order $m_{3/2}$: $\langle X^{(1)}\rangle \propto \langle f_H\rangle$, which induces $\mu \sim O(m_{3/2})$. Even if this doesn't occur, μ can arise radiatively at the GUT threshold. For example, suppose

$$f_0 \supset GAB + G'A'B' + AA' + BB' \qquad (20)$$

giving the one loop radiative contribution to μ shown in Figure 3. In fact the most elegant supersymmetric SU(5) GUT theories possess such

FIGURE 3: RADIATIVE CONTRIBUTIONS TO μ

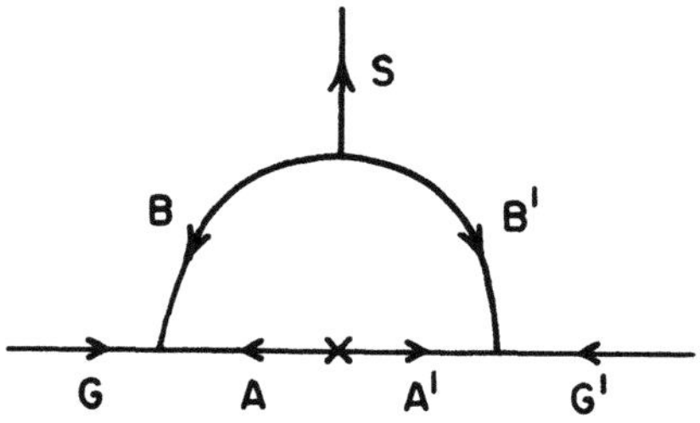

S is a spurion field representing supersymmetry breaking.

terms with representations G(5), $B(\overline{50})$, $G'(\bar{5})$, B'(50), A = A'(75) [11].

One of the things I like about this model is that at tree level it simply doesn't work. It is obvious that none of the squarks or sleptons acquire tree level vevs since they all have positive mass squareds. I'm assuming the Fayet-Iliopoulos term to be small. The Higgs have both positive mass squared and a term $\mu BHH'$ which could drive a vev. The Higgs mass matrix is

$$(H\ H'^{*})\begin{pmatrix} m_{3/2}^2+\mu^2 & -\mu B \\ -\mu B & m_{3/2}^2+\mu^2 \end{pmatrix}\begin{pmatrix} H^* \\ H' \end{pmatrix} \tag{21}$$

There are two possibilities. Either $m_{3/2}^2+\mu^2 > \mu B$, in which case SU(2) x U(1) is unbroken, or $m_{3/2}^2+\mu^2 < \mu B$ in which case the tree level potential is unbounded. To see this, notice that

$$H = \begin{pmatrix} 0 \\ h \end{pmatrix} \quad H' = \begin{pmatrix} h \\ 0 \end{pmatrix}$$

is a flat direction of the quartic potential. This is simply telling us that if the model is to be realistic radiative corrections must renormalize the coefficients of H^*H and H'^*H' differently.

The complete set of renormalization group equations for this model are rather simple. In the limit that all Yukawa couplings are neglected except for that of the top quark: $(\lambda_U)_{33} = \lambda$, all other elements of λ_U, λ_D and λ_E are taken to zero, they are:

$$\frac{d}{dt}\left(\frac{M_n}{\alpha_n}\right) = 0 \tag{22}$$

$$\frac{d}{dt} m_a^2 = -\frac{2}{\pi}\sum_n \alpha_n T_{na}^2 M_n^2 \qquad n \neq Q_3, \bar{U}_3, H \tag{23}$$

$$\frac{d}{dt}\begin{pmatrix} m_H^2 \\ m_{\bar{U}_3}^2 \\ m_{Q_3}^2 \end{pmatrix} = -\frac{2}{\pi}\sum_n \alpha_n T_{na}^2 M_n^2 + \frac{\lambda^2}{8\pi^2}\begin{pmatrix} 3 \\ 2 \\ 1 \end{pmatrix}\left(m_H^2 + m_{\bar{U}_3}^2 + m_{Q_3}^2 + A^2\right) \tag{24}$$

$$\frac{dA}{dt} = \frac{1}{4\pi}\left(-\frac{32}{3}\alpha_3 M_3 - 6\alpha_2 M_2 - \frac{26}{15}\alpha_1 M_1 + 12\frac{\lambda^2 A}{4\pi}\right) \tag{25}$$

$$\frac{dB}{dt} = \frac{1}{4\pi}\left(-6\alpha_2 M_2 - \frac{6}{5}\alpha_1 M_1 + 6\frac{\lambda^2 A}{4\pi}\right) \tag{26}$$

$$\frac{d\lambda}{dt} = \frac{\lambda}{4\pi}\left(-\frac{16}{3}\alpha_3 - 3\alpha_2 - \frac{13}{15}\alpha_1 + \frac{6\lambda^2}{4\pi}\right) \tag{27}$$

$$\frac{d\mu}{dt} = \frac{\mu}{4\pi}\left(-3\alpha_2 - \frac{3}{5}\alpha_1 + \frac{3\lambda^2}{4\pi}\right) \tag{28}$$

where t is the logarithm of the renormalization point mass. I've written these equations down not for completeness, but because they are full of interesting physical effects. For example, in Eq. (23) we see that gaugino masses tend to increase squark and slepton masses in the infrared. The up squark will be heavier than the selectron. The gaugino masses themselves scale in a fashion identical to the gauge coupling constants; they grand unify in a similar fashion to the gauge couplings. The Majorana mass of the photino at low energies is therefore probably an order of magnitude smaller than that of the gluino. From the point of view of electroweak symmetry breaking the essential point is that H appears in a large Yukawa interaction while H' does not. Thus the last term of equation (24) renormalizes the soft mass m_H^2 splitting it from $m_{H'}^2$.

I will do the simplest thing possible: use the above renormalization group equations to scale from the boundary at M_p to the low scale $m_{3/2}$. This involves a large assumption. Suppose that no new thresholds occur between $m_{3/2}$ and M_p except for the GUT threshold at a scale $\sim 10^{-2}M_p$ or so. Even a scaling across two orders of magnitude can greatly effect our results if the GUT representations are fairly large.

The Higgs potential relevant for electroweak symmetry breaking is obtained by renormalization group scaling the relevant parameters from M_p to $m_{3/2}$, to give:

$$V = (m_H^2+\mu^2)H^*H + (m'^2_H+\mu^2)H'^*H' - \mu BHH' + \text{h.c.} \tag{29}$$
$$+ \frac{g'^2}{2}\left(\frac{1}{2}H^*H - \frac{1}{2}H'^*H'\right)^2 + \frac{g^2}{2}\left(H^*\frac{\bar{\tau}}{2}H + H'^*\frac{\bar{\tau}}{2}H'\right)^2.$$

There are two conditions for a realistic stable minimum. Firstly the mass matrix must have a negative eigenvalue so that SU(2) xU(1) breaks:

$$C(I) \qquad (m_H^2+\mu^2)(m_{H'}^2+\mu^2) < \mu^2B^2 \tag{30}$$

Secondly, the potential must be bounded from below

$$C(II) \qquad (m_H^2+\mu^2) + (m_{H'}^2+\mu^2) > 2\mu B. \tag{31}$$

These conditions are shown on Figure 4 as a straight line and hyperbola passing through $(\mu B, \mu B)$:

FIGURE 4: HIGGS PHASE DIAGRAM

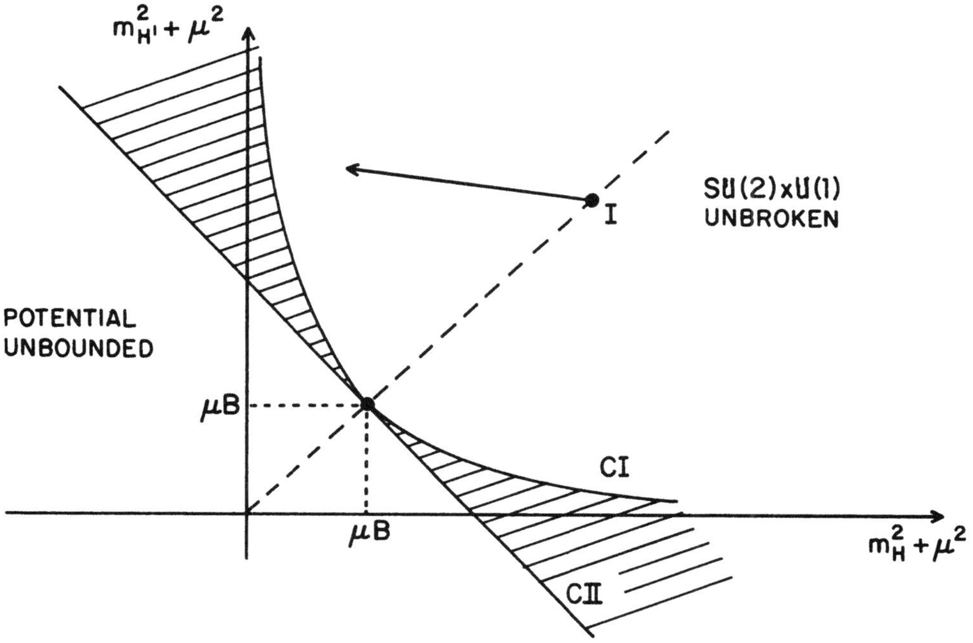

As the renormalization group equations are evolved, the parameters follow trajectories in this plot. At M_p the initial condition I lies on the dotted line $m_H^2 = m_{H'}^2$. As we saw previously, no point along this line gives a satisfactory potential at tree level. As we scale down in energy two things occur. The parameters m_H^2 and $m_{H'}^2$ move away from I as indicated by the solid line trajectory. Secondly μB also changes so that the positions of the constraints CI and CII move. A successful model must have the endpoint of the trajectory (i.e. at scale $m_{3/2}$) lying in the final position of the shaded region. For parameters lying in this shaded region it is easy to minimize V:

$$\langle H\rangle = \begin{pmatrix}0\\h\end{pmatrix} \quad \langle H'\rangle = \begin{pmatrix}h'\\0\end{pmatrix} \qquad \begin{aligned}h &= v\cos\theta\\ h' &= v\sin\theta\end{aligned}$$

$$\left(\frac{g^2+g'^2}{2}\right)v^2 = \left|\frac{m_{H'}^2-m_H^2}{\cos 2\theta}\right| - m_H^2 - m_{H'}^2 - 2\mu^2 = M_Z^2 \qquad (32)$$

$$\sin 2\theta = \frac{2\mu B}{m_H^2+m_{H'}^2+2\mu^2} \; .$$

All variants of the minimal model which have been proposed fall into two categories. Models in which the dominant scaling effect is of m_H, that is the trajectory moves rapidly from point I, I call type I [12-15]. Models in which the dominant scaling effect is that of μB (i.e. the movement of the shaded region) I call type II [18-21].

Type I models can be realized by making the top quark heavy, i.e. λ large. In these models the variation of μB and of $m_{H'}^2$ is small compared to that of m_H^2 which is driven negative by:

$$\frac{d}{dt}\begin{pmatrix} m_H^2 \\ m_{\bar{U}_3}^2 \\ m_{Q_3}^2 \end{pmatrix} = \frac{\lambda^2}{8\pi^2}\begin{pmatrix} 3 \\ 2 \\ 1 \end{pmatrix}\left(m_H^2 + m_{\bar{U}_3}^2 + m_{Q_3}^2 + A^2\right) + \cdots \qquad (33)$$

The factors of 3, 2, and 1 are crucial. The Higgs mass is driven negative first, and becoming negative it presents the colored scalars from acquiring negative masses. This rather non-trivial behaviour results from the SU(3) and SU(2) loop counting factors illustrated in Figure 5:

FIGURE 5: NUMERICS OF SCALAR MASS RENORMALIZATION

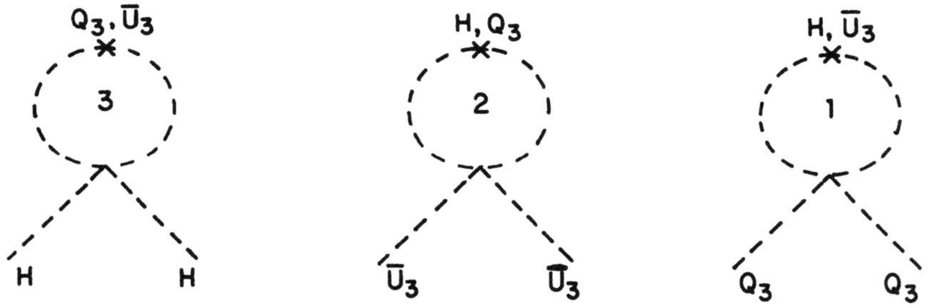

The parameter space is rather large, but following [12] we can gain considerable insight for the case of small μ and M_n. In this case the results of evolving the renormalization group equations are shown in Figure 6 [16]. A great deal can be understood from this sketch. If the top quark is too light, we violate condition (I) (Eq. 30) so that SU(2) x U(1) is unbroken. We can get away with smaller top quark masses in this regard if we are willing to put up with a large A parameter, which from Eq. (33) is seen to be equally capable of driving our trajectory in the Higgs phase diagram [15]. For too large A or λ we run the risk of overkill--the trajectory of Figure 4 runs too quickly and passes straight through the shaded region into the region of unboundedness, violating constraint II.

FIGURE 6: MODELS WITH A HEAVY TOP QUARK AND SMALL μ, M_n

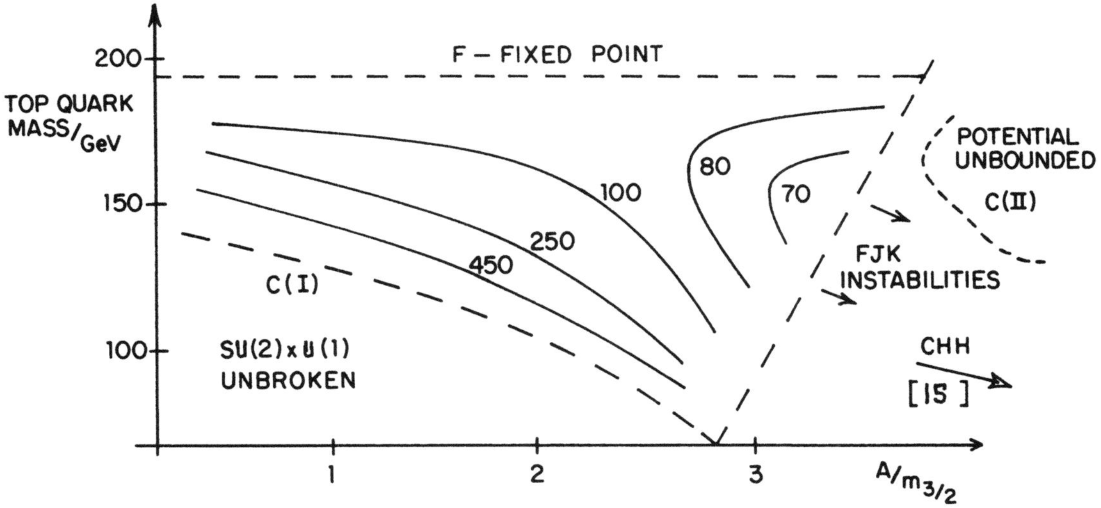

Even if A is small, the top quark cannot exceed about 180 GeV. Equation (27) has an infrared fixed point where the bracket vanishes. This occurs for a λ which gives a top quark mass less than about 180 GeV. Large values of A introduce vacuum instabilities noticed by Frère, Jones, and Raby [17]. These instabilities may be overcome however by having a theory in a long-lived false vacuum [15]. Such models may have a small top quark. Figure 6 shows a large, roughly triangular allowed region in which the desired vacuum is the true one. In this region the solid contours connect points with equal values of $m_{3/2}$. These contours show that, generally, lighter top quarks require larger values of $m_{3/2}$. For μ small equations 32 show that $\cos\theta \sim 1$ amd $M_z^2 \sim -2m_H^2$. As the top quark is made ligher, m_H^2 is driven less negative so for the Z mass to take its measured value requires an increase in the overall scale of the soft operators, i.e. an increase in $m_{3/2}$. From Figure 6 one can see that in these type I models with μ and M_n small, and with a stable vacuum, $100 \text{ GeV} \lesssim m_t \lesssim 180 \text{ GeV}$. If M_n is increased, the gluinos increase the squark masses thus helping to decrease m_H^2, and we can get away with a lighter top mass: $m_t \gtrsim 55 \text{ GeV}$.

The trajectory for models of type II is shown in Figure 7. In Figure 7a the initial values of the Higgs masses at scale M_p are shown, together with the constraints and domain of symmetry breaking. For simplicity suppose that λ, A, and M_n are all taken small. In this case none of the $z_a^* z_a$ coefficients scale very much. However, while λ is small, it's still the biggest Yukawa coupling so we get masses at

FIGURE 7: SCALING μB FOR ELECTROWEAK BREAKING

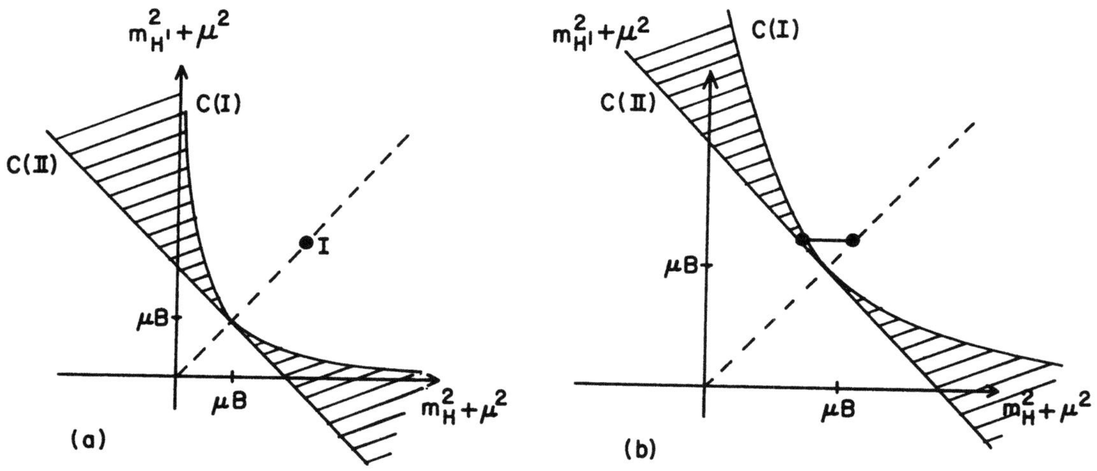

scale $m_{3/2}$

$$m_{H'}{}^2 = m_{3/2}^2, \quad m_H^2 = m_{3/2}^2 - \varepsilon \tag{34}$$

where ε is the small scaling due to λ and is shown as the short trajectory in Figure 7b. The model has been made realistic by having μB increase according to

$$\frac{d}{dt}(\mu B) = -\left(3\alpha_2 + \frac{3}{5}\alpha_1\right)\frac{\mu B}{4\pi} \tag{35}$$

such that the endpoint of the trajectory ends in the shaded region, which has moved up to meet the endpoint. Since μB scales a fixed amount, and since the width of the shaded region is very narrow at the endpoint, the initial value of μB must be chosen rather judiciously for this scheme to work.

This careful choice for μB at M_p can be seen by analyzing Eqs. (32) in this case. We find

$$M_z^2 = \frac{\varepsilon}{\cos 2\theta} - 2(m_{3/2}^2+\mu^2) \tag{36}$$

which can only be positive if $\cos 2\theta$ is close to zero. This requires θ near $\pi/4$, or

$$\sin 2\theta = \frac{\mu B}{m_{3/2}^2+\mu^2} \sim 1. \tag{37}$$

This requires that μ and B be chosen at M_p so that after scaling they satisfy $\mu B \sim m_{3/2}^2 + \mu^2$. It is clear that as the top quark is made heavier we can construct models which incorporate features of both

Type I and Type II models.

It is worth comparing general features of the sparticle spectrum for models of Type I and II. In Type I models, if the top quark is heavy, then very little can be said about the spectrum. However, we have seen that as the top quark is made lighter, so generally $m_{3/2}$ must increase. In this limit we have the disappointing result that most of the sparticles are very heavy. The exceptions are the Wiggsino [12], providing M_n and μ are both small, and possible one of the top squarks. In type II models the scale of M_W is related more to the initial choice of μB than it is to $m_{3/2}$. Thus even with a light top quark, $m_{3/2}$ can be low and thus the great interest in these models is fired by the accessibility of the sparticles.

One interesting scheme, which always has a low value of $m_{3/2}$, has M_W given by a dimensional transmutation [22]. Suppose that at a scale M_p, $m_{3/2}$, μ, and B are such that condition I is satisfied. This tells us that $SU(2)\times U(1)$ is not broken at this huge scale and that we should scale the parameters until the trajectory reaches the shaded region of Figure 4. Suppose that after scaling m_H, $m_{H'}$, μ, and B down to the scale $m_{3/2}$, we find that condition II is violated. The potential is unbounded, and it is clear that we scaled too far, passing right across the shaded region. The unboundedness is telling us that the vev is large on the scale of $m_{3/2}$. We have argued that $m_{3/2} < v < M_p$, but how do we determine v? The answer is that we must add the Coleman-Weinberg logarithms to the low energy potential, which give it the improved form

$$V(H,H') = m_1^2(H)H^*H + m_2^2(H)H'^*H' - m_3^2(H)HH' + \text{h.c.} + \frac{D^2}{2} \tag{38}$$

Now when we found unboundedness at scale $m_{3/2}$ the unbounded direction was $h = h'$. We will therefore assume that (38) has a minimum at

$$h = v+\delta, \quad m_{3/2} < v < M_p, \quad \delta \sim \frac{m_{3/2}^2}{v} \qquad h' = v-\delta \tag{39}$$

Substituting this ansatz into (38) gives

$$V = (m_1^2+m_2^2-2m_3^2)v^2 + (m_1^2-m_2^2)2v\delta + \left(\frac{g^2+g'^2}{8}\right)(2v\delta)^2 \tag{40}$$

where $m_i^2 = m_i^2(h)$. $\partial V/\partial\delta = 0$ gives $\delta \sim m_{3/2}^2/v$, while v is determined by order m^2v^2 terms:

$$V(v) = m^2(v^2)v^2, \quad \frac{dV}{dv^2} = m^2(v^2) + v^2\frac{dm^2}{dv^2} = 0 \tag{41}$$

with $m^2 = m_1^2 + m_2^2 - 2m_3^2$. Assume that the minimum occurs for v near v_0, where v_0 is defined by

$$m^2(v_0) = 0. \tag{42}$$

Near v_0 the renormalization group equation for m can be written as

$$\frac{dm^2}{dt} = \gamma m_0^2 \tag{43}$$

where m_0^2 is the right hand side of the true renormalization group equation evaluated at scale v_0. This equation has a solution

$$m^2(v) = \gamma m_0^2 \ \ln \frac{v}{v_0} + (m^2(v_0) - \gamma m_0^2) \tag{44}$$

valid for v near v_0. Substituting (43) and (44) into (41) gives

$$\ln \frac{v}{v_0} = \frac{1}{2} - \frac{m^2(v_0)}{\gamma m_0^2} \sim 0(1).$$

Thus we have demonstrated that the minimum of the potential (38) occurs at $\langle H \rangle \sim 0(v_0)$ where v_0 is defined by (42). It is the scale at which there is no quadratic term in the potential. This scale was not present in the original Lagrangian, it is a dimensional transmutation scale, which has been traded for the ratio $m_1^2 + m_2^2/m_3^2$. In principle this scale could be anything between $m_{3/2}$ and M_p. However, phenomenology requires it to be M_W, so that it is actually rather close to its lower bound.

I will just mention some recent work which explains why the scale v_0 turns out to be close to $m_{3/2}$ [23]. If the hidden sector superpotential possesses a certain global symmetry, then the potential for the hidden sector fields may actually vanish. This leaves $\langle z_i \rangle$ and therefore $m_{3/2} \sim \langle f_H \rangle$ undetermined. Thus the scale of the soft operators in the low energy theory is not fixed by the hidden sector, but will be fixed by minimizing the radiatively corrected observable sector potential. We can rewrite the potential of Eq. (41) as

$$V(v, m_{3/2}) = m_{3/2}^2 \xi^2 h^2 \tag{42}$$

where $m = \xi m_{3/2}$, and we are now considering $m_{3/2}$ and h as fields whose vacuum expectation values we wish to compute, and ξ is a dimensionless parameter which scales according to a renormalization group equation. At the Planck scale $\xi > 0$ and the potential has a minimum at $m_{3/2}$, $h \sim 0$; or at least much less than M_p. As we scale ξ down it will go negative at some scale v_0. As soon as it is negative, it is favorable to have h and $m_{3/2}$ both non-zero. In fact V looks unbounded from below. Of course it isn't, the situation is like the previous case. As $m_{3/2}$ and h become larger than v_0 we should evaluate ξ at the large scale and not at v_0. To obtain a minimum which we can believe in, assume $m_{3/2} \sim h \sim v$ and minimize $V = \xi^2(v)v^4$. This isn't really different from minimizing Eq. (41), and the same argument shows that the minimum occurs at v of order v_0. The vitally important new ingredient is that we understand why $m_{3/2}$ and M_W should turn out to be so similar: they are the same dimensional transmutation scale. Even more exciting is the realization that these theories need no scale other than M_p appearing in the Lagrangian. This is remarkable progress, and tells us that it is worth understanding the origin of the symmetry which makes the hidden sector potential vanish [23].

I have described in detail the various mechanisms which allow SU(2) xU(1) to break in these theories. The phenomenology of these models has been considered, together with other models, by Pran Nath [4]. However, before finishing, I would like to mention one aspect of the phenomenology which I find rather exciting: CP violation.

The minimal model of N=1 supergravity has four CP violating phases; the two usual ones $\bar{\theta}$ and δ, and two new ones ϕ_A and ϕ_B. Global symmetries allow us to shuffle these new phases from one interaction term to another, but they can never be eliminated. For example rotating the phases of H and H' fields in the same direction can make μ real or μB real but not both. An R rotation can make one of M_n, A and B real. I find it useful to work in a basis in which μB is real (so the Higgs vevs are real) and the gaugino masses are real. A then has an observable phase ϕ_A, and B an observable phase ϕ_B. For $m_{3/2} \sim$ 100 GeV, the limit on the neutron electric dipole moment [24] constrains these phases to be $\lesssim 10^{-2}$, from the diagram of Figure 8.

FIGURE 8: ONE LOOP CONTRIBUTIONS OF ϕ_A AND ϕ_B TO NEUTRON ELECTRIC DIPOLE MOMENT

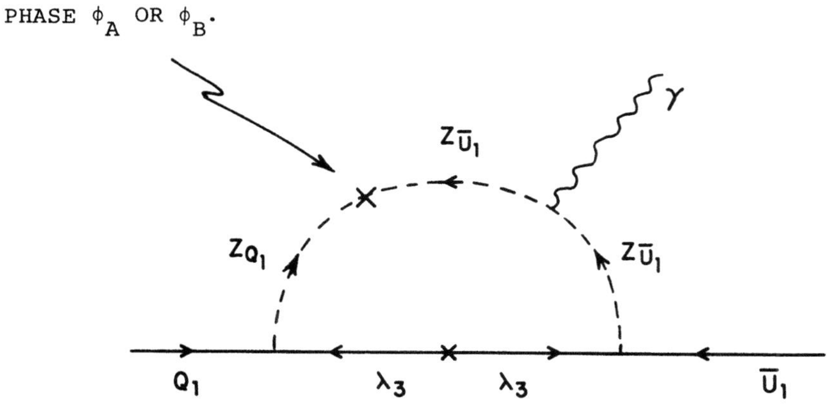

There are many other places to search for effects of these new phases. It seems that they give negligible contributions to ε, but may be interesting for ε' [25].

Of course, in the Polonyi model we know that $\phi_A = \phi_B = 0$ at the Planck scale. Even in this rather special case one can enquire as to how ϕ_A and ϕ_B will be effected by scaling. If $\bar{\theta} = 0$ for some reason, there is only one independent CP violating phase (δ), but it enters in new ways because it generates ϕ_A and ϕ_B through scaling [25]. Another possibility is that CP violation is spontaneously broken in the hidden sector of the theory. Then $\bar{\theta} = \delta = 0$ but ϕ_A and ϕ_B are non-zero.

In these lectures I have argued that certain assumptions must be made to obtain a low energy theory with high predictive power. I have shown that the low energy theory with fewest fields can have realistic electroweak symmetry breaking in a variety of non-trivial ways. Furthermore, the phenomenology of this theory has a great deal of structure and interest, without being arbitrary and cumbersome.

Acknowledgements

I would like to thank my many collaborators in this field; especially Mike Dugan, Ben Grinstein, and Joe Polchinski at Harvard, and Stuart Raby at this school. I congratulate the organizers of the school, and thank the participants whose lively interest and questions made life enjoyable.

This research is supported in part by the National Science Foundation under Grant No. PHY-82-15249.

REFERENCES

[1] P. Fayet, Phys. Lett. 69B (1977) 489, and 84B (1979) 416; S. Weinberg, Phys. Rev. D26 (1982) 287; L. Hall and I. Hinchliffe, Phys. Lett. 112B (1982) 351.

[2] L. E. Ibanez and G. G. Ross, Phys. Lett. 110B (1982) 215; L. Alvarez-Gaumé, M. Claudson, and M. B. Wise, Nucl. Phys. B207 (1982) 96; M. Dine and W. Fischler, Phys. Lett. 110B (1982) 277; C. R. Nappi and B. Ovrut, Phys. Lett. 113B (1982) 175; J. Ellis, L. E. Ibanez, and G. G. Ross, Phys. Lett. 113B (1982) 283.

[3] S. Dimopoulos and H. Georgi, Nucl. Phys. B193 (1981) 150; N. Sakai, Z. Physik C11 (1981) 153.

[4] For a review of these theories, and for references see: P. Nath, these proceedings.

[5] S. Ferrara, these proceedings.

[6] J. Polonyi, Budapest preprint KFKI-1977-93 (1977).

[7] R. Barbieri, S. Ferrara, and C. A. Savoy, Phys. Lett. 119B (1982) 343; H. P. Nilles, M. Srednicki, and D. Wyler, Phys. Lett. 124B (1983) 337.

[8] A. H. Chamseddine, R. Arnowitt, and P. Nath. Phys. Rev. Lett. 49 (1982) 970.

[9] L. Hall, J. Lykken, and S. Weinberg, Phys. Rev. D27 (1983) 2359.

[10] S. Soni and A. Weldon, Phys. Lett. 126B (1983) 215.

[11] S. Raby, these proceedings.

[12] L. Alvarez-Gaumé, J. Polchinski, and M. B. Wise, Nucl. Phys. B221 (1983) 499.

[13] J. Ellis, D. V. Nanopoulos, and K. Tamvakis, Phys. Lett. 121B (1983) 123.

[14] L. E. Ibáñez and C. Lopez, Phys. Lett. 126B (1983) 94 and CERN preprint TH-2650 (1983).

[15] M. Claudson, L. J. Hall, and I. Hinchliffe, LBL preprint 15948 (1983).

[16] J. Polchinski, Proceedings of the 4th Workshop on Grand Unification (1983).

[17] J.-M. Frère, D. Jones, and S. Raby, Michigan preprint UMHE-82-58.

[18] J. Ellis, J. S. Hagelin, D. V. Nanopoulos, and K. Tamvakis, Phys. Lett. 129B (1983) 275.

[19] L. Ibáñez, Madrid preprint FTUAM/83/4 (1983).

[20] C. Kounnas, A. Lahanas, D. Nanopoulos, and M. Quiros, Phys. Lett. 132B (1983) 95.

[21] K. Inoue, A. Kakuto, S. Takeshita, Kyushu preprint 83-HE-6 (1983).

[22] J. Ellis, A. B. Lahanas, D. V. Nanopoulos, and K. Tamvakis, CERN preprint TH-3734 (1983).

[23] J. Ellis, C. Kounnas, and D. V. Nanopoulos, CERN preprint TH-3768.

[24] J. Polchinski and M. Wise, Phys. Lett. 125B (1983) 393; D. Nanopoulos and M. Srednicki, Phys. Lett. 128B (1983) 61; W. Buchmüller and D. Wyler, Phys. Lett. 121B (1983) 321; F. del Aguila, M. B. Gavela, J. A. Grifois, and A. Mendez, Harvard preprint HUTP-83/A037 (1983).

[25] M. Dugan, B. Grinstein, and L. Hall, Harvard preprint (1984).

GEOMETRIC HIERARCHY - GLOBAL AND LOCAL SUSY

Stuart Raby
Theory Division, T-8
Los Alamos National Laboratory
Los Alamos, NM 87545

GEOMETRIC HIERARCHY [GH]

I. GLOBAL SUPERSYMMETRY [SUSY][1]

What good is SUSY? The following is a list of problems for which SUSY may provide some answers.

(i) It may provide a unified field theory of all forces. Certainly this is the goal for N=8 SUSY.[2]
(ii) It may predict three families of quarks and leptons. This possibility has been suggested on several occasions.[3]
(iii) It may solve the problem of the cosmological constant. Unfortunately, at the present time there is no real solution in sight.
(iv) It can provide an explanation for the small ratio $M_W/M_{pl} \simeq 10^{-17}$. A verifiable consequence of such an explanation is a SUSY GAP [the mass splitting between fermions and their scalar partners] of order 1 TeV or less.
(v) Finally, we may further hope that SUSY would provide some clue to quark and lepton mass matrices and hence, reduce the number of independent parameters, of order 20.

This talk shall include the following topics.

(A) Motivation for GH
(B) Generating the Grand Unification [GUT] Scale
(C) The Effective Low-Energy Theory
(D) Decoupling of SUSY Breaking Effects
(E) The Second Hierarchy Problem
(F) R Invariance and the Peccei-Quinn Symmetry

A. Motivation

Perhaps the most desirable theory would not contain any explicit scales. Then all scales would be generated via dimensional transmutation. The next best possibility is a theory with one scale. In GH(4) we introduce one intermediate scale $\mu \sim 10^{11}$ GeV. We then generate the GUT scale

$$M_{GUT} \sim e^{2\pi/\alpha}\mu \sim 10^{18} \text{ GeV}$$

using Witten's "inverse" hierarchy scheme,(5) and the weak scale

$$M_W \sim \alpha \frac{\mu^2}{M_{GUT}} \sim 10^2 \text{ GeV} -$$

utilizing the property of decoupling of SUSY breaking effects.(4,6,7)

B. Generating M_{GUT}

Consider an SU_5 GUT with the following set of superfields.

SU_5 gauge multiplet:

$(W_\alpha)_j{}^i$ (24)

Chiral Superfields:

Witten-O'Raifeartaigh Sector

$Z_j{}^i \; , \; A_j{}^i$ (24)

X (1)

Higgs

$H_i \; , \; H_{1i}$ (5) [B.1]

$\bar{H}^i \; , \; \bar{H}_1^i$ ($\bar{5}$)

Quarks and Leptons

10^a_{ij} (10)

$\bar{5}^{ia}$ ($\bar{5}$)

where α=1,2 (Lorentz spinor index), i,j=1,...,5 (SU_5 index) and a=1,2,3 (generation index). (The numbers in parentheses indicate the SU_5 representation.)

The SUSY gauge field strength W_α contains the gauge field strength ($F_{\mu\nu}$), the gauge fermion (χ_α) and the real auxiliary field (D); (group indices are suppressed). A chiral superfield (for example, Z) includes a complex scalar (Z), a left-handed Weyl fermion (ψ_z) and a complex auxiliary field (F_z).

The superspace potential W is given by the expression

$$W = \lambda_1 \mathrm{Tr}(ZA^2) + \lambda_2 X(\mathrm{Tr}A^2 - \mu^2)$$

$$+ \lambda_3(\bar{H}_1 AH + \bar{H}AH_1) + \mu'(\bar{H}_1 H + \bar{H}H_1) \qquad [B.2]$$

$$+ g^u_{ab} H 10_a 10_b + g^d_{ab} \bar{H} 10_a \bar{5}_b \quad .$$

The second line we shall discuss later [Sec. IE] in connection with the second hierarchy problem, the third line will then be discussed in the next talk with regard to proton decay. For now, let's ignore λ_3, μ', g^u_{ab} and g^d_{ab} and consider only the consequences of

$$W = \lambda_1 \mathrm{Tr}(ZA^2) + \lambda_2 X(\mathrm{Tr}A^2 - \mu^2) \quad . \qquad [B.3]$$

The tree level potential which follows is

$$V_{tree} = |F_X|^2 + \sum_{ij} |F_{Z_i{}^j}|^2 + \sum_{ij} |F_{A_i{}^j}|^2 + \tfrac{1}{2}\mathrm{Tr}D^2 \qquad [B.4]$$

where

$$-F^*_X = \frac{\partial W}{\partial X} \sim \mathrm{Tr}A^2 - \mu^2$$

$$-F^*_{Z_i{}^j} = \frac{\partial W}{\partial Z_i{}^j} \sim A^2 - \frac{1}{5}(\mathrm{Tr}A^2)1$$

$$-F^*_{A_i{}^j} = \frac{\partial W}{\partial A_i{}^j} \sim [(\lambda_1 Z + \lambda_2 X)A] - \frac{1}{5}(\mathrm{Tr}[(\lambda_1 Z + \lambda_2 X)A])1 \qquad [B.5]$$

$$D = -g([A^*,A] + [Z^*,Z])$$

(1 is the 5x5 unit matrix and g is the SU_5 coupling constant.)

Clearly, $F_X=0$ and $F_Z=0$ are inconsistent. Thus, SUSY is spontaneously broken at a scale μ. When we minimize V_{tree}, we obtain

$$\langle A\rangle = A_0\begin{pmatrix}2&&&&\\&2&&&\\&&2&&\\&&&-3&\\&&&&-3\end{pmatrix} \quad , \; A_0 = \frac{\lambda_2\mu}{(\lambda_1^2 + 30\lambda_2^2)^{\frac{1}{2}}}$$

$$\langle Z\rangle = Z_0\begin{pmatrix}2&&&&\\&2&&&\\&&2&&\\&&&-3&\\&&&&-3\end{pmatrix} \quad , \qquad [B.6]$$

$$\langle X\rangle = X_0 \quad , \; Z_0 = \frac{\lambda_2}{\lambda_1} X_0$$

The value of Z_0 is undetermined at the tree level. We have

$$V_{\substack{\text{tree}\\ \text{min}}} = \frac{\lambda_1^2\lambda_2^2\mu^4}{(\lambda_1^2 + 30\lambda_2^2)} = \text{constant} \qquad [B.7]$$

(independent of Z_0). The one-loop potential does however, depend on Z_0(5) and we find

$$V_{1\text{loop}} = V_{\text{tree}} + \sum_S (-1)^{2S}(2S+1) \int \frac{d^3\vec{p}E_{\vec{p}}}{(2\pi)^3} \qquad [B.8]$$

where

$$E_{\vec{p}}^{\;2} = \vec{p}^{\;2} + M_S^2(Z_0) \text{ and } M_S(Z_0)$$

is the tree level mass of a state with spin S. Equivalently, we find

$$V_{1\text{loop}} = V_{\text{tree}} + \frac{1}{64\pi^2}\sum_S (-1)^{2S}(2S + 1)M_S^4 \ln\frac{M_S^2}{\mu^2}$$

$$= \frac{\lambda_1^2\lambda_2^2\mu^4}{(\lambda_1^2 + 30\lambda_2^2)}\left[1 + \frac{\lambda_2^2}{\lambda_2^2 + \frac{1}{30}\lambda_1^2}\,\frac{29\lambda_1^2 - 50g^2}{40\pi^2}\ln\frac{Z_0}{\mu}\right] \qquad [B.9]$$

where (i) $\mu \sim 10^{11}$ GeV is also the renormalization point;

(ii) $V_{1\text{loop}}$ is valid for $Z_0 \sim \mu$;

(iii) note: if at the scale μ we have

$\lambda_1^2(\mu)/g^2(\mu) < \frac{50}{29}$, then clearly $Z_0>\mu$ is energetically favorable; and finally,

(iv) for values of $Z_0>>\mu$ we need to use a renormalization group improved one-loop potential.

The renormalization group improved one-loop effective potential is given by (8)

$$V(Z_0) \equiv V_{tree}\left[\lambda_1(Z_0),\lambda_2(Z_0),\mu(Z_0),g(Z_0)\right] = \frac{\lambda_1^2(Z_0)\lambda_2^2(Z_0)\mu^4(Z_0)}{\lambda_1^2(Z_0) + 30\lambda_2^2(Z_0)} \quad . \qquad [B.10]$$

Using

$$\frac{d\lambda_1}{d\ell nZ_0} \equiv \beta_{\lambda_1} = \lambda_1(2\gamma_A + \gamma_Z)$$

$$\frac{d\lambda_2}{d\ell nZ_0} \equiv \beta_{\lambda_2} = \lambda_2(2\gamma_A + \gamma_X) \qquad [B.11]$$

$$\frac{d\mu}{d\ell nZ_0} = -\mu\gamma_A \quad ,$$

we find

$$\frac{dV}{d\ell nZ_0} = 2V(Z_0)\left(\frac{\lambda_1^2\gamma_X + 30\lambda_2^2\gamma_Z}{\lambda_1^2 + 30\lambda_2^2}\right) \qquad [B.12]$$

where

$$\gamma_Z \equiv \frac{1}{8\pi^2}\left(\frac{21}{10}\lambda_1^2 - 5g^2\right)$$

and [B.13]

$$\gamma_X = \frac{1}{8\pi^2}(24\lambda_2^2)$$

are determined via the supergraphs of Fig. 1 (a,b), respectively. Substituting Eq. [B.13] into Eq. [B.12], we have

$$\frac{dV}{d\ell nZ_0} = V(Z_0)\left(\frac{\lambda_2^2(Z_0)}{\frac{\lambda_1^2(Z_0)}{30} + \lambda_2^2(Z_0)}\right)\frac{29\lambda_1^2(Z_0) - 50g^2(Z_0)}{40\pi^2} \quad . \qquad [B.14]$$

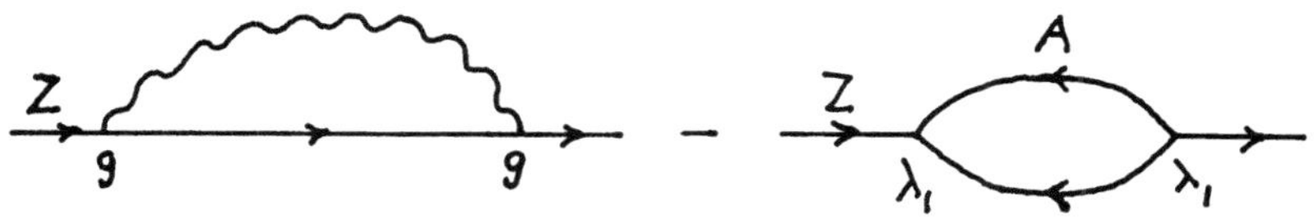

Fig. 1(a)

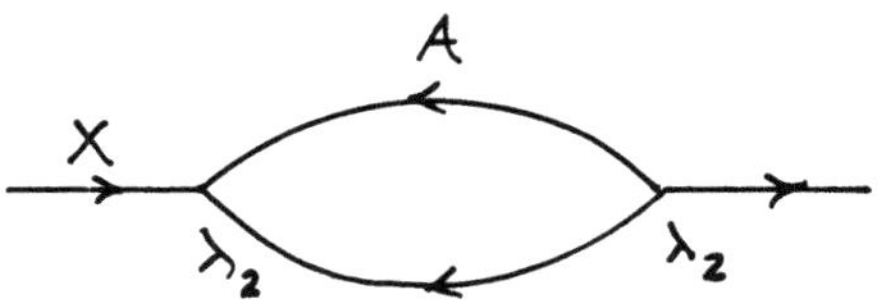

Fig. 1(b)

Hence

$$\frac{\partial V}{\partial Z_0} = 0$$

at the value of Z_0 which satisfies

$$29\lambda_1^2(Z_0) = 50g^2(Z_0) \quad . \qquad [B.15]$$

We must now use the renormalization group equations for λ_1, λ_2, and g to find the value of Z_0 satisfying Eq. [B.15] with initial conditions given by

$$\frac{\lambda_1^2(\mu)}{g^2(\mu)} < \frac{50}{29} \quad .$$

In Fig. 2 we give the renormalization group (RG) flows for the parameters $q \equiv \lambda_2^2/g^2$ and $p \equiv \lambda_1^2/g^{2(8)}$.

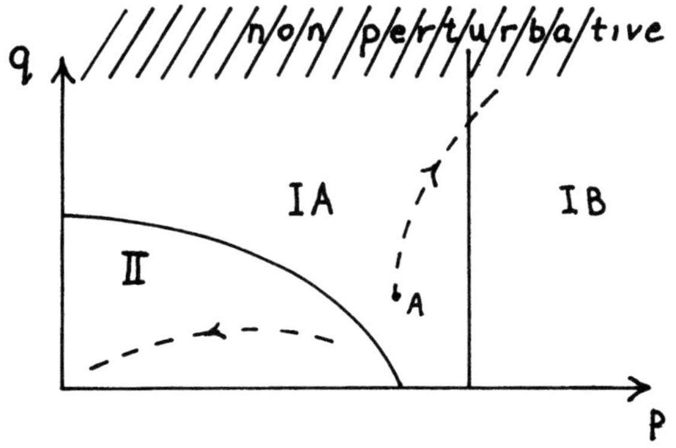

Fig. 2

[The relevant equations to derive these flows are presented for the reader in Eqs. [B.16] and [B.17].

$$\frac{d\lambda_1}{d\ell nZ_0} \equiv \beta_{\lambda_1} = \frac{\lambda_1}{8\pi^2}\left(\frac{21}{2}\lambda_1^2 + 4\lambda_2^2 - 15g^2\right) \ ,$$

$$\frac{d\lambda_2}{d\ell nZ_0} \equiv \beta_{\lambda_2} = \frac{\lambda_2}{8\pi^2}\left(\frac{42}{5}\lambda_1^2 + 28\lambda_2^2 - 10g^2\right) \ , \qquad \text{[B.16]}$$

$$\frac{dg}{d\ell nZ_0} \equiv \beta_g = -\frac{g^3}{16\pi^2}(5 - C) \quad ,$$

$C \equiv 2N_g + N_H$, N_G = number of generations of quarks and leptons and N_H = number of $5+\bar{5}$ Higgs pairs. We then find

$$\frac{dq}{d\tau} = q\left(\frac{84}{5}\, p + 56q - 15 - C\right)$$

$$\frac{dp}{d\tau} = p(21p + 8q - 25 - C) \qquad [B.17]$$

and

$$\tau \equiv \int d\ell n Z_0 \, \frac{g^2(Z_0)}{8\pi^2} = \frac{1}{5 - C}\, \ell n\left(1 + \frac{\alpha}{2\pi}\,(5 - C)\ell n\frac{Z_0}{\mu}\right)$$

is a monotonically increasing function of Z_0.]

There are three relevant regimes of RG flow. Region IB (see Fig. 2) has $p > \frac{50}{29}$ and is thus uninteresting. Regions IA and II have $p \leq \frac{50}{29}$. Any point in IA will eventually cross the axis $p = \frac{50}{29}$ (assuming the parameters remain small) at some value of Z_0 (or τ) and thus provide a minimum for the potential $V(Z_0)$. Unfortunately, in order to obtain a value for $M_{GUT} \sim 10^7 \mu$, one needs to fine-tune the initial values of p and q at the scale μ to about one part in 10^7 of the available parameter space ([see Hall et al., ref. 8). Thus, although a large hierarchy is possible, it is not natural. Finally in Region II, all RG flow approaches an ultraviolet fixed point at the origin. Thus, in this approximation, $V(Z_0)$ has no extremum at any finite value of Z_0.

C. The Effective Low-Energy Theory

In order to determine L_{eff}, we first shift all the fields by their vacuum expectation values and then integrate out (i.e. use the classical equations to eliminate) the heavy states with mass >> μ.[9] We define

$$A = \langle A\rangle + A'$$
$$Z = \langle Z\rangle + Z'$$
$$X = \langle X\rangle + X' \qquad [C.1]$$

with

$$\langle A\rangle = A_0 \begin{pmatrix} 2 & & & & \\ & 2 & & & \\ & & 2 & & \\ & & & -3 & \\ & & & & -3 \end{pmatrix}$$

$$\langle Z\rangle = \frac{\lambda_2}{\lambda_1}\,\langle X\rangle \begin{pmatrix} 2 & & & & \\ & 2 & & & \\ & & 2 & & \\ & & & -3 & \\ & & & & -3 \end{pmatrix}$$

and

$$\langle X \rangle \equiv X_0 + F_{X_0}\theta^2 \quad .$$

Note that $X_0 \sim M_{GUT}$ and

$$F_{X_0} = \frac{\lambda_1^2\lambda_2\mu^2}{\lambda_1^2 + 30\lambda_2^2} \quad .$$

The light states (with mass << μ), include the color octet and SU_2 triplet components of Z (denoted by Z_8 and Z_3 respectively) and a linear combination of the hypercharge component of Z and X (denoted by $\tilde{X}$) which contains the goldstino, a pseudoscalar axion and a scalar scalon. In addition, we have the SUSY gauge sector of $SU_3 \times SU_2 \times U_1$. We obtain

$$L_{eff} = \int d^4\theta \mathrm{Tr}\left[Z_8^* e^{g_3V_3} Z_8 e^{-g_3V_3}\right] + \int d^4\theta \mathrm{Tr}\left[Z_3^* e^{g_2V_2} Z_3 e^{-g_2V_2}\right]$$
$$+ \int d^4\theta\, \tilde{X}^*\tilde{X} + \int d^2\theta (W_3^\alpha W_{3\alpha} + W_2^\alpha W_{2\alpha} + W_1^\alpha W_{1\alpha})$$
$$+ [\int d^2\theta W_{eff} + \mathrm{h.c.}] \quad , \qquad [C.2]$$

where

$$W_{eff} = \frac{\lambda_1 A_0^2}{Z_0}\left(1 - \frac{F_{Z_0}}{Z_0}\theta^2\right)\left(- \frac{4}{3}\mathrm{Tr}Z_8^2 + \frac{9}{2}\mathrm{Tr}Z_3^2 + \text{similar term for } \tilde{X}\right)$$

+ terms of higher order in an expansion in μ/M_{GUT}.

Note: (1) L_{eff} includes explicit "soft" SUSY breaking terms ("Soft" operators preserve the renormalizability of L_{eff}[(10)]) apparent in W_{eff} of order μ^2/M_{GUT}, and (2) the goldstino $\tilde{X}$ decouples from all light states. As a result of (2), any additional radiative SUSY breaking effects must proceed via massive [$O(M_{GUT})$] intermediate states. We thus find that all these radiative effects are of order

$\alpha^n \frac{\mu^2}{M_{GUT}}$ (where α is the relevant coupling constant).

D. Decoupling

We have already seen an example of the decoupling of SUSY breaking effects in the effective low-energy theory. Recall (Eq. [C.2]) we found at the tree level

$$\delta L_{\text{breaking}} \sim \frac{\mu^4}{M_{GUT}^2} \, \mathrm{Tr} Z_8^2 + \ldots \qquad [D.1]$$

i.e. a "soft" SUSY breaking scalar mass of order μ^2/M_{GUT}, instead of μ. It is possible to show that decoupling is true to all orders in perturbation theory.(4,6,7)

Table 1 provides a list of operators which can be generated via radiative corrections. Dimension 4 operators have been omitted. The operator

$$\int d^4\theta \tilde{X}^* L \to \mu^2 F_\ell \qquad [D.2]$$

with L a light $SU_3 \times SU_2 \times U_1$ gauge singlet could, in principle, be generated. If L also couples to Higgs i.e. there exists in L_{eff} a term of the form

$$\int d^2\theta L\bar{H}H \quad , \qquad [D.3]$$

then these operators result in a scalar Higgs mass of the form

$$\mu^2 \bar{H}H \quad . \qquad [D.4]$$

This term violates decoupling and must be avoided(6,11). This is just one of the problems which can arise with light singlets. [Operator b (Table 1) is also dangerous and also requires a light singlet.] For a more complete discussion of the problems with light singlets (see ref. 6,11).

The remaining operators are all consistent with decoupling. (To prove decoupling, one must show that the correct dimensions be supplied by the large scale M_{GUT}.) Operators a, c, d, e, h, and i are all "soft" operators; f, g, j, etc. are "hard" operators which are suppressed by additional powers of m/M (where $m \equiv \mu^2/M$, $M \equiv M_{GUT}$). "Hard" operators are only important if (a) there are $SU_3 \times SU_2 \times U_1$ singlets in the low-energy theory. This is because these operators can induce large linear terms in the singlet fields which may cause problems(11) or (b) they break an exact symmetry of the renormalizable part of L_{eff}. Otherwise, they can safely be neglected. All divergences (resulting from "hard" operators) are naturally cut off at M and the only effect is to change the value of some renormalizable parameters at the scale M.

Two important "soft" operators which are generated in GH are (1) gaugino masses and (2) scalar masses. The one loop graph of Fig. 3 leads to the gaugino mass.[12]

$$\tilde{M}_i(M) = \frac{\alpha_{GUT}}{4\pi}\,[C(SU_5) - C(SU_i)]\,\frac{F_{\tilde{X}}}{\langle\tilde{X}\rangle}$$

where $\tilde{M}_i$ is the mass for the gaugino corresponding to the unbroken subgroup SU_i of SU_5, $C(SU_i) \equiv i$, $i=2, 3, 5$ and $C(U_1)=0$, and $\tilde{X}$ is the linear combination of X and Z which contains the goldstino. Finally, the two-loop graph of Fig. 4 results in a scalar mass squared of order

$$m_s^2 \sim \left(\frac{\alpha_{GUT}}{4\pi}\right)^2 \frac{\mu^4}{M_{GUT}^2} .$$

Kaplunovsky[13] has obtained a negative scalar mass squared which is a serious problem for global GH.

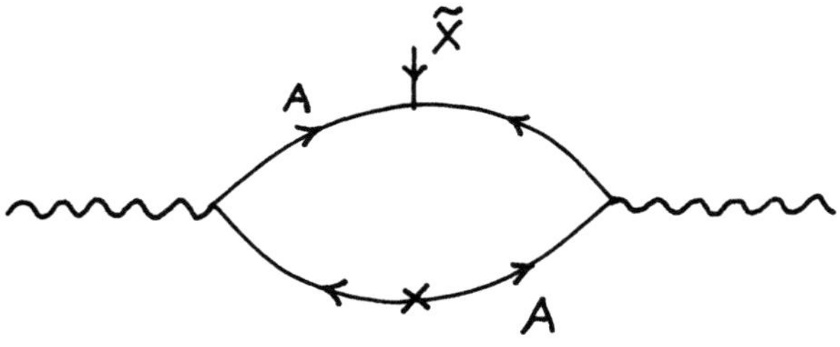

Fig. 3

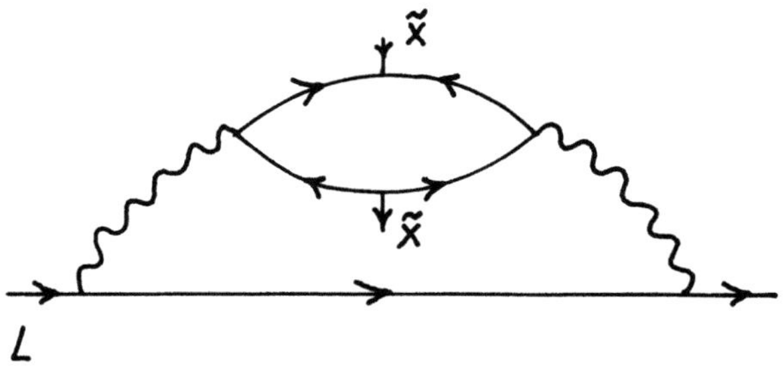

Fig. 4

E. The Second Hierarchy Problem

In SUSY GUTS, another hierarchy problem arises. In SU_5, a Higgs 5 and $\bar{5}$ are necessary to break $SU_2 \times U_1$ to U_1 at M_W, and also to give mass to quarks and leptons. Although the Higgs doublets must have mass of order M_W, the color triplet components must be heavier than about 10^{16} GeV for the fermion and about 10^{10} GeV for the scalars. This is necessary in order to avoid the rapid proton decay processes which we shall discuss in the next talk. Splitting the doublet and triplet components of a Higgs 5 is the second hierarchy problem. [Bowick and Ramond(14) have noted that in a SUSY GUT it may be possible to have light color triplets.] There are three mechanisms for splitting H.

(1) Fine tuning(15)

In this case the superspace potential contains the terms

$$W \supset \bar{H}^i A_i{}^j H_j + \bar{H}^i H_i M \qquad [E.1]$$

where the adjoint A receives an expectation value

$$A_j{}^i = A_0 \begin{pmatrix} 2 & & & & \\ & 2 & & & \\ & & 2 & & \\ & & & -3 & \\ & & & & -3 \end{pmatrix} \qquad [E.2]$$

such that

$$3A_0 = M \quad . \qquad [E.3]$$

This requires fine tuning of the bare parameters. Nevertheless, as a result of the nonrenormalization theorems, the Higgs doublets will remain massless to all order in perturbation theory as long as SUSY is preserved. Once SUSY is broken, the doublets will obtain mass m which, as a consequence of decoupling, is only of order

$$m = \frac{\mu^2}{M} \quad .$$

(2) Sliding singlet(15,16)

Replace M (Eq. [E.1]) by a chiral superfield S. The scalar potential which results has several flat directions at the tree level. In particular, if $\langle\bar{H}\rangle=0$, then the potential is flat along the directions depicted in Fig. 5. One-loop corrections will lift the degeneracy.(17) On dimensional grounds alone, we see that the energy at point A has two components

$$V \sim m^2 |S|^2 + m^2 |H^0|^2 \quad . \qquad [E.4]$$

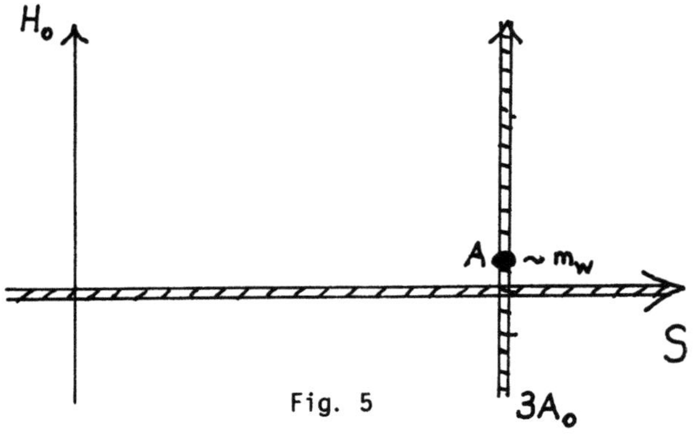

Fig. 5

Since S>>H° we can first check for the extremum in the S direction. Typically, the extremum is not located at $3A_0$.(17) As a result, all five components of H obtain large mass.

(3) <u>Missing partner</u>(18)

We describe the version which is relevant for SU_5. Consider the superspace potential which includes the following SU_5 multiplets:

$75,\ 50,\ \overline{50},\ H,\ \text{and}\ \bar{H}.$

$$W \supset \overline{50}\,75\,H + \bar{H}\,75\,50 + M\,\overline{50}\,50 \qquad [E.5]$$

$M \sim M_{GUT}$ and the expectation value of 75 is assumed to break $SU_5 \rightarrow SU_3 \times SU_2 \times U_1$. Under $SU_3 \times SU_2 \times U_1$, the 50 contains:

$$50 \supset \underline{(3,1,-\tfrac{2}{3})} + (\bar{3},2,-\tfrac{7}{3}) + (\bar{6},3,-\tfrac{2}{3}) + (6,1,\tfrac{8}{3}) + (8,2,1) + (1,1,-4) \qquad [E.6]$$

compared to the decomposition of $\bar{H}=\bar{5}$:

$$\bar{5} \supset \underline{(\bar{3},1,\tfrac{2}{3})} + \boxed{(1,2,-1)}\,.$$

All the states in the 50 obtain mass at the scale M. In addition, the color triplet states obtain a SUSY mass given by the 2 × 2 matrix:

$$\begin{array}{c|cc} & 50_3 & H_3 \\ \hline \overline{50}_{\bar{3}} & M & \langle 75\rangle \\ \bar{H}_{\bar{3}} & \langle 75\rangle & 0 \end{array} \qquad [E.7]$$

All states are massive. If <75>/M<<1, then the lightest state is predominantly a Higgs triplet with mass term

$$\frac{<75>^2}{M} \bar{H}_{\bar{3}} H_3 \quad . \qquad [E.8]$$

Finally, the Higgs doublets are "naturally" massless at the tree level.

F. R Invariance and the Peccei-Quinn Symmetry

The model of Eq. [B.2] has a global chiral symmetry which distinguishes scalars and their fermionic partners. Such a global symmetry has been called an R-Symmetry.(19) Under the R symmetry, a chiral superfield transforms as:

$$\Phi'(X_\mu,\theta_\alpha) = e^{iR_\Phi\beta}\,\Phi(X_\mu, e^{-\frac{i}{2}\beta}\theta) \qquad [F.1a]$$

and the components transform as:

$$\Phi'(X_\mu) = e^{iR_\Phi\beta}\,\Phi(X_\mu)$$

$$\psi'(X) = e^{i(R_\Phi-\frac{1}{2})\beta}\,\psi(X) \qquad [F.1b]$$

$$F'(X) = e^{i(R_\Phi-1)\beta}\,F(X) \quad .$$

Similarly, the gauge field strength transforms as:

$$W'_\alpha(X,\theta) = e^{i\beta/2}\,W_\alpha(X, e^{-\frac{i}{2}\beta}\theta) \qquad [F.2a]$$

and the components as:

$$A'_\mu(X) = A_\mu(X)$$

$$\chi'(X) = e^{i\beta/2}\chi(X)$$

$$D'(X) = D(X) \qquad [F.2b]$$

Under this R symmetry, the fields of Eq. [B.1] have the charges:

$$R = 1 \text{ for } Z,\ X,\ H,\ \bar{H}$$

$$R = 0 \text{ for } A,\ H_1,\ \bar{H}_1,\ 10^a,\ \bar{5}^a \quad . \qquad [F.3]$$

It is then a simple exercise to show that the SU_3 color anomaly of R is nonvanishing. Since R is a U_1 chiral symmetry of quarks which has a nonvanishing color anomaly, we may identify it as the Peccei-Quinn symmetry.[20] Hence, this theory has no strong CP problem.

R is spontaneously broken at M_{GUT}. As a result, there is an invisible axion with mass of order

$$m_a \sim \frac{\Lambda_{QCD}^2}{M_{GUT}} \sim 10^{-11}\text{eV} \quad . \tag{F.4}$$

(Note: There may be cosmological problems associated with this light axion. For more details, see ref. 21.)

II. LOCAL SUSY[22]

This talk includes the following topics.

(G) Nonrenormalizability
(H) Cosmological Constant
(I) O'Raifeartaigh or Not?
(J) A Simple SU_5 GUT
(K) Proton Decay

G. Nonrenormalizability

N=1 local SUSY is <u>not</u> renormalizable. We shall thus treat the full theory at the tree level only and consider radiative corrections only for the effective low-energy theory which is obtained by integrating out states with mass $\sim M_{pl}$.

The effective low-energy theory is renormalizable. Boundary conditions at M_{pl} can, in principle, be determined in terms of a cut-off theory. Since these corrections depend on the arbitrary ratio (Λ/M_{pl}) x couplings, it may be preferable however, to assume the most general low-energy theory consistent with the symmetries, for energy $E \ll M_{pl}$, and with decoupling. (For example, a SUSY interaction term of the form $\int d^2\theta H A_3 H$, where A_3 is a light SU_2 triplet, will, in general, be suppressed by a factor $\frac{m}{M}$ if it is not present at the tree level.) Finally, the effective cut-off Λ will perhaps, be fixed in the future in the framework of a more complete theory.

H. Cosmological Constant

The observed vacuum energy density

$$\Lambda_{obs} \leq (10^{-3}eV)^4 \quad .$$

If there is no SUSY, then the natural theoretical value for Λ is of order M_{pl}^4. Hence, $\Lambda_{theor.}/\Lambda_{obs} \cong 10^{120}$. Global SUSY lowers the theoretical value to $\Lambda_{theor-SUSY} \sim \mu^4$ which gives

$$\Lambda_{theor-SUSY}/\Lambda_{obs} \simeq 10^{92} \quad .$$

Finally, in local SUSY, we may fine-tune Λ to zero. However, this is not very satisfactory.

Hawking[23] has suggested that the Lagrangian value of the cosmological constant may be irrelevant. Perhaps space-time foam provides an effective cosmological constant which is naturally zero. The main point of this discussion is that the cosmological constant is the most poorly understood parameter in nature. As a result, there are two possible ways to go:

(1) one can ignore the theoretical value of Λ, assuming that whatever solves the problem will do so naturally; or

(2) we can fine-tune Λ_{theory} to zero.

Case 1: ignore Λ[24]. Consider local GH. The tree potential is given by

$$V_{tree} = e^k(|D_iW|^2 - \frac{3|W|^2}{M^2}) + \tfrac{1}{2}TrD^2 \qquad [H.1]$$

where

$$M \equiv \frac{M_{pl}}{\sqrt{8\pi}} \simeq 2.4 \times 10^{18} \text{ GeV} \quad ,$$

$$K \equiv k(\phi_i,\phi_j^*) = \sum_i \frac{|\phi_i|^2}{M^2} \quad ,$$

$$D_iW \equiv \left(\frac{\partial}{\partial\phi_i} + \frac{\partial k}{\partial\phi_i}\right) W = \left(\frac{\partial}{\partial\phi_i} + \frac{\phi_i^*}{M^2}\right) W \quad ,$$

ϕ_i are the scalar components of superfields and $W = W_{W-O'R}$ (see Eq. [B.3]).

Minimizing V_{tree} we obtain

$$\langle A\rangle = \left[1 - \left(\frac{\lambda_1^2}{\lambda_1^2 + 30\lambda_2^2}\right)\frac{\mu^2}{4M^2} + O(\varepsilon^4)\right] \frac{\lambda_2\mu}{\left(\lambda_1^2 + 30\lambda_2^2\right)^{\frac{1}{2}}} \begin{pmatrix} 2_{2_{2_{-3_{-3}}}} \end{pmatrix}$$

$$\langle Z\rangle = \left[1 + \frac{\mu^2}{2M^2} + O(\varepsilon^4)\right] \frac{\lambda_2}{\lambda_1} \langle X\rangle \begin{pmatrix} 2_{2_{2_{-3_{-3}}}} \end{pmatrix} \qquad [H.2]$$

with $\varepsilon \equiv \mu/M$. If $\langle X\rangle \equiv X_0 \gg \langle A\rangle$, then SU_5 breaks to $SU_3 \times SU_2 \times U_1$ at the scale X_0. In addition, SUSY is spontaneously broken since $D_Z W \sim D_X W \sim \mu^2$. The value of X_0 cannot be determined at the tree level since for $X_0 \ll M$, $V_{tree} \propto X_0^4$ and radiative corrections are important. At one-loop, in renormalizable interactions,

$$V_{eff} = \exp\left(\frac{\mathrm{Tr}\langle A^2\rangle}{M^2}\right)\left(\frac{\lambda_1^2\lambda_2^2\mu^4}{\lambda_1^2 + 30\lambda_2^2} + O(\varepsilon^2)\right)$$

$$\times\left\{\left[1 + \frac{3}{4\pi^2}\frac{\lambda_2^2}{\lambda_1^2 + 30\lambda_2^2}(29\lambda_1^2 - 50g^2)\ln\frac{X_0}{\mu}\right]\right. \qquad [H.3]$$

$$\left. + \frac{1}{2M^4}\left(\frac{\lambda_1^2 + 30\lambda_2^2}{\lambda_1^2}\right)^2 |X_0|^4 + O(M^{-6})\right\}$$

For $X_0 \gg \mu$, we should use the RNG improved V_{eff}. We find[24]

$$\frac{\partial V_{eff}}{\partial \ln X_0} = \left[\frac{\lambda_1^2(X_0)\lambda_2^2(X_0)\mu^4(X_0)}{(\lambda_1^2(X_0) + 30\lambda_2^2(X_0))^2}\,\frac{87\lambda_2^2(X_0)}{4\pi^2}\right]\left(p(X_0) - \frac{50}{29}\right) + \frac{2\mu^4}{M^4}\frac{\lambda_2^2}{\lambda_1^2}(\lambda_1^2 + 30\lambda_2^2)X_0^4 \equiv 0$$

where

$$p(X_0) \equiv \frac{\lambda_1^2(X_0)}{g^2(X_0)} \quad . \qquad [H.4]$$

It is then clear from Fig. 2 that for any point in Region II, V_{eff} will have a minimum. We find

$$X_0 \simeq \frac{1}{10} M \text{ for } \mu \sim 10^{11}\text{ GeV} \quad .$$

Thus

$M_{GUT} \sim M$ is natural. [H.5]

<u>Case 2</u>: fine-tune $\Lambda_{theor.}$ to zero.[25] Consider $W = W_{W-O'R} + \beta\mu^2 M$ where β is a dimensionless constant which is chosen such that at the minimum $V(X_0) \equiv 0$. With the addition of β, the tree potential is given in Fig. 6. The minimum is at

$$\langle A \rangle = \frac{\lambda_1 \mu}{(\lambda_2^2 + 30\lambda_1^2)} \begin{pmatrix} 2 & & & & \\ & 2 & & & \\ & & 2 & & \\ & & & -3 & \\ & & & & -3 \end{pmatrix}$$

$$\langle Z \rangle = \frac{\lambda_1}{\lambda_2} \langle X \rangle \begin{pmatrix} 2 & & & & \\ & 2 & & & \\ & & 2 & & \\ & & & -3 & \\ & & & & -3 \end{pmatrix} \qquad [H.6]$$

$$\langle X \rangle = X_0 = \frac{\lambda_2}{(\lambda_2^2 + 30\lambda_1^2)^{\frac{1}{2}}} (\sqrt{3}-1)\, M$$

$$\beta = \frac{-\lambda_1 \lambda_2}{(\lambda_2^2 + 30\lambda_1^2)^{\frac{1}{2}}} (2 - \sqrt{3}) \quad .$$

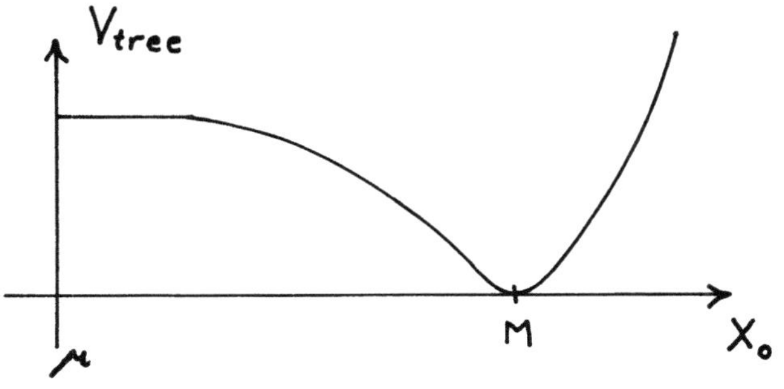

Fig. 6

<u>Note</u>:

(i) M_{GUT} is fixed already at the tree level.

(ii) The term $\beta\mu^2 M$ explicitly breaks the R symmetry (Sec. F). This is unfortunate since we then lose the solution to the strong CP problem. There is possibly one way out of this dilemma. Claudson et al.[25] have shown that it is possible to fine-tune Λ to zero by changing the function $K(\phi_i, \phi_j^*)$. For example, for a singlet X, they showed that the parameter β in $K = |X|^2 + \beta|X|^4$ may be chosen so that at the minimum $V(X)=0$. In this method, R invariance can be preserved. It would be interesting to use this method for local GH.

I. O'Raifeartaigh or Not?

In global SUSY there are two ways of spontaneously breaking SUSY. Either by using an O'Raifeartaigh sector[26] or via Fayet-Iliopoulos D term.[27] Either way, SUSY breaking requires very special choices of potentials. Global GH makes use of an O'Raifeartaigh sector. Local GH, as discussed so far, also uses the same O'Raifeartaigh sector to spontaneously break SUSY.

We now want to show that in local SUSY one can, in fact, break SUSY without ever having to use either an O'Raifeartaigh sector or a Fayet-Iliopoulos sector. We give two examples.[28]

(1) Consider the superspace potential

$$W(A) = \mu^2 M\left[\tfrac{1}{2}\frac{A^2}{M^2} + \frac{\lambda}{3}\frac{A^3}{M^3} + \frac{\lambda_4}{4}\frac{A^4}{M^4}\right] \qquad \text{[I.1]}$$

with a singlet A. We can find a value of A for which

$$-F_A^* \equiv \frac{\partial W}{\partial A} = \mu^2\left(\frac{A}{M} + \frac{\lambda A^2}{M^2} + \lambda_4\frac{A^3}{M^3}\right) = 0$$

Thus, SUSY is unbroken in the global limit. The tree potential for local SUSY is given in Fig. 7. In this case, we can always find a value of λ and λ_4 such that at the minimum $\langle D_A W\rangle \neq 0$ (SUSY is spontaneously broken) and $V(A)=0$ (the cosmological constant is fine-tuned to zero).

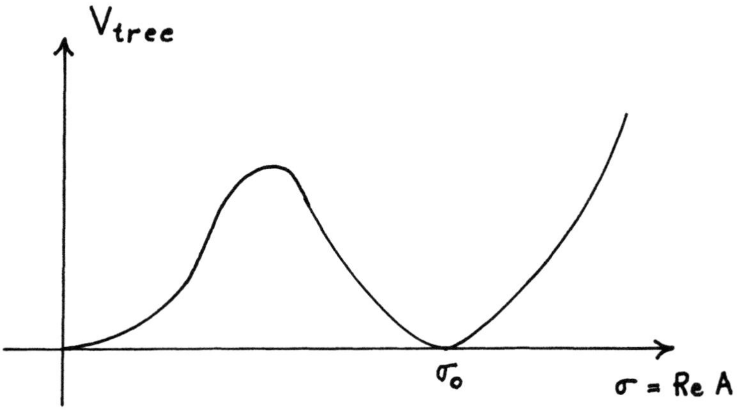

Fig. 7

(2) Consider the superspace potential

$$W(B,Z) = \mu^2 Z + BZ^2 + \lambda Z^3 \quad . \qquad [I.2]$$

The minimum of the tree potential for global SUSY is at

$$B \cong -\frac{\mu^2}{2Z} \to \infty$$

$$Z \to 0 \quad . \qquad [I.3]$$

This minimum is supersymmetric. For local SUSY, the potential is given in Fig. 8. We find

$$B \sim M$$

$$Z \sim \frac{\mu^2}{M}$$

$$D_B W \sim \left(\frac{\mu^2}{M}\right)^2 \qquad [I.4]$$

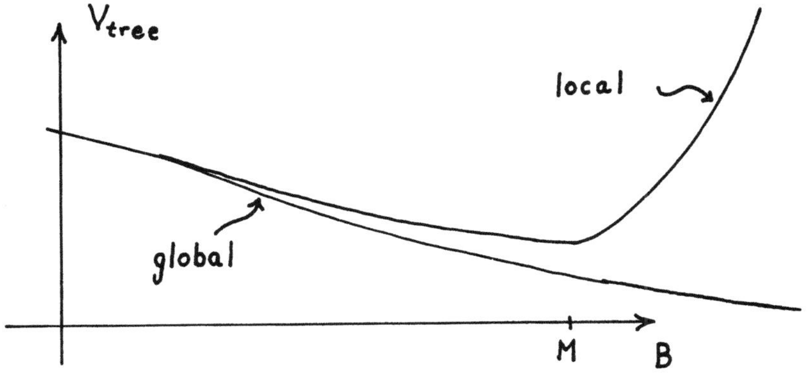

Fig. 8

λ is chosen to fine-tune the cosmological constant to zero. Thus, SUSY is spontaneously broken at the scale μ^2/M. In this case, both scales in the Lagrangian $\mu \sim 10^{15}$ GeV and $M \sim 10^{18}$ GeV are much larger than the SUSY breaking scale $\mu^2/M \sim 10^{11}$ GeV. Oakley and Ross[28] have even given examples where the SUSY breaking scale is of order

$$\mu \left(\frac{\mu}{M}\right)^{n+1} \quad , \; n = 0,\, 1,\, 2,\, \ldots \quad .$$

We conclude that it is not necessary to have an O'Raifeartaigh or Fayet-Iliopoulos sector in local SUSY in order to break SUSY or fine-tune λ to zero.

J. A Simple SU_5 GUT[29]

Consider

$$W = \frac{m_1}{2}\,TrA^2 + \frac{\lambda_1}{3}\,TrA^3 + \frac{\lambda_4}{4M}\,TrA^4 + M_{50}\overline{50}50 + \frac{\lambda_2}{M}(\overline{50}A^2H) + \frac{\lambda_3}{M}(\bar{H}A^2 50)$$

$$+ M_{40}\overline{40}40 + \frac{\lambda_2'}{M}(\overline{40}A^2 10) + \frac{\lambda_3'}{M}(\overline{10}A^2 40) \qquad [J.1]$$

$$+ g^u_{ab}H10_a 10_b + g^d_{ab}\bar{H}10_a\bar{5}_b \quad ,$$

where A is an SU_5 adjoint. The constants

$$m_1 \sim m_g \quad ; \quad \lambda_1 \sim \frac{m_g}{M} \quad ; \quad \lambda_4 \sim \frac{m_g}{M};\ m_g \text{ is the gravitino mass;}$$

$$\lambda_2,\ \lambda_3,\ \lambda_2',\ \lambda_3' \sim 1 \text{ and } M_{40} \sim M_{50} \sim M \quad .$$

We find

$$\langle A\rangle = A_0\begin{pmatrix} 2 & & & & \\ & 2 & & & \\ & & 2 & & \\ & & & -3 & \\ & & & & -3 \end{pmatrix}$$

with $A_0 \sim M$ and $D_AW \sim m_gM$. Thus, SU_5 is broken at M and SUSY is broken at $\sim 10^{11}$ GeV.

Let's now consider the effective low-energy theory. At energies E<<M, we have the states of Table 2. The Higgs doublets are naturally light because we have used the missing partner mechanism of Sec. E. In addition, we have a pair of light charged $SU_3 \times SU_2$ singlets ($E^{\pm}$). These states are also naturally light as a result of a simple generalization of the missing partner mechanism involving the states 40, $\overline{40}$, 10, and $\overline{10}$ (see ref. 30). These states are necessary for reasonable phenomenology.

The effective low-energy potential contains soft SUSY breaking terms which result in $SU_2 \times U_1$ breaking at a scale of order m_g. For a detailed discussion of this phenomenon, see Hall and Nath in this school.

Let's now evaluate the quantities:

$$\alpha_{GUT},\ \sin^2\theta(M_W),\ M_{GUT} \text{ and } m_b/m_\tau \qquad [J.2]$$

using RNG analysis in the effective low-energy theory. For the gauge couplings, we have

$$\frac{d\tilde{\alpha}_i}{dt} = -\tilde{\alpha}_i^2 \left[b_i + \sum_{j=1}^{3} b_{ij}\tilde{\alpha}_j \right] \qquad [J.3]$$

where

$$\tilde{\alpha}_i \equiv \alpha_i/4\pi$$

$$t = \ell n\mu^2$$

$$\alpha_i(M_G) \equiv \alpha_G$$

The one-loop coefficients b_i are given by

$$b_3 = 6 - 2N_G$$

$$b_2 = 4 - 2N_G - N_H$$

$$b_1 = -2N_G - \frac{3}{5} N_H - \frac{6}{5} N_E \ , \qquad [J.4]$$

where

N_G = # of generations

N_H = # $(H_2 + \bar{H}_2)$

N_E = # $(E^+ + E^-)$.

(Note: Perturbative unification requires $N_G \leq 3$.)

The two-loop coefficients are given in Table 3.(31) The solutions are well known.(31)

$$\frac{1}{2\pi}\, \ell n \frac{M_{GUT}}{\bar{\mu}} = \frac{1}{D}\left\{ \frac{1}{\alpha} - \frac{8}{3\alpha_3} - \frac{1}{4\pi} \sum_j \frac{5/3 b_{1j} + b_{2j} - \frac{8}{3} b_{3j}}{b_j}\ \ell n X_j \right\}$$

$$\sin^2\theta = \frac{1}{D}\left\{ b_3 - b_2 + \frac{5}{3}(b_2 - b_1)\frac{\alpha}{\alpha_3} + \frac{5\alpha}{24\pi}\sum_{\ell}\sum_{ijk} \varepsilon_{ijk}(b_i - b_j)\frac{b_{k\ell}}{b_\ell}\ln X_\ell \right\}$$

$$\alpha_G^{-1} = \alpha_3^{-1}(\bar{\mu}) + \frac{b_3}{2\pi}\ln\frac{M_G}{\bar{\mu}} - \frac{1}{4\pi}\sum_j \frac{b_{3j}}{b_j}\ln X_j \qquad \text{[J.5]}$$

where

$$X_j \equiv \left[1 - \frac{b_j}{2\pi}\alpha_G \ln\frac{M_G}{\bar{\mu}}\right]$$

$$D \equiv b_3 - b_2 + \frac{5}{3}(b_3 - b_1) = 12 + 2N_H + 2N_E$$

and X_j is evaluated using one-loop results for α_G, M_G.

Note:

(1) increasing D, decreases M_G; and

(2) $\sin^2\theta \sim \dfrac{2 + N_H}{12 + 2N_H + 2N_E}$.

Thus increasing N_H or N_E, decreases M_G. Increasing N_H, increases $\sin^2\theta$, however, increasing N_E, decreases $\sin^2\theta$. The one-loop results are given in Table 4. Note that $\sin^2\theta$ is very stable under small changes in the boundary conditions at $\bar{\mu} \sim$ 80 GeV. M_{GUT} however, changes by a factor of 3.

Two-loop results are given in Table 5. Once more, $\sin^2\theta$ is extremely stable, whereas M_{GUT} has changed by an order of magnitude from one loop to two loops. α_G has grown to $\sim \frac{1}{6}$ since at two loops α_3 is nonasymptotically free. To summarize, the value of M_G is very sensitive to small changes in the boundary conditions at $\bar{\mu}$. Since we have completely ignored all the new thresholds at $\bar{\mu}$ due to the SUSY zoo, we cannot trust our value of M_G to within one or two orders of magnitude. $\sin^2\theta$ is however, stable under all changes. We thus prefer the solution with $N_H = N_E = 1$. We have

$$M_G \sim 10^{19\pm2}\ \text{GeV}$$

$$\sin^2\theta \sim .225 \pm .003 \qquad \text{[J.6]}$$

$$\alpha_G^{-1} \sim 6.3 \ .$$

The errors quoted are only good guesses. It is pleasing to note that we have been able to retain perturbative unification at M_{GUT}. Moreover, M_{GUT} is of order M as required by self-consistency with the classical solution.

Let's now evaluate the ratio m_b/m_τ. We shall use the renormalization group equation

$$\frac{d}{dt}\,\ell n\,\frac{\lambda_b}{\lambda_\tau} = -\frac{8}{3}\,\tilde{\alpha}_3 + \frac{2}{3}\,\tilde{\alpha}_1 + \tfrac{1}{2}Y_t \qquad [J.7]$$

where

$$Y_t = \frac{\lambda_t^2}{(4\pi)^2} \quad ; \quad \lambda_t,\ \lambda_b,\ \lambda_\tau$$

are the standard Higgs Yukawa couplings ($g^u_{tt} \equiv \lambda_t$, $g^d_{bb} \equiv \lambda_b$ and $g^e_{\tau\tau} \equiv \lambda_\tau$). $t = \ell n\,\mu^2$ and $\tilde{\alpha}_i$ were defined previously (see Eq. [J.3]). The solution to Eq. [J.7] is given by

$$\left(\frac{\lambda_b}{\lambda_\tau}\right)_{M_W} = \left(\frac{\lambda_b}{\lambda_\tau}\right)_{M_G} \left(\frac{M_G}{M_8}\right)^{\frac{4\alpha_G}{3\pi}} \left(\frac{\alpha_G}{\alpha_3(M_W)}\right)^{\frac{8}{3b_3}} \qquad [J.8]$$

$$\times \left(\frac{\alpha_1(M_W)}{\alpha_G}\right)^{\frac{2}{3b_1}} \exp\left(-\tfrac{1}{2}\int_0^{\ell n \frac{M_G^2}{M_W^2}} Y_t\,dt\right) .$$

In Eq. [J.8] we have allowed for the possibility that the states Z_8 have mass $M_8 \geq M_W$. Then b_3=3 for the range $M_8 \geq \mu \geq M_W$. We find (for $(\lambda_b/\lambda_\tau)_{MG} \equiv 1$)

$$\frac{m_b}{m_\tau} = 4.7 \quad (\text{for } N_H = N_E = 1,\ M_8 \equiv 80\ \text{Gev}). \qquad [J.9]$$

This result neglects the effect of the top quark Yukawa coupling Y_t.

Clearly a heavy top quark can lower this result. Note:

$$\left.\frac{m_b}{m_\tau}\right|_{\substack{\text{experimental} \\ \text{at } M_W}} \sim 2.0 \quad .$$

If we now increase M_8, we obtain

$$\frac{m_b}{m_\tau} = 2.95 \text{ (for } N_H = N_E = 1,\ M_8 = 10^5 \text{ Gev).} \qquad [J.10]$$

Note: for $N_H = N_E = 1$, $M_8 = 10^5$ Gev we find

$$M_{GUT} = 1 \times 10^{18} \text{ Gev}$$

$$\sin^2\theta = 0.226 \qquad [J.11]$$

$$\alpha_G^{-1} = 13.4 \quad .$$

K. Proton Decay

As we have discussed in the preceding section, M_{GUT} is of order $M_{p\ell}$. As a result proton decay, via the standard dimension 6 operators resulting from gauge exchange, is highly suppressed (see ref. 32).

It has been shown, however, that in SUSY GUTS there are additional mechanisms leading to proton decay via both dimension 4 and dimension 5 operators.(33) Dimension 4 operators, with a dimensionless coupling of order 1, lead to rapid proton decay and must be avoided. In the simple model presented here, there is a discrete family symmetry, given by

$$10_a \rightarrow -10_a$$

$$\bar{5}_a \rightarrow -\bar{5}_a$$

which forbids dangerous dimension 4 operators.

Dimension 5 operators, on the other hand, are not necessarily dangerous and do, in fact, make interesting predictions for the dominant decay modes.(34,35) The following dimension 5 operators may be generated via the exchange of a color triplet Higgs fermion.

$$\int d^2\theta\ U_i^a\ U_j^b\ D_k^c\ E^d\ \varepsilon^{ijk}$$

$$\int d^2\theta\ U_i^a\ D_j^b\ D_k^c\ V^d\ \varepsilon^{ijk} \qquad [K.1]$$

$$\int d^2\theta\ \bar{U}_a^i\ \bar{U}_b^j\ \bar{D}_c^k\ \bar{E}_d\ \varepsilon_{ijk} \quad .$$

i,j,k=1,2,3 are color indices and a,b,c,d,=1,2,3, are generation indices. An important consequence of the SUSY structure of [K.1] is that if there is only one generation (i.e., a=b=c=d) then all three operators vanish. As a result there must be at least one second-generation particle in the decay mode.(34)

In order to determine the dominant decay modes, we must consider the graphs of Fig. 9. We find the dominant modes

$$p \to K^{+}\bar{\nu}$$
$$p \to K^{0}\mu^{+} \qquad [K.2]$$
$$n \to K^{0}\bar{\nu}$$

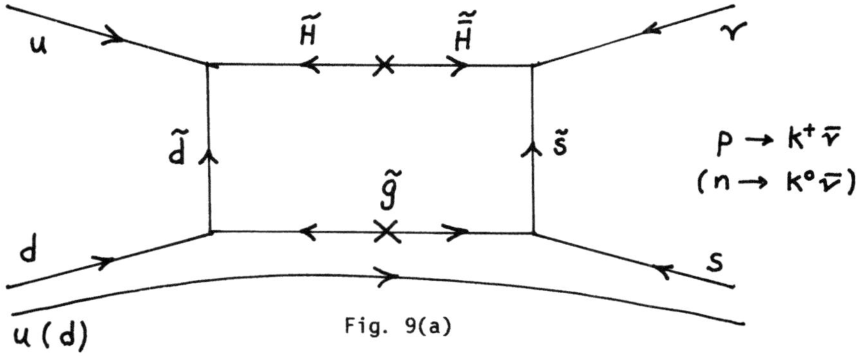

Fig. 9(a)

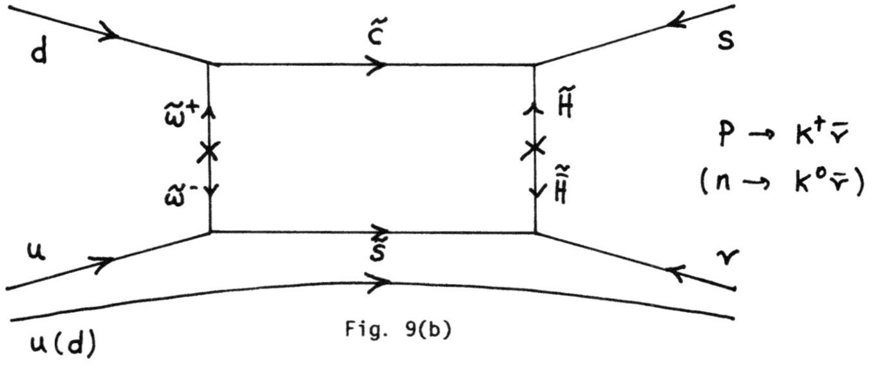

Fig. 9(b)

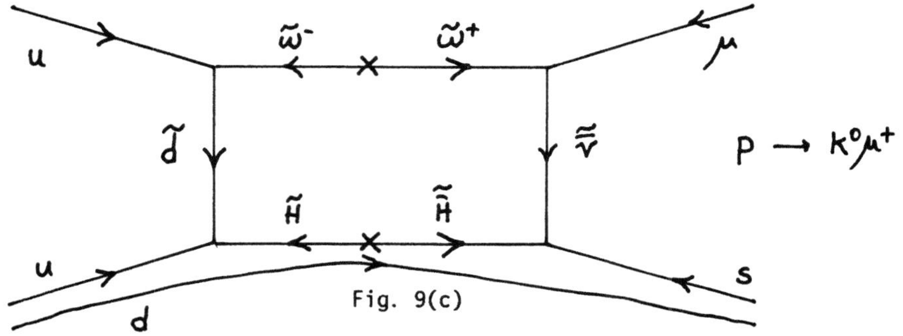

Fig. 9(c)

(i) Note: Gluino exchange graphs vanish for $\tilde{m}_u^2 = \tilde{m}_d^2$.(36) This relation among scalar quark masses is <u>not</u>, however, satisfied in supergravity. Instead we find

$$\tilde{m}_d^2 - \tilde{m}_u^2 \sim \left(\frac{\langle H_0 \rangle^2 - \langle \bar{H}_0 \rangle^2}{\langle H_0 \rangle^2 + \langle \bar{H}_0 \rangle^2} \right) M_Z^2 \sin^2\theta_W$$

which is not necessarily small compared to $\tilde{m}_d^2 + \tilde{m}_u^2$ (see ref. 37).

(ii) In order to obtain proton decay via dimension 5 operators, we must have a Higgs mass term of the form

$$M \int d^2\theta \, \bar{H}H \qquad [K.3]$$

where both H and $\bar{H}$ couple to light quarks and $M \gtrsim 10^{16}$ Gev (in order to avoid rapid proton decay).

Light color triplet Higgs scalars can lead to dimension 6 operators contributing to proton decay. If the Higgs scalars have mass of order $\sim 10^{10}$ Gev, then they may be the dominant decay mechanism.(38) Light-color triplet scalars appear naturally in global GH (see Eq. [B.2]). They are not necessary in general and for example, they are not contained in the model of sect. J. The graphs of Fig. 10 lead to the dominant modes

$$\begin{aligned} &p \to k^o \mu^+ \\ &p \to k^+ \bar{\nu}_\mu \\ &n \to k^o \bar{\nu}_\mu \end{aligned} \qquad [K.4]$$

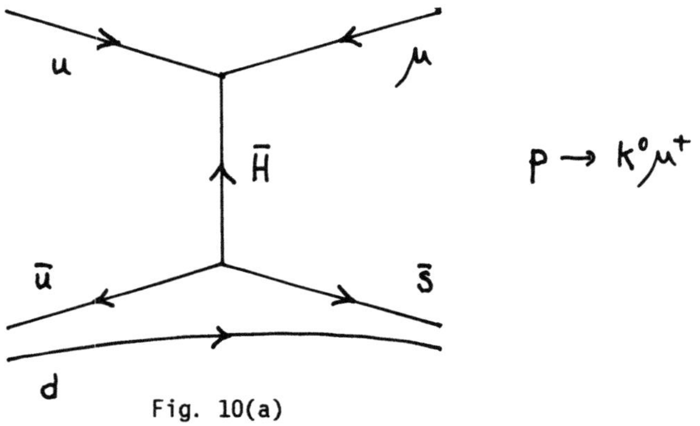

Fig. 10(a)

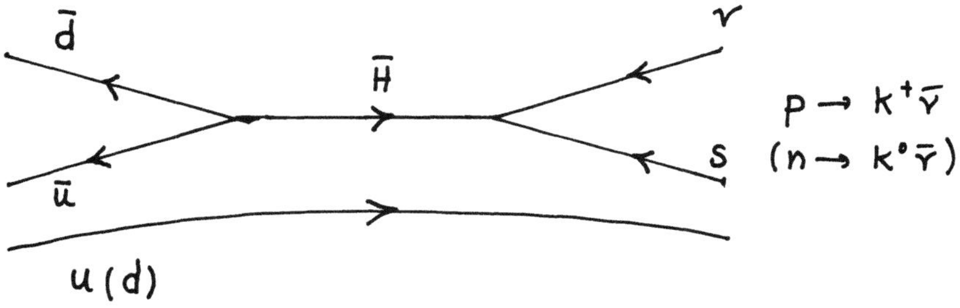

Fig. 10(b)

TABLE 1

	Operator Dimension	Operator	Effective Operator	Description
a.	5	$\frac{1}{M}\int d^4\theta \tilde{X}^* L^2$	$m\int d^2\theta L^2$	SUSY mass term
b.	5	$\frac{1}{M}\int d^4\theta \tilde{X}^* \tilde{X} L$	$m^2 M \Phi_\ell$	linear soft breaking term
c.	6	$\frac{1}{M^2}\int d^4\theta \tilde{X}^* \tilde{X} L^2$	$m^2 \Phi_\ell^2$	soft breaking mass term
d.	6	$\frac{1}{M^2}\int d^4\theta \tilde{X}^* \tilde{X} L^* L$	$m^2 \lvert\Phi_\ell\rvert^2$	soft breaking mass term
e.	6	$\frac{1}{M^2}\int d^4\theta \tilde{X}^* L^3$	$\frac{m}{M}\int d^2\theta L^3$	SUSY interaction term
f.	7	$\frac{1}{M^3}\int d^4\theta \tilde{X}^* \tilde{X} L^3$	$\frac{m^2}{M}\Phi_\ell^3$	
g.	7	$\frac{1}{M^3}\int d^4\theta \tilde{X}^* \tilde{X} D^\alpha L D_\alpha L$	$\frac{m^2}{M}\psi_\ell^\alpha \psi_{\ell\alpha}$	
h.	5	$\frac{1}{M}\int d^2\theta \tilde{X} W^\alpha W_\alpha$	$m\chi^\alpha \chi_\alpha$	gaugino mass
i.	6	$\frac{1}{M^2}\int d^4\theta \tilde{X}^* \tilde{X} D^\alpha W_\alpha$	$m^2 D$	Fayet-Iliopoulos term
j.	7	$\frac{1}{M^3}\int d^4\theta \tilde{X}^* \tilde{X} D^\alpha L W_\alpha$	$\frac{m^2}{M}\psi_\ell^\alpha \chi_\alpha$	

$\tilde{X}$ is the superfield containing the goldstino, $L \supset \Phi_\ell$, ψ_ℓ are light states, and $m \simeq \mu^2/M$, $M \sim M_{GUT}$.

TABLE 2

Light Sector E<<M

$SU_3 \times SU_2 \times U_1$		Gauge States
Chiral Multiplets		
A_8	(8,1,0)	
A_3	(1,3,0)	
A_1	(1,1,0)	
H_2	(1,2,1)	H
$\bar{H}_2$	$(1,\bar{2},-1)$	$\bar{H}$
E^+	(1,1,2)	10
E^-	(1,1,-2)	$\overline{10}$
$\begin{pmatrix} \bar{u} & ud \\ & \bar{e} \end{pmatrix}_a$		10_a
$\begin{pmatrix} \bar{d} \\ e \\ \nu \end{pmatrix}_a$		$\bar{5}_a$ a = 1,2,3

TABLE 3

$$b_{11} = -N_G \frac{38}{15} - N_H \frac{9}{25} - N_E \frac{72}{25}$$

$$b_{12} = -N_G \frac{6}{5} - N_H \frac{9}{5}$$

$$b_{13} = -N_G \frac{88}{15}$$

$$b_{21} = -N_G \frac{2}{5} - N_H \frac{3}{5}$$

$$b_{22} = -N_G\, 14 - N_H\, 7$$

$$b_{23} = -N_G 8$$

$$b_{31} = -N_G \frac{11}{15}$$

$$b_{32} = -N_G\, 3$$

$$b_{33} = -N_G \frac{68}{3}$$

TABLE 4

1 Loop Results

M_G [Gev]

$N_E \backslash N_H$	0	1	2
0	8.8×10^{24}	4.5×10^{21}	1.5×10^{19}
1	4.5×10^{21}	1.5×10^{19} (4.6×10^{19})	1.8×10^{17} (4.9×10^{17})
2	1.5×10^{19}	1.8×10^{17} (4.9×10^{17})	5.4×10^{15}

$\sin^2\theta$

$N_E \backslash N_H$	0	1	2
0	0.210	0.248	0.276
1	0.191	0.227 (0.233)	0.254 (0.251)
2	0.177	0.210 (0.206)	0.233

$\alpha_G \equiv \alpha_3(\bar{\mu})$

We have used the boundary conditions $\alpha_3(\bar{\mu})=0.1$. ($\alpha_3(\bar{\mu})=0.11$), $\alpha^{-1}(\bar{\mu})=128$ ($\alpha^{-1}(\bar{\mu})=128.4$) for $\bar{\mu}=80$ Gev. The results in parantheses indicate the effect of small changes in the boundary conditions at $\bar{\mu}$.

TABLE 5

2 Loop Results

	M_G[Gev]	$\sin^2\theta$	α_G^{-1}
$N_H = N_E = 1$	1.2×10^{20}	0.226	6.28
$N_H = N_E = 2$	1.8×10^{16}	0.236	7.03

The boundary conditions at $\bar{\mu}$ = 80 Gev are $\alpha_3(\bar{\mu})$ = 0.1 and $\alpha^{-1}(\bar{\mu})$ = 128.

REFERENCES

1. Y. A. Gof'fand and E. P. Likhtam, JETP LETT. 13, 323 (1971); D. V. Volkov and V. P. Akulov, Phys. Lett. 46B, 109 (1973); J. Wess and B. Zumino, Nucl. Phys. B70, 39 (1974), and A. Salam and J. Strathdee, Nucl. Phys. B76, 477 (1974), Phys. Rev. D11, 1521 (1975).

2. B. deWit and D. Z. Freedman, Nucl. Phys. B130, 105 (1977); B. deWit, Nucl. Phys. B158, 189 (1979); E. Cremmer and B. Julia, Phys. Lett. 80B, 48 (1978), Nucl. Phys. B159, 141 (1979), and B. deWit and H. Nicolai, Phys. Lett. 108B, 285 (1981), Nucl. Phys. B208, 323 (1982).

3. J. Ellis, M. K. Gaillard, and B. Zumino, Phys. Lett. 94B, 343 (1980); J.-P. Derendinger, S. Ferrara, and C. Savoy, Nucl. Phys. B188, 77 (1981).

4. S. Dimopoulos and S. Raby, Nucl. Phys. B219, 479 (1983).

5. E. Witten, Phys. Lett 105B, 267 (1981).

6. J. Polchinski and L. Susskind, Phys. Rev. D26, 3661 (1982).

7. R. Barbieri, S. Ferrara, and D. V. Nanopoulos, Phys. Lett. 116B, 16 (1982); M. Dine and W. Fischler, Nucl. Phys. B204, 346 (1982); see also T. Banks and V. Kaplunovsy, ref. 8.

8. M. B. Einhorn and D. R. T. Jones, Nucl. Phys. B211, 29 (1982); H. Yamagishi, Princeton preprint (1982); T. Banks and V. Kaplunovsky, Nucl. Phys. B211, 529 (1982), and L. Hall and I. Hinchliffe, Phys. Lett. 119B, 128 (1982).

9. L. Hall, J. Lykken, and S. Weinberg, Phys. Rev. D27, 2359 (1983).

10. L. Girardello and M. T. Grisaru, Nucl. Phys. B194, 65 (1981).

11. H. P. Nilles, M. Srednicki, and D. Wyler, Phys. Lett. 124B, 337 (1983); A. B. Lahanas, Phys. Lett. 124B, 341 (1983); S. Ferrara, D. V. Nanopoulos, and C. A. Savoy, Phys. Lett. 123B, 214 (1983); L. Alvarez, J. Polchinski, and M. Wise, Nucl. Phys. B221 495 (1983).

12. J. Polchinski, Phys. Rev. D (1982); V. Kaplunovsky, Nucl. Phys. B (1983).

13. V. Kaplunovsky, Nucl. Phys. B (1983).

14. M. J. Bowick and P. Ramond, University of Florida preprint, UF-TP-83-1 (1983).

15. E. Witten, Nucl. Phys. B185, 513 (1981).

16. L. E. Ibañez and G. G. Ross, Phys. Lett. 110B, 215 (1982); E. Witten, Phys. Lett. 105B, 267 (1981).

17. N. Dragon, N. Marinescu, and M. G. Schmidt, Univ. of Heidelberg preprint HD-THEP-83-11 (1983); D. Nemeschansky, Princeton Univ. preprint (1983).

18. S. Dimopoulos and F. Wilczek, Proceedings of the 1981 Erice Summer School on Particle Physics, A. Zichichi, ed.; B. Grinstein, Nucl. Phys. B206, 387 (1982); A. Masiero, D. V. Nanopoulos, K. Tamvakis, and T. Yanagida, Phys. Lett. 115B, 380 (1982).

19. P. Fayet, Nucl. Phys. B90, 104 (1975); L. O'Raifeartaigh, Lecture notes on Supersymmetry, Comm. Dias. Series A, No. 22.

20. R. D. Peccei and H. R. Quinn, Phys. Rev. Lett. 38, 1440 (1978); Phys. Rev. D 16, 1791 (1977); S. Weinberg, Phys. Rev. Lett. 40, 223 (1978); F. Wilczek, Phys. Rev. Lett. 40, 279 (1978); W. A. Bardeen and S.-H. H. Tye, Phys. Lett. 74B, 229 (1981).

21. L. Abbott and P. Sikivie, Phys. Lett. 120B, 139 (1983); J. Preskill, M. Wise, and F. Wilczek, Phys. Lett. 120B, 127 (1983); M. Dine and W. Fischler, Phys. Lett. 120B, 137 (1983).

22. For a review, see P. Van Nieuwenhuizen, Phys. Rep. 68 189 (1981).

23. S. W. Hawking, Phys. Lett. 134B, 403 (1984).

24. B. A. Ovrut and S. Raby, Phys. Lett. 121B, 381 (1983).

25. B. A. Ovrut and S. Raby, Phys. Lett. 125B, 270 (1983); M. Claudson, L. J. Hall, and I. Hinchliffe, Phys. Lett. 130B, 260 (1983).

26. L. O'Raifeartaigh, Nucl. Phys. B96, 331 (1975).

27. P. Fayet and J. Iliopoulos, Phys. Lett. 51B, 461 (1974).

28. H. P. Nilles, Phys. Lett. 115B, 193 (1982); C. Oakley and G. G. Ross, Phys. Lett. 125B, 59 (1983); B. A. Ovrut and S. Raby, Phys. Lett. 134B, 51 (1984).

29. B. A. Ovrut and S. Raby, to be published in Phys. Lett. B (1984).

30. L. E. Ibañez, Phys. Lett. 126B, 196 (1983).

31. D. R. T. Jones, Nucl. Phys. B87, 127 (1975); T. J. Goldman and D. A. Ross, Nucl. Phys. B171, 273 (1980); L. J. Hall, Nucl. Phys. B178, 75 (1981); C. H. Llewellyn Smith, G. G. Ross, and J. R. Wheater, Nucl. Phys. B177, 263 (1981).

32. S. Dimopoulos, S. Raby, and F. Wilczek, Phys. Rev. D 24, 1681 (1981); M. B. Einhorn and D. R. T. Jones, Nucl. Phys. B196, 475 (1982); W. J. Marciano and G. Senjanovic, Phys. Rev. D25, 3420 (1982); L. E. Ibañez and G. G. Ross, Phys. Lett. 105B, 439 (1981).

33. N. Sakai and T. Yanagida, Nucl. Phys. B197, 533 (1982); S. Weinberg, Phys. Rev. D26, 257 (1982).

34. S. Dimopoulos, S. Raby, and F. Wilczek, Phys. Lett. 112B, 133 (1982).

35. J. Ellis, D. V. Nanopoulos, and S. Rudaz, Nucl. Phys. B202, 43 (1982).

36. V. M. Belyaev and M. I. Vysotsky, Phys. Lett. 127B, 215 (1983).

37. J. Milutinovic, P. B. Pal, and G. Senjanovic, NSF-ITP-84-05 preprint (1984).

38. E. Golowich, Phys. Rev. D24, 2899 (1981); M. Machacek, Nucl. Phys. B159, 37 (1979).

SUPERSYMMETRY, COMPOSITENESS AND A NEW APPROACH TO GENERATION PROBLEM

Rabindra N. Mohapatra
Department of Physics and Astronomy,
University of Maryland,
College Park, Maryland 20742

Abstract

After a brief review of the various aspects of composite models of quarks and leptons, we present the quasi-Nambu-Goldstone phenomena as the basis for supersymmetric composite model building. Using this approach, we propose a new way to understand the generation problem and fermion mass hierarchies.

1. Introduction

Despite the spectacular successes of the $SU(2)_L \times U(1)_Y \times SU(3)_c$ model of electroweak interactions, many aspects of quark-lepton physics remain unexplained. Outstanding among them is the observed proliferation of quarks and leptons known as the generation problem, which has following three aspects to it:

(i) Why are there three generations with no visible differences among them except for their masses?

(ii) There is a mass hierarchy among the generations, which can be expressed in terms of the average mass per generation as follows: i.e. $\langle m\rangle_1 \approx 0$, $\langle m\rangle_2 \approx 1$ GeV and $\langle m\rangle_3 \approx 10$-$20$ GeV, where the subscript denotes the generation label.

(iii) Finally, there appears to be a definite pattern in the mixing among the generations i.e. $V_{aa} \gg V_{a,a\pm1} \gg V_{a,a\pm2} \approx 10^{-2}$.

Earlier approaches to this problem based on unified gauge theories, where smaller masses and mixings are supposed to arise as radiative corrections have been interesting although no convincing realistic model has emerged. Recent works in supersymmetric composite models appear to have raised a tantalizing possibility where various dynamical and group theoretical constraints may not only lead to a prediction for the number of generations but also to an acceptable pattern for fermion mass hierarchy. It is the purpose of this article to review the basic ideas leading to this approach and illustrate it with two examples.

The basic strategy of composite models[1] building is to assume the existence of a new layer of constituents, called preons, which interact via non-abelian gauge interactions[2] analogous to QCD (to be called hypercolor). The quarks and leptons are assumed to be singlets under this gauge group, much the same way that nucleons are singlets under $SU(3)_c$. The characteristic scale of these new interactions corresponds to the scale of compositeness and is assumed to be of order $\sim$ 1-10 TeV[3] to be compatible with the precise measurements of g-2 of the muon, upper limits on $\mu \to e\gamma$ decay etc.

An immediate problem with the above approach is to understand the approximate masslessness of the quarks and leptons, i.e. $m_{q,\ell}/\Lambda_H << 1$ in contrast with the situation in QCD where $m_N/\Lambda_C \not\approx 1$. One way to understand the masslessness of fermions is to require the theory to possess an exact chiral symmetry. But, as was pointed out by 't Hooft, the existence of chiral symmetry, via Ward inequalities necessarily leads to zero mass singularities of the three point function of the chiral currents. These singularities are dynamical features of the model and must therefore be present both in the composite as well as in the preonic levels of the theory with identical coefficients. This requirement of matching the coefficients of the singularities at the preonic and composite level will be called 't Hooft anomaly matching constraints and will fix the massless fermionic spectrum of the composite theory. Such constraints on composite models have been studied extensively.[4]

Recently, Buchmuller, Peccei and Yanagida[5] have proposed a different solution to the light fermion problem in composite models. They propose to start with a supersymmetric hypercolor theory of preons and have the associated global flavor symmetry undergo spontaneous symmetry breakdown. Supersymmetry then requires that there be massless fermions (to be called quasi-Nambu-Goldstone fermions), which are superpartners of the Nambu-Goldstone bosons. The proposal of ref.5 is that these quasi-Nambu-Goldstone fermions be identified with quarks and leptons. The advantage of this approach is that, even for unbroken global symmetries which are pure vector, massless fermions can arise and they are not subject to any additional constraints due to 't Hooft anomaly matching. It is this approach, which has recently been pursued by us[6] at Maryland in trying to understand the generation puzzle and we will discuss our results here.

This paper is organized as follows: in sec.2, we discuss a non-supersymmetric preon model where flavor degrees of freedom are carried by fermions and color and leptonic degrees by bosons; in sec.3, we study the general phenomenon of quasi-Nambu-Goldstone particles and the counting rules, sec.4 is devoted to supersymmetric preon models where quarks and leptons arise as quasi-Nambu-Goldstone fermions and a resolution of the generation puzzle is proposed; in sec.5, we conclude with a brief enumeration of general problems with this approach.

2. Color-flavor factorization and a non-supersymmetric preon model

In this section, we introduce the basic nature of the preon model we will be considering throughout this article. The idea is based on the observation that for each generation of fermion written as

$$\psi = \begin{pmatrix} u_1 & u_2 & u_2 & \nu \\ d_1 & d_2 & d_3 & e^- \end{pmatrix} \tag{1}$$

the flavor gauge interactions act vertically whereas the color and any possible lepto-quark interactions act horizontally. Furthermore, the color interactions are known to be pure vector whereas flavor interactions must necessarily be chiral. This suggested a preonic structure[7], where there are two fermionic preons (F_u, F_d) that carry flavor degrees of freedom and four scalar preons $(\phi_{i=1,2,3}, \phi_\ell)$ which carry the color lepton number degree of freedom. The quarks and leptons are composites of $F\phi$ as follows:

$$\psi \equiv \begin{pmatrix} F_u \\ F_d \end{pmatrix} \times (\phi_1 \phi_2 \phi_3 \phi_\ell) \tag{2}$$

The binding occurs due to an $SU(N)_H$ interaction operating on F and ϕ, such that $F\phi$ is $SU(N)_H$ singlet. This separation of color and flavor degrees of freedom at the preonic level will form the basis of all the models we will discuss in this paper.

To illustrate how the near masslessness of quarks leptons is understood[8] in this picture, we start with mass less fermionic preons and $SU(4)_H$ as the hypercolor gauge group, which enables us to add a renormalizable interaction of the form $\lambda\phi^4$. The global chiral symmetry of the preonic theory is then given by $SU(2)_L \times SU(2)_R \times U(1)_F \times SU(4)_c$, when color and $U(1)_{em}$ gauge interactions, which are also assumed to be present at the preonic level, are switched off. If this chiral symmetry remains exact at the composite level, this will keep the quarks and leptons massless. To see if this is true, we have to look at the only three point function of currents that has anomaly at the preonic level ie. $U(1)_F[SU(2)_L]^2 + L \to R$. It is clear that for $SU(4)_H$, precisely one generation of quarks and leptons match the preonic anomaly, this keeps $m_q = m_\ell = 0$.

While this model captures the basic spirit of our approach and provides a consistent picture for one generation of quarks and leptons, to go beyond to explain the generation structure and fermion mass hierarchies becomes a difficult task. Therefore, in the next section, we consider supersymmetric generalizations of this model in the hope that they will throw light on the other questions.

3. Supersymmetry and Quasi-Nambu-Goldstone (QNG) Particles

As a prelude to our study of the supersymmetric composite models, let us first discuss the question of spontaneous breakdown of global symmetries in supersymmetric models. Let G be the global symmetry associated with a supersymmetric theory and let the symmetry of the vacuum be a subgroup H of G. Goldstone theorem then tells us that associated with the broken generators of G, there will be Nambu-Goldstone bosons and supersymmetry of the theory will then demand that there be massless chiral fermions as well as additional massless non-Goldstone bosons in the theory to complete the supersymmetric multiplet. We will call these additional massless particles of the theory, the quasi-Nambu-Goldstone particles. Since we would like to build realistic models of particle physics using this phenomenon, we first discuss a method of counting the quasi-Nambu-Goldstone bosons and fermions. This has been discussed in recent papers.[9] We will illustrate the procedure with the following simple example:

Consider a superpotential invariant under a U(1) global symmetry, involving fields $\Phi\Phi'$ and X which carry U(1) charges +1, -1 and zero respectively i.e.

$$W = \lambda X(\Phi\Phi' - M^2) \tag{3}$$

The minimum of this potential preserves supersymmetry and breaks U(1) global symmetry by the following vacuum expectation values (VEV)

$$\text{i.e. } \langle A_\phi \rangle = \langle A_{\phi'} \rangle = M \tag{4}$$

The mass spectrum of the field theory with W as its superpotential is as follows:

$$\text{Nambu-Goldstone boson:} \quad \chi \equiv \frac{1}{\sqrt{2}}(\mathrm{Im}A_\phi - \mathrm{Im}A_{\phi'})$$

$$\text{QNG Boson:} \quad \sigma \equiv \frac{1}{\sqrt{2}}(\mathrm{Re}A_\phi - \mathrm{Re}A_{\phi'})$$

$$\text{QNG Fermion:} \quad \psi_- \equiv \frac{1}{\sqrt{2}}(\psi_\phi - \psi_{\phi'}) \tag{5}$$

Rest of the particles are massive. We note that, $\sigma + i\chi$ and ψ_- are members of a supermultiplet. $\psi_+ = \frac{1}{\sqrt{2}}(\psi_\phi + \psi_{\phi'})$ pairs with ψ_χ to form a Dirac particle with mass $\sqrt{2}\lambda M$ and has as its superpartners $\frac{1}{\sqrt{2}}(A_\phi + A_{\phi'})$ and A_χ. The important point to note is that corresponding to each massless chiral fermion there must be two real massless scalar fields.

These considerations can be generalized to arbitrary groups. In counting the massless particles, however, one has to be careful. For instance, let us consider the example where SU(N) breaks down to SU(N-1) x U(1) subgroup. The number of massless fermions is either 2(N-1) which equals the number of broken generators (as in the U(1) case above) or (N-1) which is half the number of broken generators depending on the number of representations of SU(N) employed in breaking the symmetry (2(N-1) if two fields transforming $N+\bar{N}$ are employed to break the symmetry and (N-1) if only one field transforming as either N or $\bar{N}$ breaks the symmetry). General methods to infer the number of massless fermions have been discussed by Lerche[9], Lee and Sharatchandra[9] and Kalara[9] from two different points of view.

The point of view advocated by Lerche et al. is based on an observation due to Ovrut and Wess[10]. The basic idea is that in supersymmetric theories, the superpotential W depends only on fields and not on their complex conjugates. This implies that if G is an invariance group of W, its complexification group $\bar{G}$ is also an invariance group. Similarly, the invariance group of the vacuum, becomes $\bar{H}$, the complexification of H. Thus, the actual number of massless bosons correspond to the generators in the Coset space $\bar{G}/\bar{H}$. If this number is $\bar{N}$, the number of massless fermions is $\bar{N}/2$. To illustrate, in the U(1) example mentioned above, $\bar{G} \equiv \overline{U(1)}$, which is nothing but U(1) x scale transformation; therefore, the quasi-Nambu-Goldstone particle, σ is really the NG boson corresponding to the breaking of dilatation symmetry. From this point of view, the name QNG bosons may be somewhat misleading. All the bosons are really Nambu-Goldstone bosons. It is worth pointing out that, even though the kinetic energy term in the superaction does not respect $\bar{G}$-invariance, the non-renormalization theorems protect the masslessness of the QNG bosons.

The point of view advocated by Kalara[9] may be more powerful in the sense that, it does not assume the property of the superpotential W. It starts from the definition of supercurrent corresponding to the global symmetry G (with generators denoted by θ_a) realized in the space of superfields Φ_i:

$$J_a = \sum_i \bar{\Phi}\, \theta_a\, \Phi \qquad (6)$$

and using the supersymmetric field equations, one can write its conservation law as

$$D^2 J_a = 0 = \bar{D}^2 J_a \qquad (7)$$

where

$$D = \left(\frac{\delta}{\delta\theta} - i\sigma_\mu \bar{\theta}\partial_\mu\right)$$

and $$\bar{D} = \left(\frac{\delta}{\delta\bar{\theta}} + i\theta\sigma_\mu\partial_\mu\right) \qquad (8)$$

Eqn.(7) leads to the conservation of the usual vector current associated with G but it also implies following additional equations involving fermionic fields:

$$A_i(\theta_a)_{ij}\not{\partial}\psi_j = 0$$

and $$\not{\partial}\bar{\psi}_i(\theta_a)_{ij}A_j = 0 \qquad (9)$$

Eqn.(9) inform us that, if $\langle A_i\rangle \neq 0$ and if $(\theta_a)_{ij}\ \langle A_j\rangle \neq 0$, then there are massless chiral fermions. Kalara's counting method involves looking at the transformation of θ_a and ψ_i under subgroup H; since $\langle A\rangle$ is a singlet under H, the product $\bar{\psi}_i$ and θ_a must be a singlet under H. It is claimed that, this will provide a reliable way to calculate the spectrum massless fermions. One advantage of this method may be in its application to supersymmetric effective field theories at low energies resulting from a theory with a heavy superfield. The point is that in such theories the low energy Lagrangian will contain D-type terms (non-renormalizable) contributing to the potential in addition to the W-term; because of this arguments based on the complexification of G do not apply.

4. Supersymmetric composite models for quarks and leptons

The basic idea of this section is based on the suggestion[5] that the quarks and leptons are the composite quasi-Nambu-Goldstone fermions corresponding to spontaneously broken global symmetry of an underlying supersymmetric hypercolor theory of preons. The masslessness of quarks and leptons in these theories does not require a chiral symmetry and therefore, the whole procedure of anomaly matching can be bypassed.

We, therefore, have two mechanisms to obtain composite massless quarks and leptons: (i) 't Hooft anomaly matching and (ii) Quasi-Nambu-Goldstone phenomena in a supersymmetric theory. Before presenting detailed models, we wish to point out a tantalizing possibility to get fermion mass hierarchies in this model[6,11].

To see this, let us assume that the global symmetry group of the preonic theory is G, which, in general, will be a chiral symmetry and let the symmetry group of vacuum be H i.e. $G \to H$. There are then two classes of possibilities:

(a) H is an anomaly free subgroup of G. In this case, only massless fermions are those associated with quasi-Nambu-Goldstone mechanism. We will denote such particles by $\{A_\chi, \chi\}$, where A_χ are the scalar partners of the QNG fermions χ.

(b) H is not anomaly free. In this case, in addition to the massless QNG fermions, $\{A_\chi, \chi\}$, there may be new massless fermions needed to match 't Hooft anomalies: if we denote them by the set $\{A_\psi, \psi\}$, in general

$$\{A_\psi, \psi\} \supset \{A_\chi, \chi\} \tag{10}$$

This implies that the masslessness of the set of fermions χ is doubly protected, whereas the masslessness of the remaining fermions in the set ψ (ie. $\{\psi'\} \equiv \{\psi\} - \{\chi\}$) is protected only by chiral symmetry. So, if the chiral symmetry in question is broken at a scale m_c, then

$$m_{\{\psi'\}} \approx m_c \tag{11}$$

whereas

$$m_{\{\chi\}} \approx \frac{m_c M_s}{\Lambda_H} \tag{12}$$

where M_s is the supersymmetry breaking scale and Λ_H is the scale of hypercolor interactions. If $M_s \ll \Lambda_H$, then we have a natural mechanism for $m_{\{\psi'\}} \ll m_{\{\chi\}}$.

We now present a realistic model[6] where precisely such a situation arises. The preons are supersymmetric generalization of the color flavor factorized model in sec.2, with $SU(N)_H$ hypercolor gauge interaction. The column vector for six preonic chiral superfields are given below:

$$\begin{matrix}\text{Flavor}\\ \\ \text{Color}\\ \\ \text{leptonic}\end{matrix}\begin{pmatrix}\Phi_u\\ \Phi_d\\ \Phi_1\\ \Phi_2\\ \Phi_3\\ \Phi_\ell\end{pmatrix} \equiv \Phi \qquad \begin{matrix}\text{Flavor}\\ \\ \text{Color}\\ \\ \text{leptonic}\end{matrix}\begin{pmatrix}\bar{\Phi}_u\\ \bar{\Phi}_d\\ \bar{\Phi}_1\\ \bar{\Phi}_2\\ \bar{\Phi}_3\\ \bar{\Phi}_\ell\end{pmatrix} \equiv \bar{\Phi} \tag{13}$$

Here Φ denotes the left-handed preonic superfields and $\bar{\Phi}$ denotes the charge conjugate left-handed superfields, each member in $\Phi(\bar{\Phi})$ transforms as N_H (N_H^*) representations under $SU(N)_H$ gauge group.

The global symmetry of this preonic theory is $U(6)_L \times U(6)_R \times U(1)_Y$ where $U(1)_Y$ is the R symmetry defined by

$$\Phi \to e^{im_L\alpha}\Phi(x,\, e^{-i\alpha}\theta)$$

$$\bar{\Phi} \to e^{im_R\alpha}\bar{\Phi}(x,\, e^{i\alpha}\bar{\theta})$$

$$V_H \to V_H(x,\, e^{-i\alpha}\theta,\, e^{i\alpha}\bar{\theta}) \tag{14}$$

The hypercolor instantons reduce this symmetry. To isolate the residual symmetry, we note that the anomaly current is generated by

$$A = 6Q_L + 6Q_R + 2N_H\eta_\lambda \tag{15}$$

where $Q_{L,R}$ generate the $U(1)_L \times U(1)_R$ in $U(6)_L \times U(6)_R$; η_λ is the number operator for the $SU(N)_H$ L gauginoes. The true global symmetry G of this preonic theory is then

$$G \equiv SU(6)_L \times SU(6)_R \times U(1)_V \times U(1)_X \tag{16}$$

with

$$Q_V = Q_L - Q_R$$

$$\text{and}\quad Q_X = \left(\frac{m_L+m_R}{2} + \frac{N_H}{6} - 1\right)(Q_L+Q_R) - Q_Y \tag{17}$$

Below the hypercolor scale Λ_H, we expect hypercolor singlet composites to form and the effective theory will be given in terms of an effective Lagrangian involving these composite fields. In the composite theory if none of the symmetry is broken by vacuum, then G also remains a global symmetry of the theory. However, let us assume that spontaneous symmetry breaking does occur by formation of hypercolor singlet condensates and the symmetry is reduced to H. We assume a specific form for H i.e.

$$H = SU(2)_L \times SU(4)_L \times SU(2)_R \times SU(4)_R \tag{18}$$

It is possible to write down condensates[6] so that the H is indeed given by eqn.(18). The Nambu-Goldstone bosons then belong to the following representation of H:

$$GB = (2,4,1,1) + (1,1,2,\bar{4}) + 4(1,1,1,1) \tag{19}$$

The associated QNG fermions transform as:

$$\chi = (2,4,1,1)_L + (1,1,2,\bar{4})_L + 4(1,1,1,1)_L \tag{20}$$

Note that in eqn.(20), the first two sets have precisely the same quantum numbers as the quarks and leptons of one generations, with SU(4) denoting group combining color and lepton number. We will identify these as quarks and leptons. They can be expressed in terms of their preonic components as follows: To do this, let us write

$$\Phi_i = (A_i, \psi_i) \quad (i = u,d,\text{ color, lepton number})$$

$$\equiv (A_f, \psi_f) + (A_c, \psi_c) \quad f = u,d \text{ and } c = \text{color, lepton number} \tag{21}$$

The QNGF fermions are then given by $A_f^*\psi_c + \bar{A}_f^*\bar{\psi}_c$. They remain massless due to the quasi-Nambu-Goldstone phenomena. However, since the subgroup H is not anomaly free, we have to look at 't Hooft anomaly matching constraints and any additional massless fermions implied by them.

Due to the choice of H_1 we have to match only one anomaly i.e. to $[SU(4)_{L,R}]^3$. This can be matched only by the fermions belonging to the same representation as the QNG fermions or the quarks and leptons. If we take N_g sets of QNG fermions corresponding to N_g families, 't Hooft anomaly matching implies

$$N_g = \frac{N_H}{2} \tag{22}$$

Thus, we predict that in our theory, the hypercolor degrees of freedom, N_H must be even and the number of generations must be $N_H/2$. Now we will argue that N_H is most likely to be six. The reason for this is that, when $SU(6)_{L,R}$ breakdown to $[SU(2) \times SU(4)]_{L,R}$, the residual U(1)'s must be broken and also $U(1)_V$ and $U(1)_X$ must be broken. In particular, to break $U(1)_V$ that is consistent with the above symmetry breaking pattern, we need to form a condensate of type:

$$\varepsilon^{i_1 i_2 i_3 i_4 i_5 i_6} < A_{i_1}^{\alpha_1} A_{i_2}^{\alpha_2} A_{i_3}^{\alpha_3} A_{i_4}^{\alpha_4} A_{i_5}^{\alpha_5} A_{i_6}^{\alpha_6} > ,$$

where i is preonic flavor index and α's are hypercolor index. Note that for this condensate to be a hypercolor singlet, N_H must be 2 or 6. This implies that we have either one or three generations of quarks and leptons whose masses are negligible.

5. Conclusions and Outlook

We have learned the following lessons from the exercise reported in this article: first that, quasi-Nambu-Goldstone phenomena provides an alternative mechanism for understanding the near masslessness of quarks and leptons in a composite picture, as suggested in ref.5; second point noted by us in ref.6 by Barbieri et al in ref.11 is that, this mechanism combined with 't Hooft anomaly matching constraints hold great promise in understanding the fermion mass hierarchies. More examples of this kind need to be worked out, however, before one can conclude that this approach has any connection with the real world.

There are several shortcomings of this approach, which must be removed before this leads to a complete understanding of quark lepton physics. First, we must introduce supersymmetry breaking into the theory that will give mass to the massless scalar bosons and also mass to the first generation fermions. We envision doing this with the help of N=1 local supersymmetry. We have checked that, with a Polony-like breaking of local supersymmetry, some of the scalar bosons can pick up mass. Furthermore, the rest of them will acquire mass from radiative gauge loop effects of order $\sqrt{\alpha} m_W$ and could therefore lie in the tens of Gev range. The massless scalars, which are gauge singlet will not pick up mass this way but they do not have direct coupling to matter and therefore appear harmless from a phenomenological point of view. Another point is that in realistic models, gauge interactions may be introduced from the beginning. If the composite fermion spectrum is determined from an effective superpotential of F-type all our considerations will remain unaffected by the presence of supersymmetric gauge interactions. But more quantitative implications of this idea remain to be investigated.

Finally, a very unsatisfactory aspect of our model (and indeed all composite models) is the adhoc postulate concerning condensate formation. One needs more theoretical guidance regarding the symmetry breakdown in supersymmetric QCD-like theories. This has been extensively discussed in literature[14] using the so-called complementarity principle. But, it is not clear, whether all composite models must satisfy this principle if the requirements such as the scalar particles being in the vector representation are met. This is, of course, a difficult question to answer. On the whole, this approach has opened up new possibilities and a paradise (temporary!!) for model builders.

I wish to thank O.W. Greenberg and M. Yasue in whose collaboration the work reported here has been carried out and also to acknowledge several useful discussions with S. Kalara on the counting of ONGF states. I also wish to thank Prof. Probir Roy and other colleagues at the Tata Institute of Fundamental Research for kind hospitality at the Winter School. This work has been supported by a grant from the National Science Foundation.

References

1. For a recent review and references, see L. Lyons, Oxford report 52/82 (1982). R.N. Mohapatra, Neutrino mass mini-conference, Telemark, edited by V. Barger and D. Cline, A.I.P. publication (1983).
2. G. 't Hooft, in Recent Developments in Gauge Theories ed. by G. 't Hooft et al (Plenum press, N.Y. 1980), p.135. R. Casalbuoni and R. Gatto, Phys. Lett. 93B, 47 (1980). O.W. Greenberg and J. Sucher, Phys. Lett. 99B, 339 (1981).
3. S.J. Brodsky and S.D. Drell, Phys. Rev. D22, 2236 (1980); G.L. Shaw, D. Silverman and R. Slansky, Phys. Lett. 94B, 57 (1980); M. Abolins et al, Proc. of 1982 DPF Summer Study, edited R. Donaldson, et al (Fermi Lab, 1983) p.274.
4. T. Banks, S. Yankielowicz and A. Schwimmer, Phys. Lett. 96B, 67 (1980); I. Bars, Nuc. Phys. B198, 269 (1982); J.M. Gérard, J. Govaerts, Y. Meurice and J. Weyers, Phys. Lett. 116B, 29 (1982).
5. W. Buchmüller, R.D. Peccei and T. Yanagida, Phys. Lett. 124B, 67 (1983). Nuc. Phys.
6. O.W. Greenberg, R.N. Mohapatra and M. Yasue, Phys. Lett. 128B, 65 (1983); Phys. Rev. Lett. 51, 1737 (1983); Nuc. Phys. B237, 185 (1984).
7. J.C. Pati in Neutrino Physics and Astrophysics, edited by E. Fiorini (Plenum, New York, 1982) p.275; O.W. Greenberg and J. Sucher, ref.2.
8. R. Barbieri, R.N. Mohapatra and A. Masiero, Phys. Lett.105B, 369 (1981).
9. C. Lee and H. Sharatchandra, Max Planck Preprint (1983); W. Lerche, Max Planck Preprint, (1983); T. Kugo, T. Yanagida, and T. Ojima; Max Planck Preprint; S. Kalara, University of Rochester Preprint (1983); O.W. Greenberg et al, in preparation; D. Lust and W. Lerche,

Max Planck Preprint (1983).
10. B. Ovrut and J. Wess, Phys. Rev. D25, 1608 (1982).
11. R. Barbieri, A. Masiero and G. Veneziano, Phys. Lett. 128B, 179 (1983).
12. Note that $SU(4)_{L,R}$ gauge symmetry must be broken at a higher scale ($\sim 10^2$ TeV) to be consistent with the upper limit for the branching ratio of $\Gamma(K_L \to \mu\bar{e})/\Gamma(K \to \text{everything})$. Such broken gauge symmetries with weak couplings $g_c^2 \ll 4\pi$ may be present in the theory. They will not effect our considerations, since we let all such couplings go to zero, in considering both quasi-Nambu-Goldstone phenomena and 't Hooft anomaly matching.
13. Here we may envisage embedding the theory in an N=1 supergravity with supersymmetry breaking scale being at 10^{12} GeV. The effective supersymmetry breaking scale at low energies is given by the gravitino mass m_g which can be $10^2 - 10^3$ GeV.
14. H.P. Nilles, Phys. Lett. 112B, 455 (1982); G. Veneziano and S. Yankielowicz, Phys. Lett. 113B, 231 (1982); T.R. Taylor, G. Veneziano and S. Yankielowicz, Nuc. Phys. B218, 493 (1983).
15. For an attempt in this direction, see, M. Yasue, R.N. Mohapatra and J.C. Pati, in preparation.

SEARCH FOR SUPERSYMMETRIC PARTICLES

Rohini M. Godbole
Department of Physics
University of Bombay
Kalina Campus, Bombay 400098, INDIA

ABSTRACT

Present limits on the masses of supersymmetric partners of the known particles (leptons, quarks, etc.) are reviewed. Many suggestions for the (future) experimental search for these states are also discussed. Special emphasis has been put on tests of supersymmetry and supergravity possible at the $\bar{p}p$ collider with reference to data which has become available recently.

INTRODUCTION

The discovery of W and Z bosons at the collider [1] has been one of the most convincing demonstration of the correctness of the standard model [2] of the electro-weak interactions. However, in the standard model the electroweak scale is explained in terms of the vacuum expectation value of a fundamental scalar field. We know that the scalar masses receive large, quadratically divergent corrections in perturbation theory. Hence the validity of the standard model can not be pushed much beyond the Fermi scale given by $\sqrt{G_F} \simeq 300$ GeV. In the grand unified theories, the simultaneous presence of a large scale other than the weak scale $\sim O(m_W)$ and the fundamental Higgs scalars gives rise to the hierarchy problem.

Supersymmetric theories [3,4] provide a neat solution to these problems. This is done by introducing a symmetry between fermions and bosons. Not only do these theories give a possible solution to the hierarchy problem and explain the weak scale, but they also relate particles to gravity. All this makes supersymmetry a very attractive idea.

As a result, one would like to find experimental evidence for supersymmetry. However none exists at present. It clearly implies that supersymmetry is badly broken if it is a symmetry of nature at all. Fortunately it is possible to think of a number of ways to find evidence for supersymmetry, though none are very obvious. A lot of effort has gone into this enterprise for the past years and some very good reviews [4,5,6,7] exist.

In this talk I will review different suggestions for the search for supersymmetry. The discussion will be limited to search possible in high energy physics experiments only. Low energy effects of extra particles present in supersymmetric theories and the constraints they receive thereby [8] will not be discussed.

Here I will discuss:

(a) Possible particle spectrum in most models, i.e. *where* to find these particles.

(b) *How* to find them - viz. the characteristic signatures.

(c) The present experimental *limits* on their masses.

(d) Searches likely to yield results in the near future.

Supersymmetric theories predict that all particles must appear in pairs of identical masses and quantum numbers differing in spin by 1/2. Since we do not see such particle pairs supersymmetry must be broken if it is a symmetry of nature. The expected particle spectrum depends on the model considered. In the minimal picture, corresponding to all the observed fermions (bosons) there are bosonic (fermionic) partners. Table I below summarises the spectrum in the minimal picture.

PARTICLE SPECTRUM

Particles	Sparticles
$u_L, d_L, u_R, \ldots$	$\tilde{u}_L, \tilde{d}_L, \tilde{u}_R, \ldots$
$e_L^-, e_R^-, \ldots$	$\tilde{e}_L^-, \tilde{e}_R^-, \ldots$
$\nu_e, \bar{\nu}_e, \ldots$	$\tilde{\nu}_e, \tilde{\bar{\nu}}_e, \ldots$
g	$\tilde{g}$
$W^\pm$	$\tilde{W}^\pm$
Z	$\tilde{Z}$
$H_1^0 \quad H_2^-$ $H_1^+ \quad H_2^0$	$\tilde{H}_1^+, \tilde{H}_2^- \quad \tilde{H}_1^0, \tilde{H}_2^0$

TABLE I

In the limit of exact supersymmetry the ordinary particles and their partners have the same masses. However, in practice, it is broken and the superpartners must be heavy enough to have avoided detection so far. Experimental observation of these will be evidence for supersymmetry. The idea of supersymmetry will be not so interesting if we do not find any signal for them upto the Tev range. This underscores the importance of precisely knowing what to look for in present-day particle physics experiments. It has been noted [5] that a signal for superparticles may have been missed because of inappropriate cuts used in the experiments. Their nonobservation so far can be used to put lower limits on their masses. Only associated production of superpartners is possible, and their nonobservation therefore excludes certain regions in two dimensional mass plots.

In Table I all the sparticles above the line are the scalar superpartners. They are squarks, sleptons and sneutrinos. The sparticles below the line (the fermionic partners) are 'inos'. The Majorana particles $\tilde{W}^{\pm}$, $\tilde{H}_1^+$, $\tilde{H}_2^-$, mix to give two Dirac particles 'W-inos' $\tilde{W}_1^{\pm}$, $\tilde{W}_2^{\pm}$. For neutral gauge fermions also the isospin eigenstates mix to give mass eigenstates. In most models the lowest mass eigenstate is the photino, which has mostly gauge couplings. The states shown in Table I are the isospin eigenstates. The mass eigenstates are called photino, Ziggsino and Higgsino depicting their dominant coupling [5].

It is necessary to have some idea of the masses and coupling of the sparticles to discuss possible methods to search for them. The masses are model dependent. There exist two distinct types of phenomenology depending upon whether we consider global or local models. In local models the gravitational effects induce spontaneous breakdown of supersymmetry. The supersymmetry breaking scale is characterised by M_s = sqrt (M_P* $m_{3/2}$), where $m_{3/2}$ is the mass of the gravitino, the spin 3/2 partner of graviton. The masses of light particles such as photino, gluino are controlled by $m_{3/2}$. Experimental information available at present allows a large value of $m_{3/2}$ ($\sim m_W$), which gives a potentially interesting mass spectrum. Table II below gives an overview of possibilities given in global [9] and local [10] models. Since the spontaneous breakdown of supersymmetry (ssb) in global models is fraught with problems, I will mainly concentrate the discussion on the predictions of the supergravity models. In global models the Goldstino (the supersymmetric partner of the goldstone boson associated with the ssb) and its couplings are important. A detailed discussion of this issue is given in Ref.4.

Even if supersymmetric theories have little predictivity about the

SPECTROSCOPY IN LOCAL AND GLOBAL MODELS

Standard particles	SUSY partners	predicted masses in global models	local models
	Gravitino Goldstino	$\}$ << 1 GeV	$m_{3/2}$ > 20 GeV
Gluon g	$\tilde{g}$	$\sim$ 0 (few GeV)	$O(m_{3/2}) > \alpha_s m_{3/2} > 4$ GeV*
Photon γ	$\tilde{\gamma}$	0 + Radiative corrections	$O(\tilde{m}_g/8)$
W-boson $W^{\pm}$ Higgs H^{+}	$\tilde{W}^{\pm}$ $\tilde{H}^{+}$	$m_{\tilde{W}_{1,2}} \sim O(m_W)$	$\simeq m_W \pm O(M_{3/2})$
H^{o} Z-boson Z_o	$\tilde{H}^{o}$ $\tilde{Z}_o$	$m_{\tilde{Z}_{1,2}} \sim O(m_{Z_o})$	$\simeq m_Z \pm O(M_{3/2})$
quark q	$\tilde{q}$	$m_{\tilde{q}}$ < 40 GeV	$\lvert m^2_{3/2}+m^2_{\tilde{g}}\rvert^{1/2} \pm O(M_q)$
lepton ℓ	$\tilde{\ell}$	$m_{\tilde{\ell}}$ < 40 GeV	$m_{3/2} \pm O(m_{\ell})$

*Input from present experiments.

TABLE II

masses of the various spartners, masses of various sparticles are related in each model. More importantly the coupling strengths of the various spartners are completely determined in terms of those of the standard particles. For example, the coupling strength for $\tilde{g}\tilde{q}\bar{q}$, $g\tilde{q}\bar{\tilde{q}}$ and $gq\bar{q}$ vertices is the same. As a result, calculation of production cross section is easy. However the background from ordinary physics is large and isolation of the signal is the real problem. For the same reasons, it is possible for the earlier experiments to have missed the spartners due to inappropriate cuts.

In almost all the supersymmetry and supergravity models R parity is a good quantum number. This implies that the lowest mass spartner must be stable. There are cosmological arguments [11] why the lightest R odd particle should be neutral. The presence of this sparticle in the final state will appear as missing energy. Hence all the processes involving sparticles will dominantly give rise to events with large missing energy and therefore will be a good signature of the super particle-physics.

In most of the discussion below we will assume $\tilde{\gamma}$ to be the lightest supersymmetric particle. (There are models with $\tilde{\nu}$ as the lightest sparticle [12]. Implications of this possibility have been discussed and will be reviewed.) The dominant decay modes of the various sparticles are $\tilde{g} \to q\bar{q}\tilde{\gamma}$, $\tilde{\ell} \to \ell\tilde{\gamma}$, $\tilde{q} \to q\tilde{g}$ (if $m_{3/2} > m_{\tilde{g}}$) and $\tilde{q} \to q\tilde{\gamma}$ (if $m_{3/2} < m_{\tilde{g}}$).

The above summarises all the general information necessary to discuss the phenomenology of the search for sparticles. The important features of $\bar{p}p$ and e^+e^- machines available in the near future are given in Table III below.

Machines Available

		$\sqrt{s}$ GeV	WHEN?
e^+e^-	PETRA	45	Running
	TRISTAN	60	1986 - 1988
	LEP/SLC	80-100	" "
$\bar{p}p$	SPS	540	Running
pp	TEVATRON	2000	1986?

TABLE III

The search with e^+e^- machines is much cleaner, but $\bar{p}p$ machines have the advantage of higher energy and early availability. These overweigh the problem of signal to background ratio which is typical for the $\bar{p}p$ machines. Hence special attention has been focussed at present on the search for spartners in $\bar{p}p$ machines.

Below I discuss possible experimental signatures specific to each sparticle listed in Table II and present experimental limits. This is followed by possible improvements on these limits in view of the machines listed in Table III. This information is summarised in Table IV and V. Table VI summarises the signals possible due to various sparticles and the competing background due to ordinary physics.

(1) Supersymmetric Scalars

1.a. Charged sleptons

Scalar partners of leptons are called sleptons. Corresponding to each massive fermion there are two sleptons. The mass of the gravitino > 20 GeV (This follows from present limits on gluino masses, for those supergravity models in which gaugino masses are radiative.) This leads to the expectation $m_{\tilde{\ell}}$, $m_{\nu_{\tilde{\ell}}} \sim 20 - 80$ GeV in most models.

First consider the case of the charged sleptons. The only possible decay mode for them is $\tilde{\ell} \to \ell\tilde{\gamma}$.

Mass limits attainable with various machines alongwith the present experimental information for the sfermions

		Limits possible in GeV			
		PETRA PEP	LEP, SLC TRISTAN	SPS	FNAL
$\tilde{\ell}$	$e^+e^- \rightarrow \tilde{\ell}^+\tilde{\ell}^-$ present limit $m > 14$ GeV	15	45		
	$e^+e^- \rightarrow e\tilde{e}\tilde{\gamma}$ Present limit $m > 19.5$ GeV	30	75		
$\tilde{\ell}, \tilde{\nu}$	$\bar{p}p \rightarrow W$, $W \rightarrow \tilde{\ell}\tilde{\nu}$; $m_W > m_{\tilde{\ell}} + m_{\tilde{\nu}}$ needed. No analysis at present			~ 30	~ 30
$\tilde{\nu}$	τ decay gives $m(\tilde{\nu}_e) + m(\tilde{\nu}_\ell) > m_\tau$				
	$e^+e^- \rightarrow \tilde{\nu}\tilde{\nu}$ counting experiment at Zo pole		50		
$\tilde{q}$	$e^+e^- \rightarrow \tilde{q}\tilde{q}$ $m_{\tilde{q}} > 15$ GeV (Not very clear)	15	40		
	$\bar{p}p \rightarrow \tilde{q}\tilde{\gamma}X$			~ 35	~ 80

TABLE IV

<u>Search for charged sleptons at e^+e^- colliders</u>

Different suggestions are given in the literature for detecting $\tilde{\ell}$ in e^+e^- machines. For the expected mass range typically one expects one unit of $R = \sigma(e^+e^- \rightarrow \tilde{\ell}^+\tilde{\ell}^-)/\sigma(e^+e^- \rightarrow \mu^+\mu^-)$ at LEP, which corresponds to ~ 100 events/day with a luminosity of 10^{31} events/cm^2.

Two different suggestions [13,14] exist:

$$\text{(a)}\quad e^+e^- \xrightarrow{Zo,\gamma} \tilde{\ell}^+ + \tilde{\ell}^-$$
$$\tilde{\ell}^+ \rightarrow \ell^+\tilde{\gamma}, \quad \tilde{\ell}^- \rightarrow \ell^-\tilde{\gamma}$$

Mass limits attainable with various machines
and present information for gauginos

		Limits possible in GeV			
		PETRA PEP	LEP, SLC TRISTAN	SPS	FNAL
$\tilde{g}$	Beam Dump NC excess present limit m_g > 2-5 GeV			$\sim$10 - 12	
$\tilde{g}$	$\bar{p}p \to \tilde{g}\tilde{g}$ $\to \tilde{q}\tilde{g}(\tilde{\gamma})$ present data consistent with $15 < m_{\tilde{g}} < 45$ GeV			$\sim$40	$\sim$90
$\tilde{g}$	$e^+e^- \to Z_o \to \tilde{g}\tilde{g}$		40		
$\tilde{\gamma}, \tilde{Z}_{(-)}$	Cosmology $m_{\tilde{\gamma}}$ > 2 GeV				
$\tilde{\gamma}, \tilde{Z}_{(-)}$	$e^+e^- \to \tilde{\gamma}\tilde{\gamma}\gamma$ $e^+e^- \to \tilde{\gamma}\tilde{\gamma}$ Present limit $m_{\tilde{\gamma}}$ > 22.5 GeV or no Goldstino	22.5	45		
$\tilde{\gamma}, \tilde{Z}_{(-)}$	$e^+e^- \to \tilde{\gamma}\tilde{Z}_{(-)}$	$m_{\tilde{\gamma}}+m_{\tilde{Z}_{(-)}}$ < 40	$m_{\tilde{\gamma}}+m_{\tilde{Z}_{(-)}}$ < 100		
$\tilde{W}_1^\pm, \tilde{W}_2^\pm$	$e^+e^- \to \tilde{W}_1+\tilde{W}_2$	$\sim$ 20	$\sim$ 40		
$\tilde{W}_1^\pm, \tilde{W}_2^\pm$	$\bar{p}p \to \tilde{W}_1+\underbrace{\tilde{\gamma}(\tilde{g})(\tilde{q})}_{\tilde{y}_o}$			$m_{\tilde{W}_1}+m_{\tilde{y}_o}$ < 90	$m_{\tilde{W}_1}+m_{\tilde{y}_o}$ < 150

TABLE V

The process suggested in Ref.13 is indicated in Fig.1a, while Fig.1b contributes only to $\tilde{e}^+\tilde{e}^-$ production. Due to the large mass of the scalar quarks, $\tilde{\gamma}$ interacts weakly with matter and hence escapes detection. The typical signature for this process will be two charged,

Signatures for sparticles in pp, $\bar{p}p$ machines with competing background from nonsupersymmetric particles

Signature			Comments
1 jet on each side		$p\bar{p} \to \tilde{q}\tilde{g}$ $p\bar{p} \to \tilde{q}\tilde{\bar{q}}$	
	Background	$p\bar{p} \to q\bar{q}$ $p\bar{p} \to gg$ $p\bar{p} \to Q\bar{Q}$	Q: heavy quark, competing background. Missing energy to be used for isolation of the signal
1 jet + missing p_T		$p\bar{p} \to \tilde{g}\tilde{\gamma}(\tilde{g})$ $p\bar{p} \to \tilde{q}\tilde{\gamma}$ $p\bar{p} \to \tilde{W}\tilde{\gamma}$ $\quad \hookrightarrow h+\tilde{\gamma}$	good isolation criteria for ss signal exist
	Background	$p\bar{p} \to Q\bar{Q}$ $p\bar{p} \to Wg$ $\quad \hookrightarrow$ slow ℓ + ν $p\bar{p} \to W \to L\nu^{*}$ $\quad \hookrightarrow q_1\bar{q}_2\nu$	
missing energy + one lepton		$p\bar{p} \to W \to \tilde{e}\tilde{\nu}$ $\quad \hookrightarrow e\tilde{\gamma}$ $p\bar{p} \to W \to \tilde{W}\tilde{\gamma}$ $\quad \hookrightarrow \ell\nu\tilde{\gamma}$	
	Background	$p\bar{p} \to W \to L^{\pm}\nu$ $\quad \hookrightarrow \ell\nu\bar{\nu}$ $p\bar{p} \to Q\bar{Q}$	→strong background →Has not yet been studied.
4 jet + electron		$p\bar{p} \to \tilde{q}\tilde{\bar{q}}$ $\quad \hookrightarrow e\tilde{\nu}X$ $\quad \hookrightarrow q+\tilde{\gamma}$ $p\bar{p} \xrightarrow{Z} \tilde{W}_1^{+} + \tilde{W}_2^{-}$ $\quad \hookrightarrow h+\tilde{\gamma} \hookrightarrow \nu+\ell+\tilde{\gamma}$	
	Background	$p\bar{p} \to Q\bar{Q}$ $\quad \hookrightarrow$ jet $\quad \hookrightarrow q\ell\nu$	Quantitative study of the background necessary.

Table VI contd.

Table VI contd.

Two leptons	$p\bar{p}$, e^+e^- $\rightarrow Z_o \rightarrow \tilde{\gamma} + \tilde{Z}_{(-)}$	Rates small.
	Background $e^+e^- \rightarrow L^+L^-$	Background separation good. clear signal exists.

TABLE VI

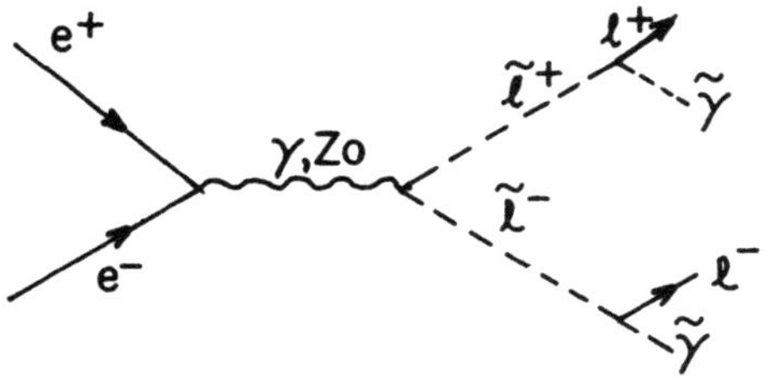

Fig. 1a

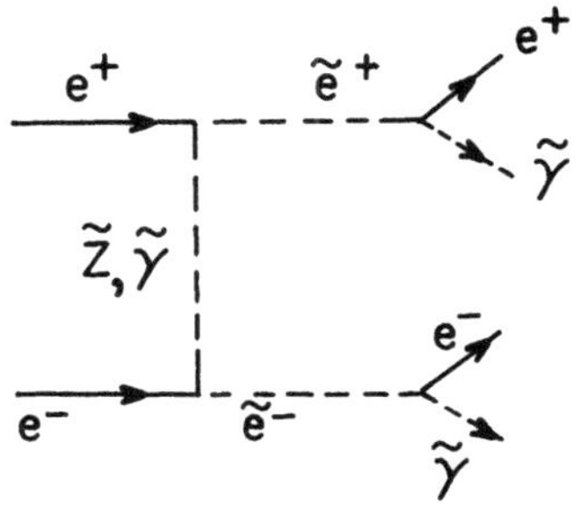

Fig. 1b.

noncoplanar leptons in the final state with half the energy missing and a $\sin^2\theta$ distribution which is characteristic of scalar production.

This signature was looked for in the present e^+e^- experiments [15]. The process has a strong background from the two photon reactions as well as heavy lepton production. The acoplanarity cut reduces the former and the latter would give equal number of $l^+l'^-$. Detailed discussion [16] of the possible cuts to improve the signal shows that the signal will be observable at LEP for $m_{\tilde{\ell}} < 40$ GeV.

(b) The mass range that can be covered by the above method is limited by $m_{\tilde{\ell}} < E_{beam} = \sqrt{s}/2$. Instead if the process [14] $e^+e^- \rightarrow e\tilde{e}\tilde{\gamma}$ shown in Fig.2 is used selectron masses upto $0.75\sqrt{s}$ can be probed [17]. Detailed calculations for the experimental signature [18], the effect of finite mass of the photino [19] exist. Two photon production of selectron in the reaction $e^+e^- \rightarrow e^+e^-\tilde{e}^+\tilde{e}^-$ [18], shown in Fig.3 may be useful if the selectron mass is very large. Search for the selectron using polarised electron beam [20] has also been discussed in the literature. Using this method bounds for observability are close to $m_{\tilde{e}}+m_{\tilde{\gamma}} < \sqrt{s}$. Limits on slepton mass can also be indirectly inferred from the photino pair production in e^+e^- collision. This is discussed

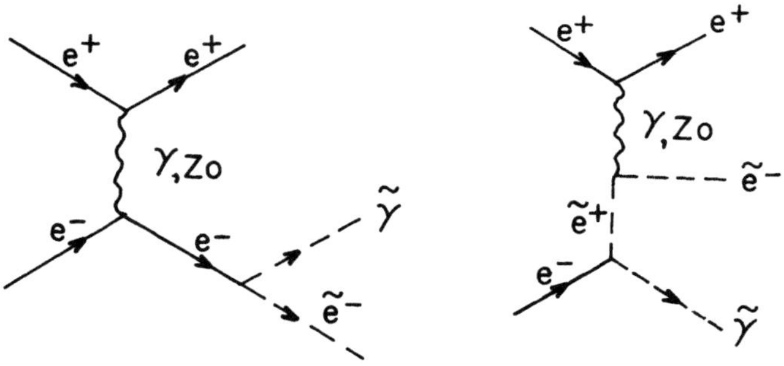

Fig. 2.

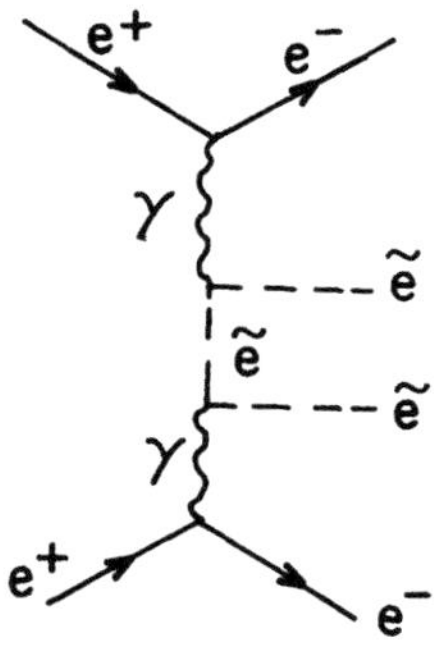

Fig. 3

in detail in the section on the photino.

Slepton signature at the collider

W can decay in a $\tilde{\ell}\tilde{\nu}$ pair just as it decays in a $\ell\nu$ pair, though the phase space effects are different. The suggested process is

$$
\begin{aligned}
p\bar{p} \to\ & WX \\
& \hookrightarrow \tilde{\ell} \qquad \tilde{\nu} \\
& \quad \hookrightarrow \ell\tilde{\gamma} \quad \hookrightarrow \nu\tilde{\gamma}
\end{aligned}
$$

for which we also expect a large p_T lepton and missing energy. However, this lepton populates a different region in the phase space and on the average, the missing energy will be greater for the $\tilde{\ell}$ process. This process, though interesting, has the disadvantage of needing $m_W > m_{\tilde{\ell}} + m_{\tilde{\nu}_\ell}$. It has also been shown [12,21] that the signature will be quite clear at the present collider energies, if the masses of the sparticles are $\sim$ 30 GeV. This will need a reasonable improvement in the statistics.

Current experimental limits for sleptons

All the detectors at PETRA and PEP viz. JADE, CELLO, TASSO, MARKII and MARKJ have been used for the slepton search [15,17]. The absence of the desired signal in all these searches excludes a $\tilde{e}$, $\tilde{\mu}$ and $\tilde{\tau}$ between 2 GeV and 15 GeV [15]. Use of the process (b) above at PEP [17] gives higher limit of $m_{\tilde{e}} > 22.4$ GeV if both the chiral partners are assumed to be degenerate. If one of them is assumed to be infinitely heavy then the limit is 19.5 GeV. In all these searches a zero photino mass has been assumed. For a heavy photino [19] the lepton spectrum will be softer. A photino heavier than 10 GeV and a slepton mass greater than 12 GeV is consistent with all the data [5].

In conclusion the best place to look for charged sleptons are e^+e^- machines. Present day machines exclude a slepton upto 20 GeV (more conservatively upto 12 GeV). One must wait for LEP, SLC and TRISTAN to probe a higher mass range.

The above discussion assumes $m_{\tilde{\gamma}} < m_{\tilde{\ell}}$. In some models it is otherwise and $\tilde{\nu}$ is the lightest neutral sparticle; the cominant decay mode for $\tilde{\ell}$ then, is $\tilde{\ell} \rightarrow \nu\tilde{\nu}\ell$. This alters the signature of the $\tilde{\ell}$ to some extent but the qualitative features remain the same.

1.b. Search for Sneutrinos

In most models we expect $m_{\tilde{\nu}} > 20 - 80$ GeV. There also exist models where it is the lightest sparticle ; this can then have important cosmological implications. The allowed decay modes of $\tilde{\nu}$ and $\tilde{\nu}$ phenomenology have been discussed in detail [12].

Decays of $\tilde{\nu}$:

The dominant decay mode of $\tilde{\nu}$ is $\tilde{\nu} \rightarrow \nu\tilde{\gamma}$ (Fig.4). However, if either $\tilde{\ell}$ or W-ino are light, very interesting signatures are predicted [5,12,21]. In this case a large decay rate through the diagrams in Fig.5 are predicted. This will give rise to large branching ratio for final states with charged particles. If $\tilde{\ell}$, $\tilde{W}$ are heavy, then $\tilde{\nu} \rightarrow \nu\tilde{\gamma}$ will be the dominant decay mode. Therefore absence of the charged decay

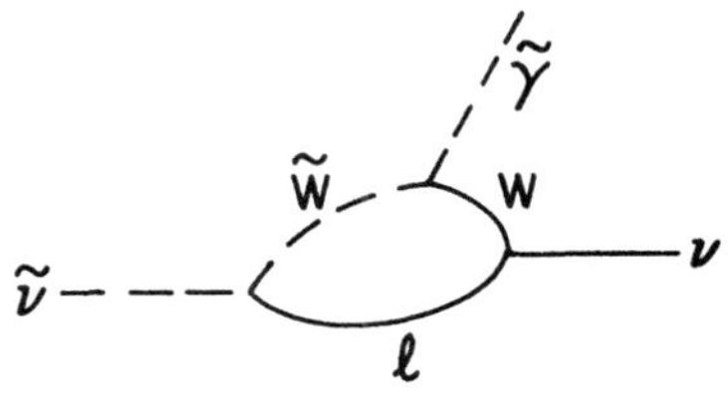

Fig. 4

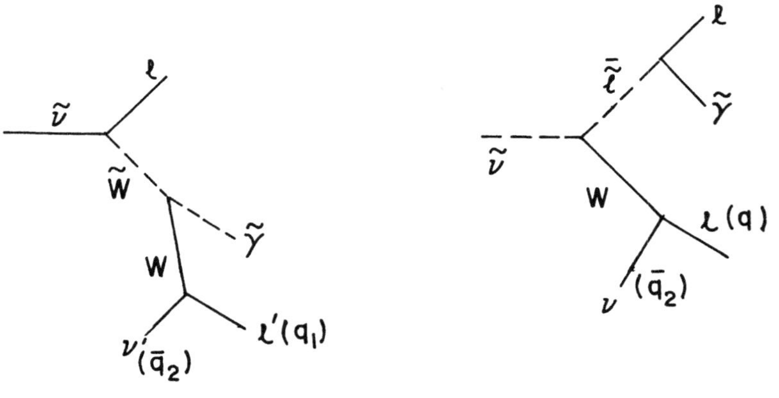

Fig. 5

mode will be useful to constrain the supersymmetric models.

Possible search for $\tilde{\nu}$ at the e^+e^- collider

Pair production of $\tilde{\nu}$ at the Z_o pole, just like the ν pair production has been suggested for the $\tilde{\nu}$ search. The diagrams pertinent to this process are shown in Fig.6. Upto $m_{\tilde{\nu}} = 40$ GeV one expects a cross section $10 \sim 50 \times \sigma(e^+e^- \to \mu^+\mu^-)$ at the Z_o pole at LEP/SLC. If the

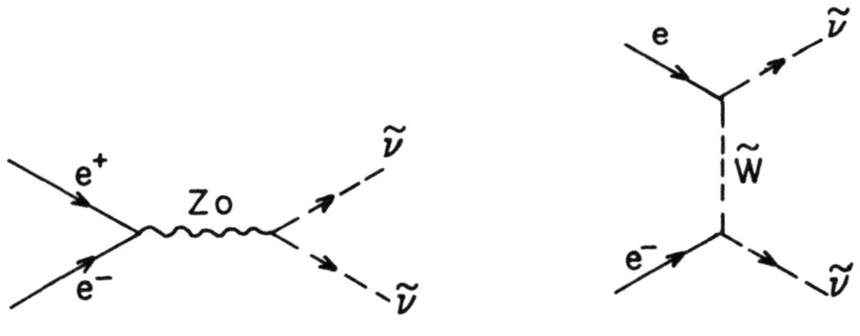

Fig. 6

two body decay mode is the dominant one then this search is just like the ν counting experiment. The process will add to the Z_o width ~ 80 MeV for each generation [6,22]. The three body decay modes will give a signature similar to the charged leptons, except with a higher missing energy. It will also have to be discriminated against the heavy lepton and the two photon background.

Search for $\tilde{\nu}$ at the $\bar{p}p$ collider

The process

$$p\bar{p} \to W \to \tilde{\ell} + \tilde{\nu}, \quad \tilde{\ell} \to \ell\tilde{\gamma}$$

can also be used to look for $\tilde{\nu}$ [12,21]. If $m_{\tilde{\ell}}, m_{\tilde{\nu}} \sim 30$ GeV, the process should be observable at the SPS collider at present energies if we can produce $\sim$ 200 events of $W \to e\nu$ type. If the two body decay mode of $\tilde{\nu}$ is the dominant one then the main signature will be $\sim$ 3/4 missing energy. If $\tilde{\nu} \to \bar{\ell}\ell'\nu\tilde{\gamma}$, then the process will have a very clear signal with 3 leptons and no jets in the final state.

Current Experimental limits for sneutrino

At present the only experimental limit on the mass comes from an analysis of τ decay [23] The process in Fig.7 modifies τ decay. Absence of any such modification of the expected decay characteristics gives $m_\tau < m_{\tilde{\nu}_\tau} + m_{\tilde{\nu}_\ell}$.

SLC and LEP are the machines to look forward to for more (and better) experimental information. For the $\bar{p}p$ collider search background studies are needed.

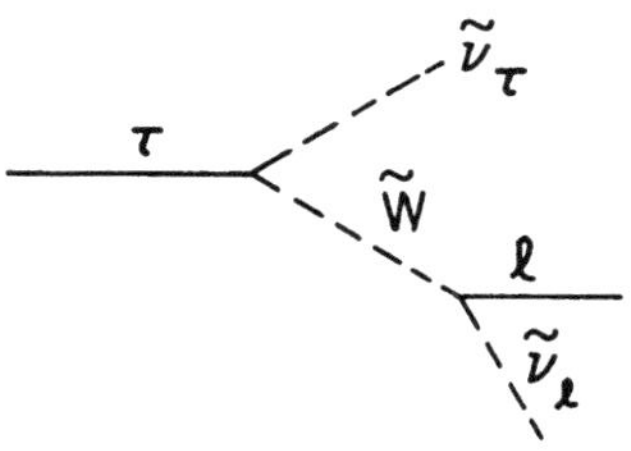

Fig. 7

1.c. Scalar Quarks (Squarks)

Most of the models predict squark masses of the same order as that of the charged sleptons. The possible decay modes are $\tilde{q} \to q\tilde{\gamma}$ if $m_{3/2} < m_{\tilde{g}}$ and $\tilde{q} \to q\tilde{g}$ if $m_{3/2} > m_{\tilde{g}}$. Squarks can be produced at e^+e^- machines as well as at the $\bar{p}p$ collider.

Squark search possible at the e^+e^- machines

For the e^+e^- machines the squark signal is similar to that for sleptons, with the charged sleptons in the final state replaced by jets. The relevant diagram is shown in Fig.8. This signal suffers from a lower rate compared with the slepton signal due to the charge factors and also the detection of jets in the final state makes the analysis

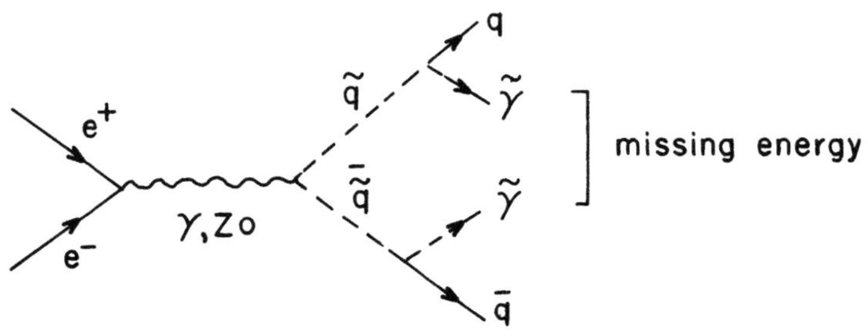

Fig.8

somewhat dependent on the model used for fragmentation. The signal will be two large p_T noncollinear jets with missing energy.

<u>Search at the $\bar{p}p$ collider for squarks</u>

$\bar{p}p$ collider seems to be a more promising place to look for squarks. The production cross section can be high due to the strong couplings, but the process will suffer from the usual problem with the $\bar{p}p$ search viz. the background. Production and decay of a squark will give rise to charged jets with missing p_T.

Two different processes have been suggested for this search. The process $qg \to q\tilde{g}(\tilde{\gamma})$ (Fig.9) will give rise to single jet (lepton) events with missing p_T. It has been shown [24,25] that this process has a detectable cross section at the Tevatron upto $m_{\tilde{q}} < 80$ GeV.

Another suggestion [26] for squark hunting is to look for events with a single jet + missing energy + lepton at the $\bar{p}p$ collider. These can come from

$$gg \to \tilde{q}\ \bar{\tilde{q}}, \quad \tilde{q} \to q\tilde{\gamma}, \quad \bar{\tilde{q}} \to \tilde{\nu}eX$$

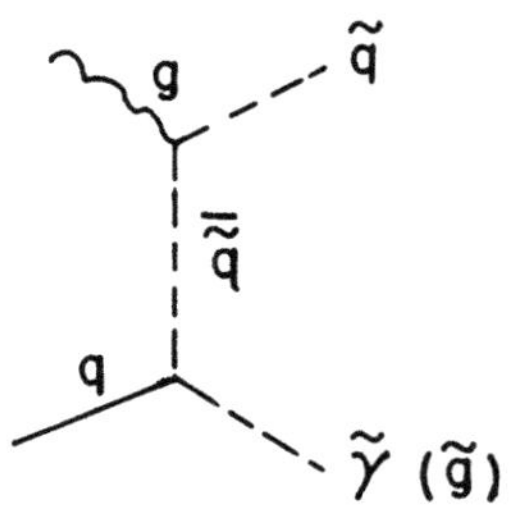

Fig.9

Similar events have been observed at the SPS and have been interpreted in terms of heavy quark production and its decay [27]. The work of Gluck and Reya [26] shows that upto squark mass $\sim$ 35 GeV the rates for the squark production can be comparable to that for heavy quark production. However, to separate the two signals from each other, comparisons of detailed distributions are necessary [28,29].

A signal for squark can also come from its contribution to the photon structure function [30]. This work shows that for $m_{\tilde{q}} \sim 40$ GeV, the effect on the photon structure function will be 5 - 10% at LEP. This is barely measurable at LEP. Absence of such a signal could be used to put limits on $m_{\tilde{q}}$.

In conclusion the best place to see a squark signal is $\bar{p}p$ collider in spite of the background problems. The signal will be events with (1) 2 jets + missing energy, (2) single jet + high p_T lepton + missing energy, (3) lepton + missing energy.

(2) Fermionic Superpartners

The discussion of the fermionic partners of various gauge bosons gets complicated because of the mixing between various isospin eigenstates mentioned earlier. However this complication does not exist for gluinos. So first I will discuss the possibilities and limitations of the gluino search.

(2.a) Gluinos ($\tilde{g}$)

Gluinos have large coupling strength characteristic of the strong interactions which gives rise to large cross sections for processes involving them. In many currently fashionable models, their mass generated through loop effects is expected to be 15 - 100 GeV [5]. The dominant decay mode of gluinos $\tilde{g} \to q\bar{q}\tilde{\gamma}$ can occur via the diagram in Fig.10. The life-time is therefore $\tau_{\tilde{g}} \simeq 10^{-6}.(m_\mu/m_{\tilde{g}})^5(m_{\tilde{q}}/m_W)^4$ - (for light $\tilde{\gamma}$). In case of nondegenerate $\tilde{q}_L$ and $\tilde{q}_R$ an additional decay mode $\tilde{g} \to \tilde{\gamma}g$ [31] shown in Fig.11 also exists. The decay width for this two body decay mode is proportional to $(m^2_{\tilde{q}_L} - m^2_{\tilde{q}_R})$. This is important in the context of the search for characteristic signatures since the existence of the two body decay mode will greatly increase the detectability of $\tilde{g}$.

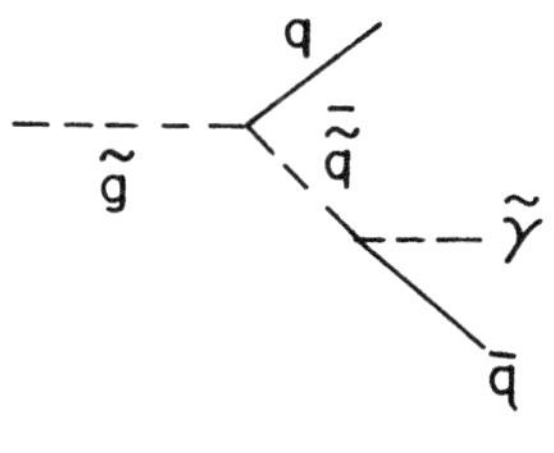

Fig. 10

Hadronic reactions are the best place to produce the gluinos. For $m_{\tilde{g}} > 1$ GeV, the perturbative calculation of the cross section for gluino production are applicable [32]. The colour factor enhances the production of gluino pair by an order of magnitude over the squark pair production of the same mass.

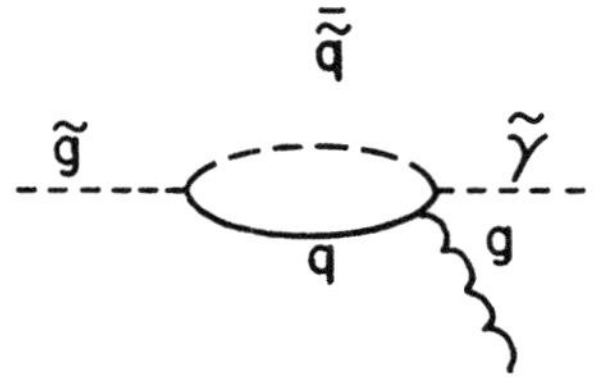

Fig.11

Limits on very light gluinos

Experimental information at present makes very light gluinos ($m_{\tilde{g}} \sim 0$ GeV) unlikely. Such light gluinos will be produced with $\sim$ mb cross section. Since they are coloured they would have to form new hadron states [32,33] which should be longlived. It is quite unlikely that such states would have escaped detection till now [32].

Gluino search at the beam dump

For heavier gluinos the new supersymmetric hadrons will decay into lighter sparticles [32,33,34]. For 1 GeV < $m_{\tilde{g}}$ < 10 GeV the search can be carried out in the beam dump experiments with beam energies~1 Tev. In these experiments the gluinos are pair produced in the dump. The $\tilde{g}$ decays to produce $\tilde{\gamma}$. The photino will travel to interact in the calorimeter. Photino interactions are shown in Fig.12 below. The masses of the various superpartners determine which of the

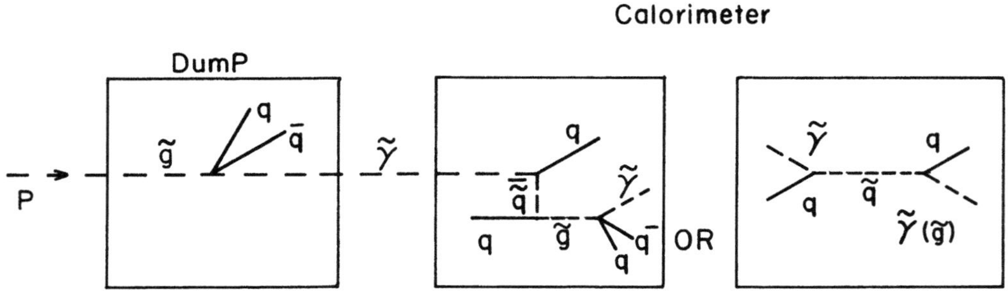

Fig.12

photino interaction will be dominant. Since no charged lepton appears in the final state these events will appear as neutral current (NC) events in the beam dump experiments. For small values of $m_{\tilde{q}}, \sigma^{int}_{\tilde{\gamma}} \sim 5\sigma^{\nu}_{cc}$

[12]. Absence of any such excess of NC events in present day experiments has put a limit on $(\sigma_{\tilde{g}}^{prod.}) \cdot (\Gamma_{\tilde{g}})(\sigma_{\tilde{\gamma}}^{int})$. It is clear from this discussion that such an experiment will not provide very clean answers, but will at least exclude certain regions in the $m_{\tilde{g}} - m_{\tilde{q}}$ plane. Such limits given by CHARM experiment [35] and FNAL experiment [36] are shown in Fig.13.

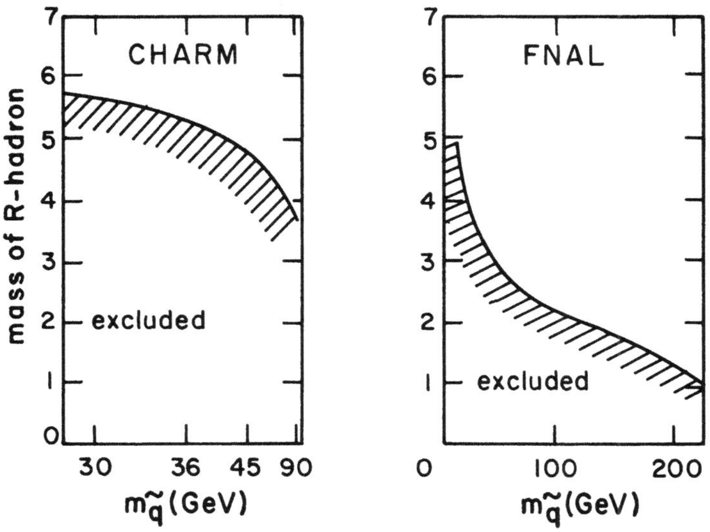

Fig. 13

Gluino search at the collider

For higher gluino masses hadron colliders are the ideal places for the search. The feasibility of such a search for the $\bar{p}p$ machine for $m_{\tilde{g}} < 40$ GeV and at the Tevatron for $m_{\tilde{g}} < 90$ GeV has been demonstrated [37,38]. Recent work [39] discusses a clean criterion for isolation of the gluino signal from the known QCD background.

The $\tilde{g}\tilde{g}$ pair production which occurs via diagrams shown in Fig.14 has a cross section ~ 1 nb for $m_{\tilde{g}} \sim 40$ GeV. At current CERN energy this corresponds to ~ 100 events, although the real problem is a good signature.

ture. Since a gluino is expected to decay in $q\bar{q}\tilde{\gamma}$, we expect

$$p\bar{p} \to \tilde{g} + \tilde{g} + X$$
$$\tilde{g} \to q\bar{q}\tilde{\gamma}, \quad \tilde{g} \to q\bar{q}\tilde{\gamma}$$

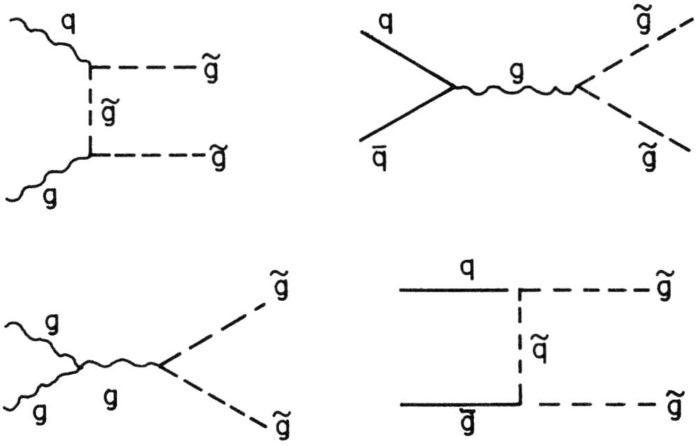

Fig.14

Hence the characteristic signature is large missing p_T, jets and no leptons. The dominant background is heavy flavour production and decay. The not-back-to-back topology of the high p_T jets and missing energy has been used to isolate the gluino signal. Two parameters $x_E = -\vec{p}_T\cdot\vec{p}_T'/|\vec{p}_T|^2$, and $p^{out} = \sqrt{|\vec{p}_T'|^2 - x_E^2|\vec{p}_T|^2}$ can be used. Here $\vec{p}_T$, $\vec{p}_T'$ are transverse momenta in opposite hemispheres. Ordinary QCD background has $x_E \sim 1$ and also has low p^{out}. A cut of $x_E < 0.5$ and $p^{out} > 5$ GeV reduces the background and a search becomes feasible at the $\bar{p}p$ collider for $m_{\tilde{g}} < 40 - 50$ GeV.

But these cuts also reduce the signal considerably; hence one must single out events with large p_T and demand that the large p_T track be isolated. This procedure reduces the background without affecting the signal [39]

$\tilde{\gamma}\tilde{g}$ production (through the reaction $q\bar{q} \xrightarrow{\tilde{q}\text{ exchange}} \tilde{g}\tilde{\gamma}$) in a $p\bar{p}$ collision can also be used for the $\tilde{g}$ search. The cross section for this process is lower by a factor α_{em}/α_s x colour factors as compared to the pair production discussed above. However the signature will be clean due to a larger p_T imbalance. The main background for this process will be

$$\bar{p}p \rightarrow gW$$
$$\quad\quad\quad \hookrightarrow e\nu$$

and where the lepton is slow. To isolate the signal one should look for large p_T jets, large missing energy and absence of leptons in the final state.

Recent events with a photon (jet) + missing energy reported by UA1 [40] can be interpreted as gluino events. They are consistent with gluino masses between 15 - 45 GeV. The optimal value of gluino mass is ∿ 35 GeV [41], although such a limit is circumspect due to the low statistics and very little predictability of the supersymmetric theories. Therefore one has to wait for a high luminosity run of the $\bar{p}p$ collider or for the Tevatron results. Gluino masses upto 80 GeV can be probed at the Tevatron.

Gluino search in e^+e^- machines

Suggestions for gluino search in e^+e^- machines have been made as well. However, in general, these are not very promising due to low branching ratios. In the continuum e^+e^- machines are not very useful, but at the Z_o pole the situation might be different. The signal due to $\tilde{g}\tilde{g}$ pair produced with $q\bar{q}$ (Fig.15) is not very clean. To the same order in strong coupling it is possible to get $\tilde{g}\tilde{g}$ pair production via Z_o decay [41] or in one loop process due to difference in masses of the two chiral squarks. The diagram is shown in Fig.16. The production cross section depends on $m_{\tilde{q}}$. For $m_{\tilde{q}} < 50$ GeV, the signal is just about detectable (O(10) events) at LEP. Therefore for $m_{\tilde{g}} < m_{Z_o}/2$, the above process can give rise to large p_T jets with missing energy. However this process will have a large background due to two-photon processes.

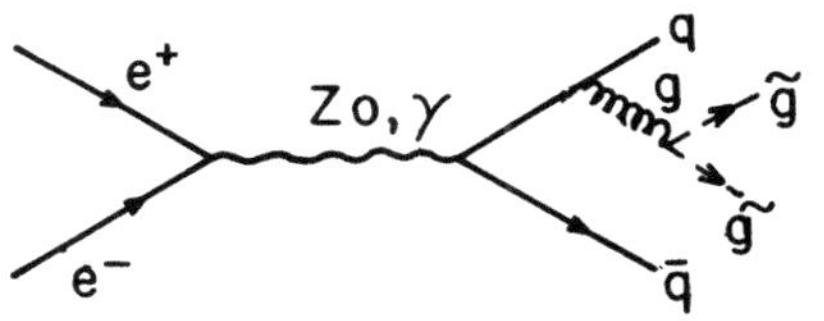

Fig. 15

For smaller gluino masses one can search in e^+e^- machines, in quarkonium decay:

$Q\bar{Q} \to gg\tilde{g}\tilde{g}$ [42], $Q\bar{Q} \to g\tilde{g}\tilde{g}$ [43], $QQ \to \tilde{g}\tilde{g}$ [44]. Either the branching ratios are small, $O(10^{-2}) - O(10^{-4})$ or the process is useful only in a limited mass range (viz.$m_{\tilde{g}}<4.5$ GeV) which is almost ruled out by other experiments [35,36]. The detection of gluino balls [45] is also not very promising. This way of looking for gluinos may be useful in the case of a heavy toponium [46].

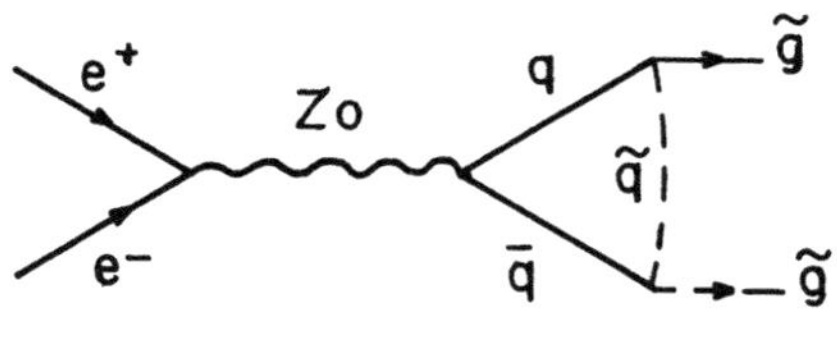

Fig.16

In conclusion the best place to look for gluinos is $\bar{p}p$ colliders. Production cross sections are large enough to give about $\sim$ 100 events for $m_{\tilde{g}}$ < 40 GeV at present collider energies. Specific selection criteria of missing energy, isolated large p_T track can be employed to separate the signal from background. The presently observed monojet events are consistent with the gluino interpretation; however it is hard to put a lower limit on $m_{\tilde{g}}$ using the present data.

(2.b) Gauginos

The search for the colour singlet, spin 1/2, supersymmetric partners of the gauge bosons $[W,Z,\gamma]$ and the Higgs bosons $[H_1^+,H_2^-,H_1^0,H_2^0]$ is perhaps the most promising. In a large class of supersymmetric theories there exists [48] at least one gaugino lighter than the gauge boson. A lot of work has been therefore done regarding the search of the charged and neutral gauginos [49,50,51,52,53,54,55]. Let us first consider the case of the neutral gauginos.

(2.bl) Neutral Gauginos

Supersymmetric partners $\tilde{Z},\tilde{\gamma},\tilde{H}_1^+,\tilde{H}_2^-$ will mix to give rise to mass eigenstates. The lightest of these will be called $\tilde{\gamma}$. In almost all models this is a pure photino. The heaviest eigenstates which has mostly gauge couplings will be called $\tilde{Z}$. The two intermediate mass states will be called $\tilde{h}^0,\tilde{Z}_{(-)}$.

Photino $(\tilde{\gamma})$

Cosmological limit on $m_{\tilde{\gamma}}$

Photino masses are expected to be small in many models [10]; $\tilde{\gamma}$ is expected to be the lightest sparticle in almost all models. Hence it will be stable. There are cosmological limits on the density of such neutral, stable particles. These give mass limits similar to the limits for the mass of a heavy stable lepton [56]. However the Majorana character of the photinos changes the annihilation cross section, and hence the mass limits are different. According to Goldberg's analysis [57], $m_{\tilde{\gamma}}$ as small as 2 GeV will be allowed for $m_{\tilde{q}} < 40$-50 GeV (Fig.17). The area to the left of the crosshatched line has been ruled by present experimental limits on the sfermion masses. A more detailed analysis of the limits on the masses of the superparticles and resulting constraints on the supersymmetric theories, coming from cosmology is given in Ref.[58].

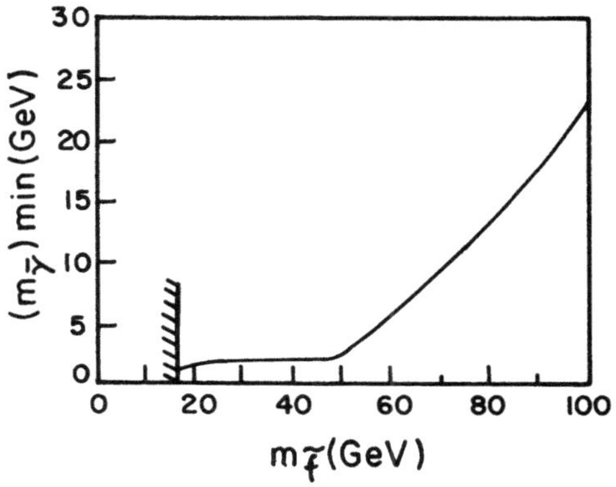

Fig. 17

Explicit search for photinos in e^+e^- experiments through reaction $e^+e^- \to \tilde{\gamma}\tilde{\gamma}$ or $e^+e^- \to \tilde{\gamma}\tilde{\gamma}\gamma$ [53] have been suggested. The relevant diagram for $e^+e^- \to \tilde{\gamma}\tilde{\gamma}$ is shown in Fig.18. Here $\tilde{G}$ is the goldstino. Absence of such a signal in an analysis carried out by CELLO [59] implies that $m_{\tilde{e}} > 20$ GeV, or $m_{\tilde{\gamma}} > 22.5$ GeV or a light Goldstino does not exist.

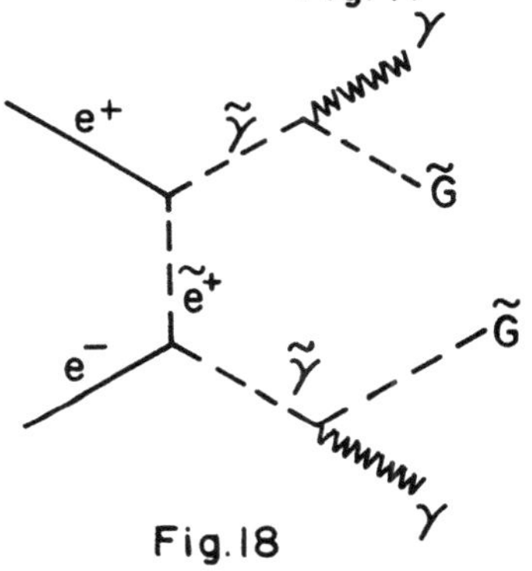

Fig.18

A more promising suggestion is to look for the recoil photon in $e^+e^- \to \tilde{\gamma}\tilde{\gamma}\gamma$ which occurs via the diagram shown in Fig.19. This process has a better signal (recoil γ) and cross sections which are large enough at LEP if $m_{\tilde{\gamma}} < 40$ GeV.

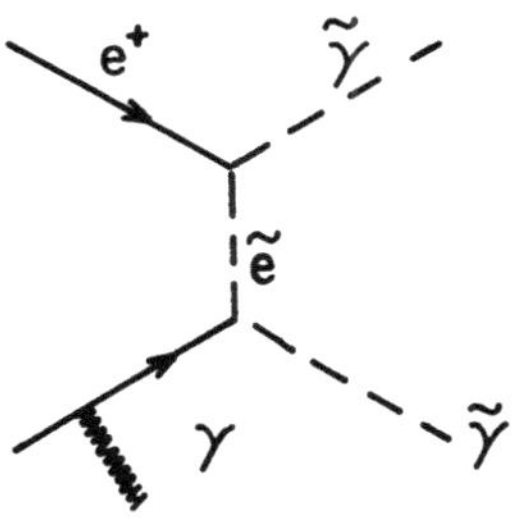

Fig.19

These discussions assume a rather small mass for the photino which is allowed by cosmological constraints, though for a heavy photino the above mentioned test would be difficult.

$\tilde{Z}$, $\tilde{Z}_{(-)}$, $\tilde{h}^o$

Among the remaining neutral gauginos $\tilde{Z}_{(-)}$ has the best chance of being detected at present in Z_o decay in e^+e^- or W decay in $p\bar{p}$ collisions. The analysis in Ref.48 shows that the existence of a $\tilde{Z}_{(-)}$ lighter than the Z_o is ensured under very general conditions. In almost all models $m_{\tilde{\gamma}} + m_{\tilde{Z}_{(-)}} < 100$ GeV [5]; hence LEP provides an ideal place to look for $\tilde{Z}_{(-)}$ in Z_o decay. In many models $m_{\tilde{\gamma}} + m_{\tilde{Z}_{(-)}} < 35$ GeV which means that the search is possible even at PETRA. The process suggested [51,54] for search is (Fig.20) $e^+e^- \to \tilde{\gamma}\tilde{Z}_{(-)}$.

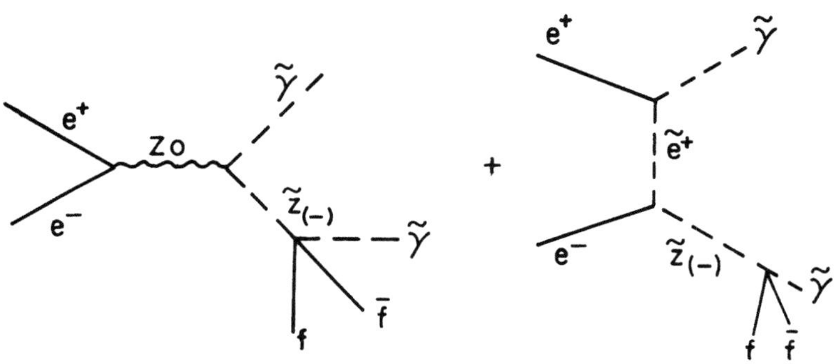

Fig. 20

The signal will be a $f\bar{f}$ pair with missing mass. Dominant background will be two photon process and heavy lepton pair production and decay. For $m_{\tilde{e}} < m_W$, at S = 1000 GeV (at PETRA) the cross section is 0.1 pb = 10^{-37} cm^2. This is about 0.1 units of R. A detailed study of cuts required to isolate the signal from the background has been done [55]. This shows that the signal can be rather clean and it could possibly be seen at LEP and SLC. $\tilde{\gamma}\tilde{Z}_{(-)}$ may be the first threshold to open up in view of the light mass expected for $\tilde{\gamma}$ in most models.

The search for neutral gauginos in $\bar{p}p$ collisions is more naturally discussed along with the case of the charged gauginos.

(2.b2) Charged Gauginos

In the minimal picture $W^{\pm}$ have four charged fermionic partners,

$\tilde{W}^{\pm}_{1,2}$, called W-inos. Since a fairly general class of theories can be shown [48] to have a Wino lighter in mass than W, the search for them is promising. As we shall show below the production and decay of a Wino can give rise to very characteristic signatures - the so-called Zen events [49,50,51]. The best place to look for these is in Z/W decay. It will be possible to produce W only in the pp or $\bar{p}p$ machines in the forseeable future. Hence I shall discuss the Wino search only at the collider machines in W/Z decay. The branching ratios for the decay modes of W into sparticles are model dependent, but various calculations show that for Wino masses upto 40 GeV there is a good chance to see the Wino signal at the collider at present. The possible processes are

(1) Wino produced in W decay: $W \to \tilde{W} + \tilde{g}(\tilde{q})$,

(2) Wino produced in W decay along with a neutral gaugino: $W \to \tilde{W}_1 + \tilde{\gamma}$,

(3) Pair production of two Winos in Z decay: $Z \to \tilde{Z}_{(\bar{\ell})} + \tilde{\gamma}$ (which has been discussed earlier). If $m_Z > m_{\tilde{W}_1} + m_{\tilde{W}_2}$ then $Z \to \tilde{W}_1 + \tilde{W}_2$ is also a possible process.

Allowed decay modes and the branching ratios for W decay do depend on models. But in general the dominant decay modes are $\tilde{W} \to h + \tilde{\gamma}$ and $\tilde{W} \to \ell\nu_\ell\tilde{\gamma}$. With these decay modes process (1) will give rise to events with missing energy + large p_T lepton + jets via

$$p\bar{p} \to W \to \underset{\hookrightarrow q\bar{q}\tilde{\gamma}}{\tilde{g}} + \underset{\hookrightarrow \ell\nu_\ell\tilde{\gamma}}{\tilde{W}_1} \qquad \text{or} \qquad p\bar{p} \to \underset{\hookrightarrow \ell\nu_\ell\tilde{\gamma}}{\tilde{W}_1} + \underset{\hookrightarrow q\tilde{\gamma}}{\tilde{q}}$$

the diagrams shown in Fig.21.

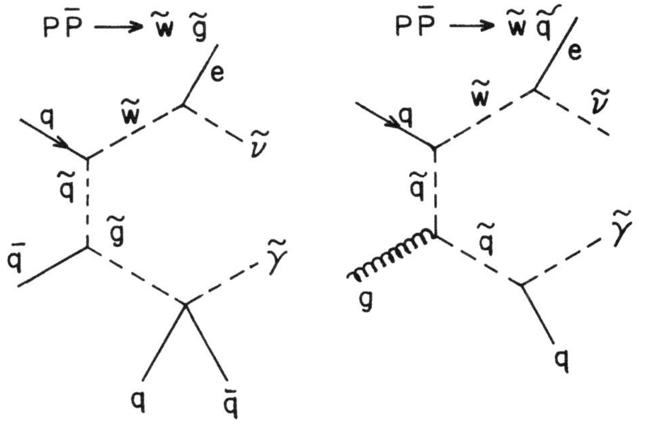

Fig. 21

For process (2) the suggested signal is

$$p\bar{p} \to W \to \tilde{W}_1 + \tilde{\gamma} \qquad \text{(a)}$$
$$\hookrightarrow h+\tilde{\gamma}$$

or

$$p\bar{p} \to W \to \tilde{W}_1 + \tilde{\gamma} \qquad \text{(b)}$$
$$\hookrightarrow \ell\nu_\ell\tilde{\gamma}$$

These two will give rise to events with single jet (lepton) + missing p_T respectively. These are the Zen events. In the case of process (b) above, the single lepton spectrum has been computed [52]. It is difficult to separate this from a heavy lepton signal. Background to this process due to heavy flavour production should be computed. Such a computation is in progress [59].

The process mentioned in (3) above can give rise to single jet + lepton events via

$$p\bar{p} \to Z \to \tilde{W}_1^+ + \tilde{W}_2^-$$
$$\hookrightarrow h+\tilde{\gamma} \quad \hookrightarrow \ell\tilde{\nu}_\ell$$

These events are similar to the events already seen in W search [1] and which have been given a possible interpretation in terms of the top production and its decay [27]. The signal can also be seen at LEP in e^+e^- reactions. Calculations of detailed backgrounds and cuts have been done only for a few of the above processes.

The single jet + missing energy + lepton events of UA2 [61] are of the type expected from process (1) above. It has been shown that the 10 observed events are consistent with a choice of $m_{\tilde{g}} \sim 70 - 100$ GeV, $m_{\tilde{q}} \sim 40$ GeV and $m_{\tilde{\nu}} < m_{\tilde{W}}$. The solution can by no means be compelling at the present level of experiments, but is tantalising enough to encourage further collider experiments with better statistics and higher energies.

In Conclusion

(1) It is possible to explain the present experimental results from UA1 [40] and UA2 [61], in terms of gluino and Wino production. Even though these are not the only explanations, these events make future experiments interesting from the point of view of the supersymmetry phenomenology.

(2) In general the e^+e^- experiments at LEP are capable of providing

signatures for sparticles of masses less than 50 GeV, but good signatures are possible for selectron masses upto 80 GeV.

(3) For the $\tilde{q}$, $\tilde{\ell}$ the nonobservation of a signal in e^+e^- experiments can be easily converted into lower mass limits.

(4) $p\bar{p}$ machines offer higher mass ranges that can be probed upto 200 GeV, and also provide very promising signatures for gauginos. However the signatures are messy: Some of the possible backgrounds have been computed but much work still remains to be done. Conversion of the nonobservation into lower limits on masses is not very straightforward as a detailed consideration of experimental cuts is necessary.

(5) Upcoming machines such as TRISTAN, TEVATRON and LEP should be able to see some evidence for sparticles if any of them have masses $\sim$ 100-200 GeV.

References

1. G. Arnison et al, Phys. Lett. 122B, 103 (1982); M. Banner et al, Phys. Lett. 123B, 476 (1982); G. Arnison et al, Phys. Lett. 126B, 398 (1983); P. Bagnaia, Phys. Lett. 129B, 130 (1983).
2. S. Weinberg, Phys. Rev. Lett., 19, 1264 (1967); S.L. Glashow, Nuclear Physics 22, 579 (1961); A. Salam, "Elementary Particle Theory", ed. N. Svartholm, p.367 (Almquist an Wikrell, Stockholm, 1968).
3. For recent reviews see, for example, P. Nath et al, Lectures at the "Summer Workshop on Particle Physics" Trieste (1983), North-eastern Univ. Nub. # 2613; J. Ellis, Invited talk at the International Symposium on Lepton and Photon Interactions, Cornell, 1983 (Ref.TH 3718 - CERN); C.A. Savoy, Invited Talk at the XVIII Recontres de Moriond, La Plagne, 1983 (Saclay SPhT/83/73); For an earlier review, see P. Fayet and S. Ferrara, Phys. Rep. 32C, 249 (1977).
4. For a review of models, motivation see, P. Fayet, 'Proceedings of the 21st International High Energy Conference; Paris 1982, ed. Petau and M. Porneuf.
5. G.L. Kane: Invited talk at the Fourth Workship on the Grand Unification, Philadelphia, PA, April 1983; See also, Invited talk at the Conference on the Physics of XXI Century, Tuscon, AZ, December 1983.
6. Proceedings of the 1982 DPF Study on future facilities, Snowmass, ed. R. Donaldson, R. Gustafson and F. Paige.
7. C.H. Llewelyn Smith, OXF. Univ. Ref.44/82; A. Savoy - Navarro, CERN - EP/83-132; Both to appear in Phys. Reports 'Supersymmetry confronts Experiments'.
8. For a review see, for example, P. Fayet, Talk at the CERN supersymmetry Workshop, To appear in Physics Reports; J. Ellis and G.R. Ross, Phys. Lett. 117B, 397 (1982).
9. P. Fayet, Phys. Lett. 86B, 272 (1979).
10. A.H. Chamseddine et al., Phys. Lett. 119B, 343 (1983); R. Barbieri et al, Phys. Lett. 119B, 343 (1983); E. Cremmer et al, Phys. Lett. 123B, 214 (1983); L. Alvarez-Gaume et al, Nucl. Phys. B221, 495 (1983); J. Ellis et al, Phys. Lett. 121B, 123 (1983).
11. S. Wolfram, Phys. Lett. 82B, 65 (1979); P.F. Smith and J.R.J. Bennet, Nucl. Phys. B149, 525 (1979).

12. J.S. Hagelin et al, Los Almos preprint LA - UR - 83 - 3711, (Submitted to Nuclear Physics B); R.M. Barnett et al, Phys. Lett. 126B, 176 (1983); Phys. Rev. Lett. 51, 176 (1983).
13. P. Fayet and G. Farrar, Phys. Lett. 89B, 191 (1982).
14. M.K. Gaillard et al, Phys. Lett. 116B, 279 (1982).
15. D.P. Barber et al, Phys. Rev. Lett. 45, 1904 (1982); H.J. Behrend et al, Phys. Lett. 114B, 287 (1982); W. Bartel et al, Phys. Lett. 114B, 211 (1982); R. Brandelik et al, Phys. Lett. 117B, 365 (1982); C.A. Blocker et al, Phys. Rev. Lett. 49, 517 (1982); D. Ritson XXI International Conference on High Energy Physics, Paris, July 1982, MAC.
16. N.Cabbibo et al, Phys. Lett. 132B, 195 (1983).
17. L. Gladney et al, Phys. Rev. Lett. 51, 2253 (1983).
18. M. Kuroda et al, Phys. Lett. 127B, 467 (1983).
19. M. Gluck and E. Reya, Phys. Lett. 130B, 423 (1983).
20. M. Gluck, Phys. Lett. 129B, 255 (1983).
21. R. Barbieri et al, Phys. Lett. 127B, 429 (1983).
22. V. Barger et al, Phys. Rev. D28, 2912 (1983).
23. G.L. Kane and W.B. Rolnick, UM-TH-83-14.
24. S.H. Aronson et al, in Ref.6.
25. N. Cabbibo et al, Phys. Lett. 132B, 199 (1983).
26. M. Gluck and E. Reya, Phys. Rev. Lett. 51, 867 (1983).
27. R.M. Godbole et al, Phys. Rev. Lett. 50, 1539 (1983).
28. R.M. Godbole, in preparation.
29. G. Altarelli and R. Ruckl, Ref. CERN TH 3814 (1984).
30. E. Reya, Phys. Lett. 124B, 424 (1983).
31. H. Haber and G.L. Kane, Nuclear Phys. B232, 333 (1984), (Preprint UM-TH-83-18).
32. G.R. Farrar and P. Fayet, Phys. Lett. 76B, 575 (1978); 79B, 442 (1976); G.L. Kane and J.P. Leville, Phys. Lett. 72B, 227 (1982).
33. T. Goldman, Phys. Lett. 78B, 110 (1983); M. Chanowitz and S. Sharpe, Phys. Lett. 126B, 225 (1983).
34. E. Franco, Phys. Lett. 124B, 271 (1983); P.R. Harrison and C.H. Llewellyn Smith, Nuclear Phys. B213, 1234 (1983).
35. F. Bergsma et al, Phys. Lett. 121B, 429 (1983).
36. R.C. Ball et al, UMHE 83-13/UW EX 83-234 (1983).
37. L. Littenberg and Hinchliffe, Berkeley preprint, LBL-15022 (1982); C. Fabjan, p.85 in the Report CERN TH 3311/EP 82-63 (1982).
38. J. Ellis and H. Kowalski REF. TH.-3843-CERN (1984).
39. E. Reya and D.P. Roy, preprint DO-TH 84/03.
40. G. Arnison et al, Phys. Lett. 139B, 125 (1983).
41. E. Reya and D.P. Roy, Dortmund preprint DO-TH 84/11.
42. B.A. Campbell et al, Phys. Lett. 126B, 376 (1983); P. Nelson and P. Osland, Phys. Lett. 115B, 407 (1982).
43. B.A. Campbell et al, Nuclear Phys. B198, 1 (1982); B.A. Campbell et al, Phys. Lett. 131B, 213 (1983); G. Eliam and M. Gluck, Technion preprint Phys-83-7 (1983).
44. W.Y. Keung, Phys. Rev. D28, 1129 (1983); M. Gluck, Phys. Lett. 129B, 257 (1983).
45. G. Eliam and A. Khare, Phys. Lett. 134B, 169 (1983).
46. J. Ellis and S. Rudaz, Phys. Lett. 128B, 248 (1983).
47. G. Eliam and A. Khare, Brookhaven preprint.
48. W. Weinberg, Phys. Rev. Lett. 50, 387 (1983).
49. P. Fayet, Phys. Lett. 133B, 363 (1983); J. Ellis et al, Phys. Lett. 127B, 233 (1983); ibid 132B, 436 (1983); Grienstein et al, Phys. Lett. 130B, 285 (1983); V. Barger et al, Phys. Lett. 131B, 367 (1983).
50. A.H. Chamseddine et al, Phys. Lett. 129B, 445 (1983); Harvard preprint HUTP-83/A077.
51. G.L. Kane and F.M. Frere, Nuclear Phys. B223, 331 (1983).

52. D. Dicus et al, Phys. Rev. Lett. 51, 1030 (1983); Phys. Lett. 129B, 457 (1983); preprint DOE ER-03992-521.
53. P. Fayet, Phys. Lett. 117B, 460 (1982); J. Ellis and J. Hagelin, Phys. Lett. 122B, 303 (1983); T. Kobayashi and M. Kuroda, Phys. Lett. 134B, 271 (1984); Phys. Lett. 139B, 208 (1984).
54. E. Reya, Phys. Lett. 133B, 245 (1983).
55. D. Dicus et al, Phys. Rev. D29, 1317 (1984); Austin preprint April 1984.
56. B. Lee and W. Weinberg, Phys. Rev. Lett. 39, 165 (1977).
57. H. Goldberg, Phys. Rev. Lett. 50, 1419 (1983).
58. J. Ellis et al, SLAC-PUB-3171.
59. H.J. Behrend et al, Phys. Lett. 123B, 127 (1983).
60. R.M. Godbole, in preparation.
61. M. Banner et al, Phys. Lett. 139B, 102 (1984).
62. G.L. Kane, Talk at the Topical Conference on Flavour Mixing in Weak Interactions, Erice, April (1984).

METHODS IN NON-PERTURBATIVE FIELD THEORY

Tohru Eguchi

Department of Physics
University of Tokyo
Tokyo, Japan 113

It has been known for some time that quantum field theories greatly simplify in the limit of large internal symmetry group O(N), SU(N) etc.. In some favorable situations in particular a field theory becomes exactly soluble in the $N=\infty$ limit and one can obtain a great deal of insight into the non-perturbative structure of the theory.

Perhaps the first known example of soluble system in the $N=\infty$ limit is the spherical model of Berlin and Kac introduced in 1952[1] (the model was later shown to be equivalent to O(N) σ-model). In this example, problem of the evaluation of path integral is reduced to that of extremizing a potential of a single variable in the $N=\infty$ limit and is thus trivially solved. Today there exists a class of 2-dimensional field theories which are solved in exactly the same manner as in O(N) σ-model and they represent a kind of showcase for 1/N expansion. These are, Gross-Neveu, CP^{N-1} etc. models which are conformally invariant at the classical level (no dimensional parameters in the Lagrangian) and asymptotic free at the quantum mechanical level. Their $N=\infty$ solutions represent a gap in the mass spectrum of the theory and hence the phenomenon of dimensional transmutation. Presence of the mass gap is the crucial aspect of these theories which is believed to persist also at finite $N(\geq 3)$ and hence the $N=\infty$ limit represents correctly the essential feature of the models also at finite N.

For the sake of illustration let us discuss (discrete)

Gross-Neveu model[2] at $N=\infty$ limit which exhibits interesting non-perturbative phenomena; dimensional transmutation and the dynamical breakdown of discrete γ_5 invariance. This is a theory of self-interacting spinors in 2-dimensions and its Lagrangian is given by

$$\mathcal{L} = \sum_i \bar{\psi}_i(x) i\gamma\cdot\partial\psi_i(x) - \frac{g^2}{2}\left(\sum_i \bar{\psi}_i(x)\psi_i(x)\right)^2 \quad . \tag{1}$$

Here $\psi_i(x)$ $(i=1,2,\dots,N)$ is an N-plet of spinor field and the partition function is given by

$$Z = \int \prod_i d\bar{\psi}_i(x)\, d\psi_i(x)\, e^{i\int \mathcal{L}(x) d^2x} \quad . \tag{2}$$

We introduce an auxiliary field $\sigma(x)$ and rewrite (1), (2) as

$$\mathcal{L}' = \sum_i \bar{\psi}_i(x) i\gamma\cdot\partial\psi_i(x) - \sigma(x)\sum_i \bar{\psi}_i(x)\psi_i(x) + \frac{\sigma^2(x)}{2g^2} \tag{3}$$

$$Z = \int d\sigma(x) \prod_i d\bar{\psi}_i(x) d\psi_i(x)\; e^{i\int \mathcal{L}'(x) d^2x} \tag{4}$$

By performing the integration over spinors we obtain an effective action of the theory in terms of σ field,

$$S_{eff} = -iN\,\mathrm{tr}\ln(i\gamma\cdot\partial - \sigma(x)) + \frac{1}{2g^2}\int \sigma^2(x) d^2x \quad . \tag{5}$$

Ground state of the theory is given by the minimum of S_{eff},

$$N\int \frac{d^2p}{(2\pi)^2}\,\mathrm{tr}\,\frac{1}{\not{p}-\sigma} = -\frac{\sigma}{g^2} \quad . \tag{6}$$

Here we assumed the translational invariance of the vacuum $\sigma(x)=\sigma$. After performing Wick rotation, the non-trivial solution ($\sigma \neq 0$) of (6) is given by,

$$2\int \frac{d^2p_E}{(2\pi)^2} \frac{1}{p_E{}^2+\sigma^2} = \frac{1}{\lambda} \quad , \tag{7}$$

where p_E is the Euclidean momentum and $\lambda=g^2N$. Integration in (7) has an ultra-violet (UV) divergence which is eliminated by the renormalization of coupling constant λ as,

$$\int_{-\Lambda}^{\Lambda} \frac{d^2p_E}{(2\pi)^2} \frac{1}{p_E{}^2+\sigma^2} \approx \frac{1}{4\pi} \ln \frac{\Lambda^2}{\sigma^2} \quad ,$$

$$\frac{1}{\lambda(\Lambda)} \equiv \frac{1}{\lambda(\mu)} + \frac{1}{2\pi} \ln \frac{\Lambda^2}{\mu^2} = \frac{1}{2\pi} \ln \frac{\Lambda^2}{\sigma^2} \quad . \tag{8}$$

Λ is the UV cut-off and an arbitrary mass scale μ is introduced. Thus the renormalized version of (7) is given by

$$\frac{1}{\lambda(\mu)} = \frac{1}{2\pi} \ln \frac{\mu^2}{\sigma^2} \quad . \tag{9}$$

Dependence of $\lambda(\mu)$ on μ is controled by the β function

$$\beta(\lambda(\mu)) = \mu \frac{d\lambda(\mu)}{d\mu} = -\frac{1}{\pi}(\lambda(\mu))^2 \quad . \tag{10}$$

Thus in fact $\lambda(\mu)$ decreases as μ is increased and the theory is asymptotically free. When we solve (9) for σ, we obtain,

$$\sigma = \mu \, e^{-\frac{\pi}{\lambda(\mu)}} \quad . \tag{11}$$

σ, the mass of the spinors ψ_i, is a physical quantity and is independent of the arbitrary parameter μ. In fact μ dependence cancels in (11) when $\lambda(\mu)$ obeys (10).

Effective potential $V_{eff}(\sigma) = S_{eff}(\sigma(x)=\sigma)/\text{volume}$ is plotted in Fig.1. We see the double-well structure and in fact the trivial vacuum $\sigma=0$ is at the top of the potential and is unstable. Having $\sigma\neq 0$ breaks spontaneously the discrete γ_5 invariance of (1),

$$\psi_i \to \gamma_5\psi_i, \quad \bar\psi_i \to -\bar\psi_i\gamma_5 \quad . \tag{12}$$

$$\sigma \to -\sigma.$$

In this respect there exists a close analogy to a system of an Ising ferromagnet. If we introduce a small explicit mass term to (1)

$$\mathcal{L} \to \sum \bar\psi_i i\gamma\cdot\partial\psi_i - \frac{g^2}{2}(\sum_i \bar\psi_i\psi_i)^2 - m\sum_i \bar\psi_i\psi_i \tag{12}$$

the gap equation (7) is modified as,

$$2\int \frac{d^2p_E}{(2\pi)^2}\,\frac{M}{p^2+M^2} = \frac{1}{\lambda}(M-m) \quad ,$$

$$M = \sigma + m \quad . \tag{13}$$

After a suitable renormalization (bare mass is renormalized like a coupling constant $m(\mu) = \frac{\lambda(\mu)}{\lambda(\Lambda)}m(\Lambda)$) we obtain,

$$M = m(\mu) + \frac{\lambda(\mu)}{2\pi} M \ln \frac{\mu^2}{M^2} \quad . \tag{14}$$

M and m correspond to the total magnetization and the applied external magnetic field of the Ising system, respectively. γ_5 invariance is the analogue of the up-down Z_2 symmetry. When we plot (14) as in Fig.2, we find a diagram similar to that of a magnetization versus external field of a ferromagnet. In Fig.2 the system is unstable in the branch $\partial M/\partial m(\mu)<0$ (σ field becomes tachyonic).

In a class of theories like Gross-Neveu model basic fields in the Lagrangian appear as the fundamental (vector) representation of the internal symmetry group and thus may be called as vector models. Vector models are usually soluble in $N=\infty$ limit since the problem is reduced to a sum over bubble diagrams which is trivial to handle. On the other hand, when one considers a model where basic fields belong to the adjoint representation of the group, the problem becomes that of summing over planar diagrams. In this case theory has been solved only in simplest cases, i.e. in the d=0 space-time dimensionality.

Let us next consider the following model discussed by Brézin, Itzykson, Parisi and Zuber[3]

$$Z = \int dA \exp\{-(\frac{1}{2}\mathrm{tr}\,A^2 + g^2\,\mathrm{tr}\,A^4)\} \quad . \tag{15}$$

Here A_{ij} $(i,j=1,\ldots,N)$ is an N x N hermitian matrix and dA means the flat measure $dA=\prod_i dA_{ii}\prod_{i>j} dReA_{ij}\; dImA_{ij}$. Model (15) may be regarded as a zero-dimensional version of a matrix ϕ^4-theory,

$$Z = \int \prod_x dA(x)\exp\{-(\frac{1}{2}\sum_{\mu=1}^{d}\mathrm{tr}(\partial_\mu A(x))^2 + \frac{m^2}{2}\mathrm{tr}\,A(x)^2 + g^2\mathrm{tr}A(x)^4)\} \tag{16}$$

obtained after eliminating the kinetic term. Then the propagator of (15) is simply

$$\langle A_{ij}A^*_{lm}\rangle = \delta_{il}\delta_{jm} \tag{17}$$

and the integrand of Feynman digram of the model becomes some constant. In fact an amplitude of a (vacuum) diagram in n-th order of perturbation theory is equal to $N^{2-2j}\lambda^n$ where j is the number of handles in the topology of the diagram and $\lambda=g^2N$. Although the model (15) appears unrealistic, one can study combinatorial aspects of planar perturbation theory making use of its analytic solution. For instance, the free energy F=ln Z of (15) has an expansion,

$$\frac{1}{N^2}F(N,\lambda) = \sum_j\sum_n\frac{1}{N^{2j}}c_{n,j}\,\lambda^n \tag{18}$$

where $c_{n,j}$ is the number of (connected) vacuum graphs with j handles in n-th order of perturbation theory. At $N=\infty$ (18) is replaced by

$$\lim_{N\to\infty} \frac{1}{N^2} F(N,\lambda) = \sum_n c_{n,0}\, \lambda^n \tag{19}$$

where $c_{n,0}$ is the number of planar graphs. Then the radius of convergence of the series (19) determines the rate of increase of $c_{n,0}$ as n becomes large.

Systems like (15) are sometimes called as random matrix models. Strategy of solving (15) is first to diagonalize A,

$$A = UDU^{+} \qquad D_{ij} = \delta_{ij}\mu_i \qquad (i=1,\dots,N) \tag{20}$$

and express the integration measure using eigenvalues

$$dA = dU \prod_{i<j} (\mu_i - \mu_j)^2 \prod_i d\mu_i \quad . \tag{21}$$

After a scale transformation $\mu_i \to x_i=\mu_i/N$, the partition function is rewritten as

$$Z = \text{const} \times \int_{-\infty}^{+\infty} \prod_i dx_i \; e^{-N^2 S_{eff}(x_i)}$$

$$S_{eff}(x_i) = \frac{1}{2}\sum x_i^{\,2} + \lambda \sum x_i^{\,4} - \frac{1}{N}\sum_{i<j} \ln(x_i - x_j)^2 \tag{22}$$

As $N \to \infty$, x integration is dominated by the saddle point,

$$\frac{\partial S_{eff}}{\partial x_i} = x_i + 4\lambda x_i^{\,3} - \frac{2}{N}\sum_{j\neq i} \frac{1}{x_i - x_j} = 0 \quad . \tag{23}$$

We then introduce a density function,

$$\rho(x) = \frac{1}{N}\sum_{i=1}^{N} \delta(x - x_i) \ , \qquad \int_{-a}^{+a} \rho(x)dx = 1 \quad . \tag{24}$$

We assume that the support of ρ is $[-a,a]$ and a is determined below. As $N\to\infty$ (23) is rewritten as

$$x + 4\lambda x^3 - 2 \times \mathcal{P} \int_{-a}^{+a} \frac{\rho(y)}{x-y} dy = 0 \tag{25}$$

($\mathcal{P}$ means the principal part). It is not difficult to solve (25). We introduce a function

$$F(x) = \int_{-a}^{+a} \frac{\rho(y)}{x-y} dy \tag{26}$$

which has the following properties;

1. $\mathrm{Re}\, F(x) = \frac{1}{2} x + 2\lambda x^3$, for $|x| \le a$

2. $\mathrm{Im}\, F(x) = 0$, for $|x| \ge a$

3. $F(x) \to \frac{1}{x}$ as $|x| \to \infty$

We then may write down a trial formula for F(x).

$$F(x) = \frac{1}{2}x + 2\lambda x^3 + (-2\lambda x^2 + b)\sqrt{x^2-a^2} \tag{27}$$

which has properties 1,3 above. a,b are determined by the condition 2,

$$F(x) \to \frac{1}{2}x + 2\lambda x^3 + (-2\lambda x^2 + b)\left(x - \frac{a^2}{2x} - \frac{a^4}{8x^3}\right) + O\left(\frac{1}{x^3}\right)$$

$$= \left(\frac{1}{2} + b + a^2\lambda\right)x + \left(-\frac{a^2 b}{2} + \frac{\lambda}{4}a^4\right)\frac{1}{x} + O\left(\frac{1}{x^3}\right) \quad . \tag{28}$$

Thus

$$\frac{1}{2} + b + a^2\lambda = 0$$

$$-\frac{a^2}{2}b + \frac{\lambda}{4}a^4 = 1 \quad . \tag{29}$$

After eliminating b, one obtains

$$a^2(\lambda) = -\frac{1}{3\lambda} \pm \sqrt{\frac{1}{9\lambda^2}+\frac{8}{3\lambda}} \tag{30}$$

The model (15) is well-defined in the region $\lambda>0$ where we choose + sign in (30). Now the crucial point is that this solution for $\lambda>0$ can be continued analytically beyond $\lambda=0$ to the region $\lambda<0$ where we take - sign in (30). The solution then exists up to $\lambda=\lambda_c$ where the discriminant vanishes

$$\frac{1}{9\lambda_c^{\,2}}+\frac{8}{3\lambda_c}=0 \qquad \text{or} \qquad \lambda_c = -\frac{1}{24}\ . \tag{31}$$

Thus the model (15), originally defined for $\lambda>0$, actually exists in a range of $\lambda<0$. In particular the series (19) has a finite radius of convergence $|\lambda| = 1/24$. Hence

$$c_{n,0} \approx (24)^n \tag{32}$$

as $n \to \infty$.

In the case of ordinary field theory the number of Feyman diagrams increases like n!. Thus

$$c_n = \sum_j c_{n,j} \approx n! \qquad n\to\infty \tag{33}$$

(we put N=1 for simplicity) and the expansion

$$F(\lambda) = \sum_n c_n \lambda^n$$

in an asymptotic series with zero radius of convergence. Rate of increase (33) reflects an instability of the vacuum at $\lambda<0$ due to tunnelling effect,

$$F(\lambda) = \frac{1}{\pi}\int_{-\infty}^{0} \frac{\lambda}{(\lambda-\lambda')\lambda'}\,\mathrm{Im}\,F(\lambda')d\lambda' \approx \frac{1}{\pi}\int_{-\infty}^{0}\frac{\lambda}{(\lambda-\lambda')\lambda'}\,e^{-\frac{S_{cl}}{|\lambda'|}}\,d\lambda'$$

$$= \frac{1}{\pi}\sum_{n=0}^{\infty}(-1)^{n+1}\left(\frac{\lambda}{S_{cl}}\right)^{n+1} n!\quad . \tag{34}$$

Here S_{cl} is the instanton action which describes the tunnelling process. The exponential behavior (32) in the case of planar diagrams implies a suppression of tunnelling (S_{cl} is proportional to N^2) and the stablization of vacua at $\lambda < 0$. As $N \to \infty$, an asymptotic series is turned into a convergent expansion.

When we go to more realistic cases of 4-dimensional field theories, analysis becomes quite complex[4], however, the basic physics is well illustrated by the 0-dimensional example.

In the examples of Gross-Neveu and random matrix model, we have used the saddle point method; fluctuations around the saddle point are suppresed at $N=\infty$ and are controled systematically by $1/N$ expansion. Closely related to the dominance by the saddle point there exists another fundamental aspect of large N field theory; factorization property of invariant physical observables.[5] In the case of gauge theory, for instance, if we consider the expectation value for a product of two Wilson loop amplitudes,

$$<W(C_1)W(C_2)>, \tag{35}$$

$$W(C_i) \equiv \frac{1}{N} P \exp ig \int_{C_i} A_\mu(x)\, dx_\mu \qquad i=1,2 \quad ,$$

the leading contribution comes from the disconnected part

$$<W(C_1)W(C_2)> = <W(C_1)><W(C_2)> + O(\frac{1}{N^2}) \quad . \tag{36}$$

Thus the expectation value of a product of loops is given by the product of expectation values for each loop. It is possible to check (36) in each order of perturbation theory.

Factorization property (36) has a profound implication on the structure of $N=\infty$ field theory. If, for instance, we take $C_1=C_2$ in (36), we find,

$$<(W(C) - <W(C)>)^2> = 0 \quad . \tag{37}$$

Therefore there exist no dispersion in the values of W(C) when it is averaged over all possible field configurations. This means that at $N=\infty$ contributions to W(C) come only from a particular class of field contigurations (possibly a unique configuration modulo gauge transformation) which give the same value for W(C). All the other configurations may be thrown away without changing the leading result in $N=\infty$ limit. Therefore the factorization condition (36) suggests the possibility of a drastic reduction in the relevant dynamical degrees of freedom in $N=\infty$ field theories. As we shall see in the following, it is in fact possible to cut down substantially the phase space of path integration and map the problem of $N=\infty$ field theory to that of a certain type of random matrices.

In order to facilitate our discussions let us now introduce a space-time lattice and use the lattice formulation of gauge theory.[6] Standard lattice action is given by

$$S = \sum_{\mu\neq\nu} \sum_{x} \mathrm{tr}\, U_\mu(x)\, U_\nu(x+\mu)\, U_\mu^+(x+\nu)\, U_\nu^+(x) \quad , \tag{38}$$

where $U_\mu(x)$ is a link variable associated with the link $(x;\mu)$ connecting x to $x+\mu$. We consider the gauge group $G=U(N)$ hereafter. Wilson loop amplitude on the lattice is defined by

$$\langle W(C)\rangle = \langle \mathrm{tr}(U_\mu(x)U_\nu(x+\mu)\ .\ .\ .\ U_\sigma(x-\sigma))\rangle \tag{39}$$

for a loop C $(x,\ x+\mu,\ x+\mu+\nu,\ \ldots,\ x-\sigma,\sigma)$.

In the following we make use a set of identities among (39) for various loops which are analogues of Schwinger-Dyson (S-D) equations in the continuum field theory. They have the following generic form,

$$\lambda\langle W(C)\rangle - \sum_{\nu\neq\mu}\langle W(C+\delta_\nu C)\rangle + \sum_{\nu\neq\mu}\langle W(C+\delta'_\nu C)\rangle = 0 \quad . \tag{40}$$

Here $\delta_\nu C$ and $\delta'_\nu C$ are infinitesimal (one lattice unit) variations of C at a link $(x;\mu)$ as depicted in Fig.3. If the varied link is traversed more than once in the contour C, additional terms appear in (40). In the case of Fig.4a (4b), for instance, the link $(x;\mu)$ is traversed twice in the opposite (same) directions. We then have a term

$$\overset{+}{(-)}\ \langle W(C_1)W(C_2)\rangle \tag{41}$$

in the right hand side of (40). These terms correspond to joining and splitting of strings and the analogues of the δ-functions in the right hand side of S-D equations in the continuum field theory. Thus they are sometimes called as source terms.

Let us now consider a lattice gauge system which is at a finite temperature T. In Euclidean formulation of field theory temperature corresponds to a periodicity in the time direction. If the temporal size of the lattice is n_T, temperature is given by

$$T = (an_T)^{-1} \quad , \tag{42}$$

where a is the lattice constant. It is known that lattice gauge theories (with non-Abelian gauge group) have two distinct phases at finite temperature[7];

1, Confined phase at low temperature,

2, Plasma phase at high temperature.

These phases are separated by a deconfining transition at $T=T_c$ which is characterized by a spontaneous breakdown of a symmetry associated with the center of the gauge group (Z_N for SU(N), U(1) for U(N)).

Now we shall argue that at $N=\infty$, as far as the gauge system stays in the confined phase, its physical properties do not depend on the temperature T or the temporal size of the lattice n_T. The arguement is based on the comparison of S-D equations at T=0 and $T\neq0$ gauge theory. In the case of finite temperature theory link variables on links separated by integer x n_T steps in the time direction are identified due to periodicity,

$$U_\mu(t + m\cdot n_T, \vec{x}) = U_\mu(t,\vec{x}) \quad . \tag{43}$$

Because of this identification new type of source terms appear in finite temperature S-D equations. If we consider a loop as shown in Fig.5, for instance, the same variable $U_\mu(x)$ appear both in links $(t,\vec{x};\mu)$ and $(t+m\cdot n_T, \vec{x};\mu)$. Then the situation is as if in the T=0 theory the link $(t,\vec{x};\mu)$ or $(t+m\cdot n_T,\vec{x};\mu)$ is traversed twice in the opposite directions and its variation gives rise to a splitting of the string as shown Fig.6. The new source term has a form

$$\langle W(L_1)W(L_2)\rangle \tag{44}$$

which is qualitatively different from the previous case (41) since L_i (i=1,2) are now open loops. At N=∞ (44) factors

$$\langle W(L_1)\rangle\langle W(L_2)\rangle \tag{45}$$

Expectation value for such an open loop $\langle W(L_i)\rangle$ serves as an order parameter for the deconfining transition. At low temperatures $T<T_c$, invariance of the action (38) under the transformation by the center of the group (U(1) for G=U(N)),

$$U_t(t,\vec{x}) \rightarrow e^{i\theta}U_t(t,\vec{x}) \quad \text{fixed } t \text{ and all } \vec{x} \tag{46}$$

is unbroken. This forces all open-loop expectation values to vanish

$$\langle W(L)\rangle = 0.$$

(Above T_c, U(1) symmetry (46) becomes spontaneously broken and we have $\langle W(L)\rangle\neq 0$). Therefore at $T<T_c$ extra source terms of $T\neq 0$ S-D equations disappear and $T\neq 0$ and $T=0$ gauge theories obey the same set of identities. Thus the physical quantities of these theories must coincide and gauge theory becomes temperature independent at N=∞ in the range $T<T_c$.

This result implies that it is possible at N=∞ to replace an infinite space-time lattice by one of finite temporal extension as far as the U(1) symmetry is unbroken. We then may repeat the same reasoning and successively shrink the size of other directions,

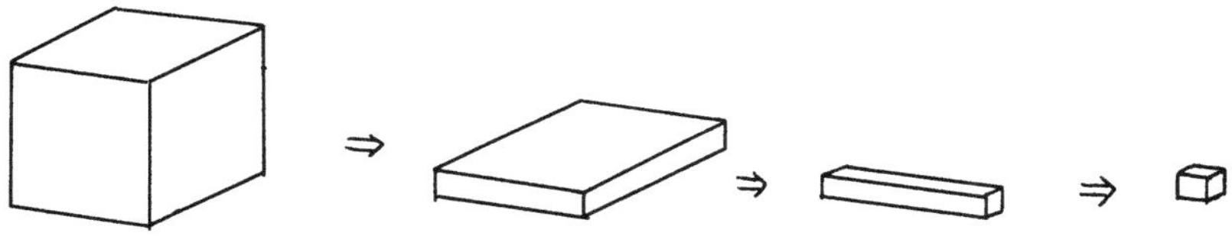

and finally arrive at a very small lattice size. During this process we must protect U(1) symmetries associated with each direction,

$$U_\mu \rightarrow e^{i\theta_\mu} U_\mu \quad . \tag{48}$$

If the symmetry (48) is left intact, it is in fact possible to show[8] that the original model of infinite lattice (38) is equivalent to a model defined on a hypercube,

$$Z_r = \int \prod_\mu dU_\mu \, e^{\frac{1}{g^2} S_r}$$

$$S_r = \sum_{\mu \neq \nu} \mathrm{tr}\, U_\mu U_\nu U_\mu^+ U_\nu^+ \tag{49}$$

and the analogue of the Wilson loop amplitude (39) defined by

$$W_r(C) = \langle \mathrm{tr}\, U_\mu U_\nu \; . \; . \; . \; U_\sigma \rangle_r \tag{50}$$

agrees with that of the original theory (in (50) the expectation value is taken with respect to weight (49)). Thus the gauge field is mapped onto a certain type of random matrix model at $N=\infty$.

As a matter of fact, according to a numerical simulation[9] U(1) symmetry (48) becomes spontaneously broken for $\lambda<\lambda_c$ (≈ 2) in the model (49) (λ_c corresponds to T_c via $T_c=(a(\lambda_c))^{-1}$ with $a(\lambda) \approx \exp(-24\pi^2/11\lambda)$) and its equivallence to the original theory (38) is lost in the weak coupling region. We need some special trick to avoid this symmetry breakdown.

At present there exist two ways to prevent U(1) breaking in the model (49); 1. quenching;[9,10] 2. twist.[11] Among these twist approach is more efficient in numerical analysis and we shall discuss its basic idea.

Since U_μ's are unitary matrices, their eigenvalues $e^{i\theta_\mu^j}$ ($j=1,\dots,N$) lie on the unit circle. Then the invariance of the vacuum under U(1) rotation (48) corresponds to a uniform distribution of these eigenvalues in the ground state configuration. As it turns out, there exists a neat way of imposing a uniform distribution on the ground state by introducing certain phase factors into the action (49),

$$S_t = \sum_{\mu\neq\nu} e^{\frac{2\pi i n_{\mu\nu}}{N}} \, \mathrm{tr}\, U_\mu U_\nu U_\mu^+ U_\nu^+ \tag{51}$$

$$n_{\mu\nu} = -n_{\nu\mu} \quad ; \quad \text{integers} \quad .$$

These integers are so-called twist factors. The ground state of the system (51), $U_\mu = T_\mu$, satisfies the equation

$$e^{\frac{2\pi i n_{\mu\nu}}{N}} T_\mu T_\nu = T_\mu T_\nu \qquad \mu \neq \nu \quad . \tag{52}$$

For the sake of simplicity we first consider 2-dimensional case where we have

$$e^{\frac{2\pi i}{N}} T_1 T_2 = T_2 T_1 \tag{53}$$

(we put $n_{12}=1$). This is the commutation relation of the twisting matrices P,Q[12]

$$P_N = \begin{pmatrix} 0 & 1 & & 0 \\ & 0 & 1 & \\ 0 & & \ddots & \ddots \\ & & & \cdot\ 1 \\ & & & 0 \end{pmatrix} \quad , \quad Q_N = \begin{pmatrix} 1 & & & 0 \\ & \omega & & \\ 0 & & \ddots & \\ & & & \omega^{N-1} \end{pmatrix} \quad , \quad \omega = e^{\frac{2\pi i}{N}} \quad . \tag{54}$$

Thus $T_1 = P$, $T_2 = Q$. It is evident from (54) that Q has uniformly distributed eigenvalues. Eigenvalues of P coincide those of Q. Furthermore they have the property,

$$\frac{1}{N} \mathrm{tr}\, P^a Q^b = \delta_{a,0} \delta_{b,0} \tag{55}$$

where δ is the mod-N Kronecker delta.

In the case of 4 dimensions, it is convenient to choose $N = L^2$ and $n_{\mu\nu} = L$ ($\mu > \nu$). Then a solution to (52) is given by

$$T_1 = \mathbb{1}_L \times P_L$$

$$T_2 = P_L \times Q_L$$

$$T_3 = P_L Q_L \times Q_L$$

$$T_4 = Q_L \times Q_L \tag{56}$$

(all the other solutions are obtained by unitary transformations.) Now the link variables are expanded around T_μ,

$$U_\mu = V_\mu T_\mu \, , \qquad V_\mu = e^{igA_\mu} \, , \tag{57}$$

and the action (51) is rewritten as

$$\begin{aligned} S_t &= \sum_{\mu\neq\nu} e^{\frac{2\pi i n_{\mu\nu}}{N}} \operatorname{tr} U_\mu U_\nu U_\mu^{+} U_\nu^{+} \\ &= \sum_{\mu\neq\nu} \operatorname{tr}(V_\mu \cdot T_\mu V_\nu T_\mu^{+} \cdot T_\nu V_\mu^{+} T_\nu^{+} \cdot V_\nu^{+}) \quad . \end{aligned} \tag{58}$$

When we compare (58) with the standard action (38)

$$S = \sum_x \sum_{\mu\neq\nu} \operatorname{tr}(U_\mu(x) U_\nu(x+\mu) U_\mu^{+}(x+\nu) U_\nu^{+}(x)) \quad ,$$

we note the correspondence

$$\begin{aligned} V_\mu &\leftrightarrow U_\mu(x) \\ T_\nu V_\mu T_\nu^{+} &\leftrightarrow U_\mu(x+\nu) \qquad \text{etc.} \end{aligned} \tag{59}$$

Thus the twisting matrix T_μ acts like a lattice translation generator in the μ-direction. Eq.(58) shows that the model (51) is in fact equivalent to the original theory (38) when its fields are restricted to lie on a special subspace (cross section) in its configuration space. In Fig.7 we have drawn a schematic picture of a configuration space and its cross section for $U(x)$ (we suppress suffix μ). When $U(x)$ lies on a cross section, a space-time translation is effected by a rotation in color space

$$U(x) = T(x) U(0) T(x)^{+} \quad . \tag{60}$$

The twist model (51) corresponds to the choice

$$T(x) = T_1^{x_1} T_2^{x_2} T_3^{x_3} T_4^{x_4} \quad , \tag{61}$$

$$\text{with} \quad T_\mu T_\nu = e^{-\frac{2\pi i n_{\mu\nu}}{N}} T_\nu T_\mu \quad .$$

Now it is not difficult to show that in general restriction of the field configurations to the cross section (60),(61) does not change the leading result in large N field theories. For the sake of illustration, let us consider a model of interacting hermitian matrices in 2-dimensions,

$$S = \frac{1}{2} \sum_x \sum_\mu \{ 2\mathrm{tr}A(x)^2 - \mathrm{tr}A(x+\mu)A(x) - \mathrm{tr}A(x-\mu)A(x) \}$$

$$+ \frac{m^2}{2} \sum_x \mathrm{tr}A(x)^2 + \frac{\lambda}{N} \mathrm{tr}A(x)^4 \quad . \tag{62}$$

Making use of the rule (60),(61), we replace (62) by

$$\begin{aligned} S_t &= 2\,\mathrm{tr}\,A^2 - \mathrm{tr}\,PAP^+A - \mathrm{tr}\,QAQ^+A \\ &\quad + \frac{m^2}{2}\,\mathrm{tr}\,A^2 + \frac{\lambda}{N}\,\mathrm{tr}\,A^4 \quad . \end{aligned} \tag{63}$$

Although there are no apparent space-time dependence left in (63), it is possible to convert color degrees of freedom A_{ij} $(i,j=1,\ldots,N)$ into those of space-time and bring the theory into a form suitable for comparison with (62). We introduce a Fourier transform

$$A_{ij} = \sum_{a,b=1}^{N} A(a,b)\, \frac{e^{\frac{\pi i a b}{N}}}{\sqrt{N}} ((P^+)^b Q^a)_{ij} \quad . \tag{64}$$

Since

$$PAP^+ = \sum_{a,b} A(a,b) e^{\frac{2\pi i a}{N}} \frac{e^{\frac{\pi i ab}{N}}}{\sqrt{N}} (P^+)^b Q^a \quad ,$$

$$QAQ^+ = \sum_{a,b} A(a,b) e^{\frac{2\pi i b}{N}} \frac{e^{\frac{\pi i ab}{N}}}{\sqrt{N}} (P^+)^b Q^a \quad , \qquad (65)$$

$0 < \frac{2\pi a}{N}, \frac{2\pi b}{N} \leq 2\pi$ have the meaning of x,y component of the momentum, respectively. The kinetic term of (63) is then rewritten as,

$$\frac{1}{2} \{ 4 \sum_{a,b=1}^{N} (\sin^2 \frac{\pi a}{N} + \sin^2 \frac{\pi b}{N}) + m^2 \} \, |A(a,b)|^2 \quad . \qquad (66)$$

Thus it has the standard form of a lattice propagator.

The interaction vertex trA^4 has an expression,

$$trA^4 = \frac{1}{N} \sum \delta_{\sum_i a_i, 0} \delta_{\sum_j b_j, 0} A(a_1,b_1)A(a_2,b_2)A(a_3,b_3)A(a_4,b_4) \, e^{i\theta(a,b)} , \qquad (67)$$

$$\theta(a,b) = \frac{\pi}{N} \sum_i a_i b_i - \frac{2\pi}{N} \sum_{i>j} a_i b_j \quad . \qquad (68)$$

Except for the phase factor (68),(67) has the form of a 4-point vertex of (color singlet) ϕ^4-theory written in momentum space. (66) combined with (67),(68) gives the Feynman rule of the model (63). Now it is easy to show that the phase factor (68) has the following properties[11]; phase factors coming from 4-point vertices in a planar diagram cancel among themselves while they do not cancel in the case of a non-planar diagram. Therefore planar Feynman amplitudes are correctly reproduced in the model (63) and at the same time non-planar amplitudes are suppressed by rapid oscillation of the phase factor (68). Thus the model (63) is in fact equivalent to (62) in the leading order in 1/N expansion. Equivalence of the random matrix model (58) with the gauge theory at N=∞ (38) can be shown in exactly the same fashion.

In the above we have discussed some examples of quantum field theories which are either exactly solved or greatly simplified in N=∞ limit. 1/N expansion appears at present to be the most promising tool to investigate non-perturbative phenomena in field theory. In particular in the case of theories where dimensional transmutation is expected to take place 1/N is the only expansion parameter of the theory (coupling constant g is traded for the mass gap as in (11) and is not an expansion parameter) and the theory will never be solved unless it is solved in N=∞ limit.

In the last year Witten[13] made an observation that baryons in N=∞ QCD are realized as solitons in chiral Lagrangian which incorporates the structure of the anomaly of underlying gauge theory. This result seems to indicate that there exists a possibility of unifying various approaches to strong interaction physics proposed in the past, chiral dynamics, dual theory, strong coupling theory etc. within the framework of 1/N expansion. Analytic solution of N=∞ gauge field will undoubtedly give us a wealth of information on strong interaction dynamics. Unfortunately, we are yet very far from achieving this goal.

References

1. T. Berlin and M. Kac, Phys. Rev. 86 821 (1952).
2. D. Gross and A. Neveu, Phys. Rev. D10 3235 (1974).
3. E. Brézin, C. Itzykson, G. Parisi and J. Zuber, Comm. Nath. Phys. 59 35 (1978).
4. G. 't Hooft, Preprints (1982),
   C. de Calana and V. Rivasseau, Comm. Math. Phys. 82 69 (1981).
5. E. Witten, Cargese Lectures (1979).
6. K. Wilson, Phys. Rev. D10 2445 (1974).
7. A. M. Polyakov, Phys. Lett. 72B 477 (1978),
   L. Susskind, Phys. Rev. D20 2610 (1978),
   L. McLerran and B. Svetitsky, Phys. Lett. 98B 195 (1981),
   J. Kuti, J. Polonyi and K. Szlanchanyi, Phys. Lett. 98B 199 (1981).
8. T. Eguchi and H. Kawai, Phys. Rev. Lett. 48 1063 (1982).
9. G. Bhanot, U. Heller and H. Neuberger, Phys. Lett. 113B 47 (1982)
10. G. Parisi, Phys. Lett. 112B 463 (1982),
    D. Gross and Y. Kitazawa, Nucl. Phys. B206 440 (1982),
    S. Das and S. Wadia, Phys. Lett. 117B 228 (1982).
11. A. Gonzalez-Arroyo and M. Okawa, Phys. Rev. D27 2397 (1983),
    T. Eguchi and R. Nakayama, Phys. Lett. 122B 59 (1983).
12. J. Groeneveld, J. Jurkiewicz and C. P. Kortals-Altes, Physica Scripta 23 1022 (1981),
    G. 't Hooft, Comm. Math. Phys. 83 267 (1981).
13. E. Witten, Nucl. Phys. B223 422,433 (1983).

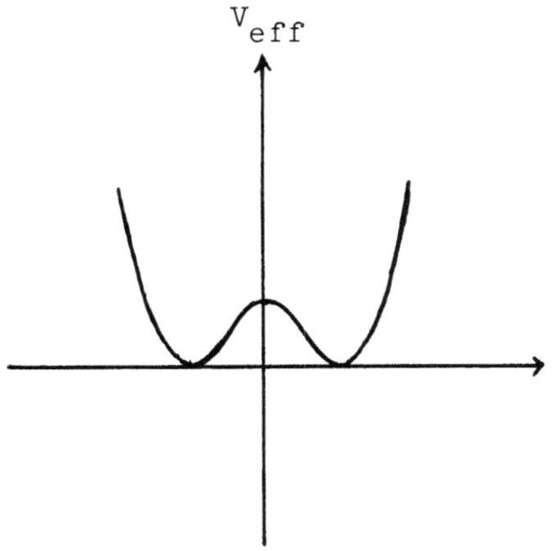

Fig. 1

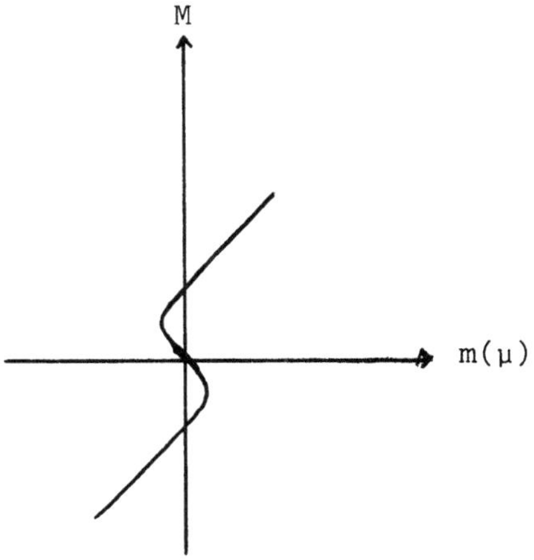

Fig. 2

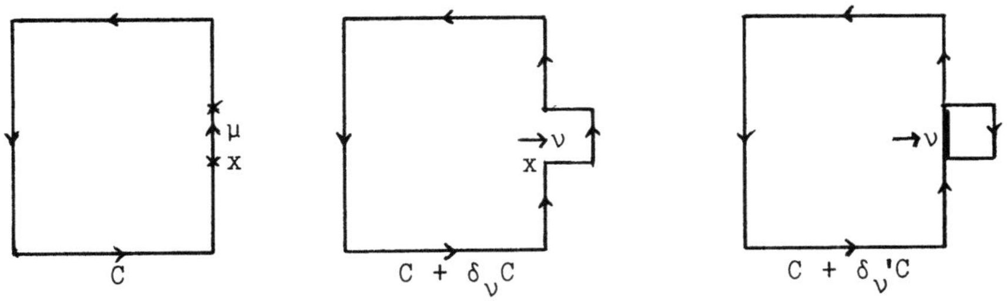

Fig. 3

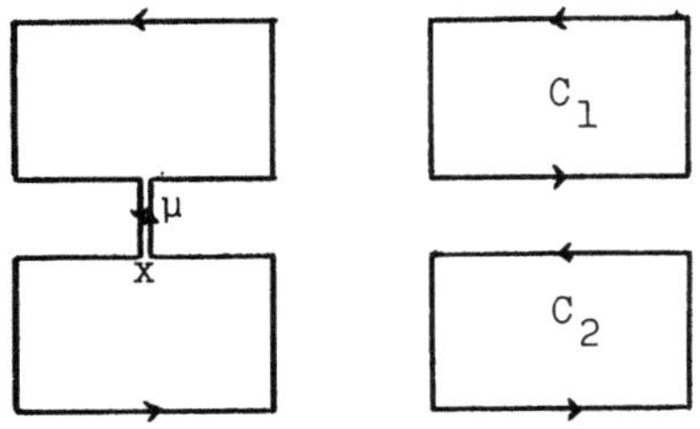

Fig. 4a

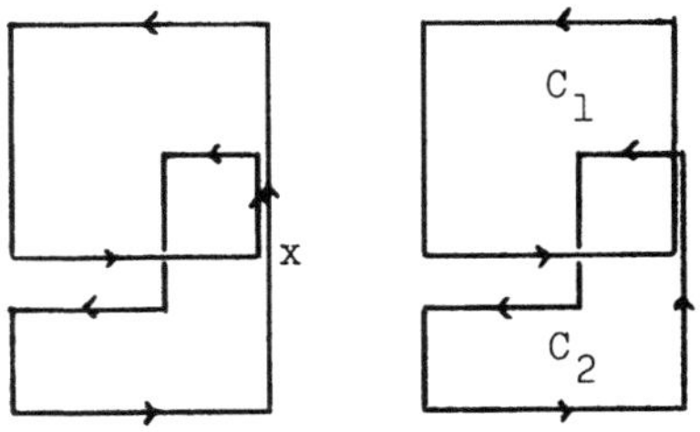

Fig. 4b

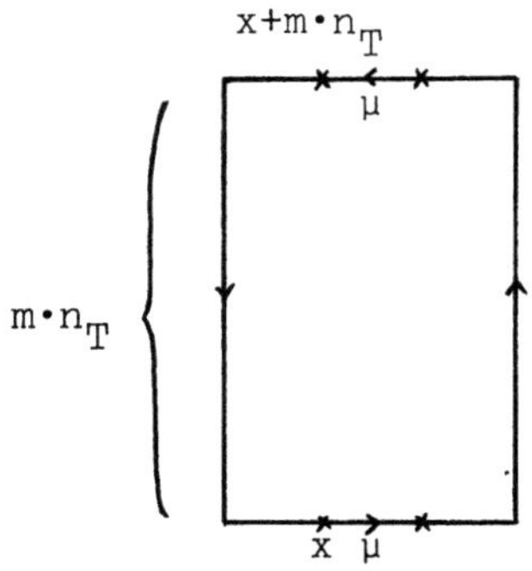

Fig. 5

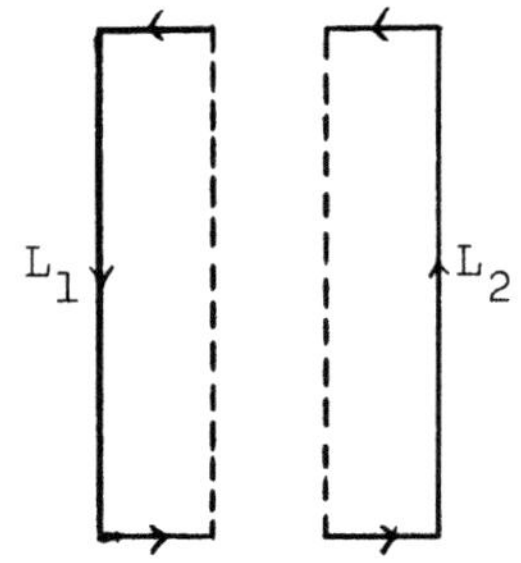

Fig. 6

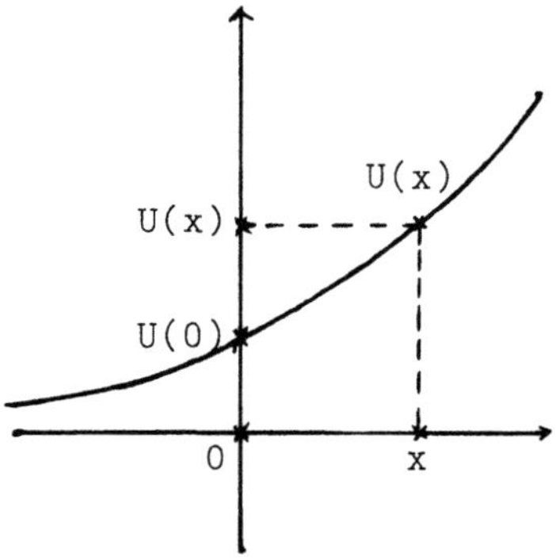

Fig. 7

ANOMALIES AND INDEX THEORY

Michael Atiyah
Mathematical Institute
Oxford OX1 3LB.

§1. Introduction

There is at present considerable interest amongst physicists in various anomalies that occur in gauge theories and their relation to the index theory of the Dirac operator. Numerous papers have appeared on the subject [6], [7], [8], [9], [10], [11], [12] and it is the purpose of this lecture to provide a general introduction and guide to the subject. I will concentrate on the main ideas since the detailed calculations are fully available in the literature.

Since this is a lecture for physicists I will begin by a brief review of the index theorem in its simplest form. This will also enable me to establish the notation and terminology I will use which (inevitably) tends to differ slightly from that used in physics. I will then go on to describe a more general version of the index theorem, dealing with parameters. It is this version which is related to the anomalies that turn up in relation to the difficulties of defining gauge invariant determinants.

After dealing with determinants I will return to the index theory explaining further refinements all of which help to throw further light on anomalies.

§2. Review of the Index Theorem

The index theorem is concerned with any elliptic differential operator but for physical (or geometric) applications we only need the special but important case of the Dirac operator. This is globally defined on any Riemannian manifold M provided M is oriented and has a spin structure (these are minor topological constraints). We assume M is compact, then the Dirac operator is elliptic and self-adjoint. In particular it has a discrete spectrum of eigenvalues and we are particularly interested in the 0-eigenspace or null-space.

The Dirac operator acts on spinor fields and there is a basic algebraic difference between spinors in even and odd dimensions.

In odd dimensions spinors belong to an irreducible representation of the spin group whereas in even dimensions spinors break up into two irreducible pieces, denoted say by S^+ and S^-. The Dirac operator interchanges these two, so that it consists essentially of an operator

$$D : S^+ \to S^-$$

and its adjoint

$$D^* : S^- \to S^+.$$

The dimensions of the null-spaces of D and D^* are denoted by N^+ and N^- and the index of D is defined by

$$\text{(2.1)} \quad \text{index } D = N^+ - N^-.$$

Although there is a formal symmetry between positive and negative spinors, and in fact they are interchanged by reversing the orientation of M, the numbers N^+ and N^- need not be equal, because of global topological effects. This is the origin of the "chiral anomaly".

The index theorem [2] gives an explicit formula for index D in terms of topological invariants of M. Moreover, using the Riemannian metric ρ and its curvature tensor $R(\rho)$, one can write down an explicit integral formula

$$\text{(2.2)} \quad \text{index } D = \int_M \Omega(R(\rho))$$

where Ω is a certain formal expression constructed purely algebraically out of $R(\rho)$. For example when $\dim M = 4$

$$\text{(2.3)} \quad \Omega = \frac{\text{Tr } \Omega^2}{96\ \pi^2},$$

where Ω is considered as a matrix of 2-forms.

The index formula (2.2) is, from the physical point of view, purely gravitational since it only involves the metric ρ. Moreover, the index is zero for the sphere or torus and one needs more complicated manifolds to exhibit a non-zero index. However, one can generalize (2.2) to deal with gauge theories. Mathematically this means we give ourselves a complex vector bundle V over M with a unitary connection A. If the fibre of V is C^N then A is a

U(N)-gauge field over M. The extended Dirac operator now acts on V-valued spinors:

$$D_A : S^{+} \otimes V \to S^{-} \otimes V$$

and as before one defines the index of D_A by a formula like (2.1). The index formula (2.2) gets generalized to a formula:

$$(2.4) \qquad \text{index } D_A = \int_M \Omega(R(\rho), F(A))$$

where $F(A)$ is the curvature of A, and $\Omega(R,F)$ stands for a certain explicit algebraic expression in R and F. For example when $\dim M = 4$ we get, as a generalization of (2.3)

$$(2.5) \qquad \Omega = \frac{\mathrm{Tr}\ \Omega^2}{96\ \pi^2} - \frac{\mathrm{Tr}\ F^2}{8\pi^2} \quad .$$

In particular, when M is the standard 4-sphere, (2.3) reduces to a form familiar in connection with instantons.

§3. The index of 2-parameter families

The important property of the index is that it is unchanged by perturbation of the operator. This is why it can be given by a topological formula. Thus when we vary the gauge field A continuously the index of D_A does not change. Nevertheless it is possible to extract more topological information from a continuous family of operators as I shall now explain. For the purpose of physical applications I will, for the present, consider only the 2-parameter case.

Let me begin for simplicity with a purely "gravitational" situation, so let X be a 2-dimensional surface (compact, connected and oriented) and suppose we are given a fibre-bundle Y over X with fibre M. For example Y could be the product manifold $M \times X$, but this will not be interesting from our point of view. The twist involved in making a non-trivial bundle will be an essential topological feature. Now by giving Y a Riemannian metric we get metrics ρ_x on all the fibres M_x, so that essentially we have a family of metrics on M parametrized by X. Correspondingly (under minor topological conditions) we get a family D_x of Dirac operators on M, parametrized by X. In this situation, in addition to the integer given by index D_x, we can define another topological invariant which may be called the degree of the family of Dirac operators. I shall now explain the definition of this degree in the special case when the index is zero (I shall return to the more general situation in §5).

Since index $D_x = 0$ the generic situation (which one can achieve by an overall perturbation of the family, i.e. by perturbing the metric on Y) is that D_x is invertible for all x in X except at a finite number of points say $x_1, x_2, \ldots, x_k$. Moreover we can assign a multiplicity ν_i to each x_i which will generically be ± 1. To do this one can, near each x_i, replace D_x by a finite matrix T_x (discarding high eigenvalues) and ν_i is the multiplicity of the zero of $\det T_x$ at $x = x_i$. More precisely $2\pi\nu$ is defined as the variation in phase of $\det T_x$ as x traverses (positively) a small circle centred at x_i.

We now define the degre ν of the family of Dirac operators D_x by putting $\nu = \Sigma\nu_i$. It is not hard to show that this is indeed a topological invariant of the family, i.e. it is invariant under perturbation. Thus it depends only on the topology of the fibre bundle Y (and is zero for the product $M \times X$).

The index theorem for families [3] gives in particular a topological formula for this degree, and I shall comment on this formula later in §5.

The preceding discussion can be extended to include gauge fields. We can now, if we wish, fix the metric ρ on X and take $Y = M \times X$, but we also take a vector bundle V on Y with a connection. This gives a family A_x of gauge fields and the corresponding Dirac family again has a degree which will now depend on the topology of V. The degree will be zero if V comes from a bundle on M.

§4. Determinants

In quantizing gauge theories one has to introduce determinants of operators and for a consistent theory these should have certain natural properties. The difficulty is of course that determinants of operators have to be regularized in some way and it may not be obvious how to retain all the required properties in the regularization. For positive definite operators of Laplace type (arising from bosons) the zeta-function method of regularization gives a fully satisfactory definition. Recall that for such an operator

Δ with eigenvalues λ one defines

$$\zeta(s) = \Sigma\lambda^{-s}$$

and then one defines

$$\det \Delta = \exp(-\zeta'(0)).$$

This makes sense since $\zeta(s)$ can be analytically continued from the half-plane of convergence $\mathrm{Re}(s) > d/_2$ and is holomorphic at $s = 0$.

For fermions however we have to define determinants of Dirac operators and these are not positive. Difficulties (or anomalies) can now arise. Consider as before an even-dimensional manifold M with Dirac operators $D_\rho : S^+ \to S^-$, depending on the metric ρ. Suppose we look for a regularized complex-valued determinant $\det D_\rho$ which should have the following properties

(i) $\det D_\rho$ is a differentiable function of ρ.

(ii) $\det D_\rho$ is gauge invariant, i.e. $\det D_{g(\rho)} = \det D_\rho$ for any diffeomorphism g of M.

(iii) $\det D_\rho = 0$ to first order if and only if $D_\rho^* D_\rho$ has exactly one zero eigenvalue.

Property (iii) is the characteristic property of determinants in finite dimensions and it is a minimum requirement for any regularized determinant.

Since $D_\rho^* D_\rho$ is a Laplace type operator there is no difficulty in defining its determinant so one can give a definition of $|\det D_\rho|$. The problem is to define the phase of $\det D_\rho$.

Now return to the situation discussed in §3. Suppose we can find a fibration over a surface X with fibre M so that the Dirac operators D_x all have <u>index</u> zero but that the <u>degree</u> of the family is non-zero, I claim that this will show that there is no function $\det D_\rho$ with the three properties listed above. Suppose in fact $\det D_\rho$ exists with these properties. Then (i) and (ii) imply that $f(x) = \det D_{\rho_x}$ is a differentiable complex-valued function on X. Property (iii) then implies that ν is the number of zeros of f (counted with multiplicities and signs), i.e. that ν is the degree of $f : X \to C$. But by elementary topology the degree of f is necessarily zero, so that $\nu = 0$, contradicting our assumption.

More insight into this argument can be obtained if we recollect that the local multiplicities ν_i at points x_i where D_x was not invertible , were defined as local phase variations (around x_i) defined by using an eigenvalue cut-off. If it were possible to use such a cut-off consistently at all points then the total phase variation ν would have to be zero. Thus a fibration with $\nu \neq 0$ implies a behaviour of the eigenvalues which prevents a good cut-off regularization.

At this point I should mention that examples exist when the degree $\nu \neq 0$ so that the above discussion applies. In fact examples exist in the lowest dimension namely when $\dim M = 2$ [1], but both M and X have to be taken as surfaces of fairly high genus (not spheres or tori). These examples are of considerable interest in algebraic geometry since it is the variation in the conformal structure of M that is essentially involved.

If we move from the pure gravitational case to a gauge theory situation examples with $\nu \neq 0$ occur when M is the 2n-sphere S^{2n} , $X = S^2$ and the vector bundle V has dimension $(n + 1)$ (or greater), so that the gauge group G is $U(n + 1)$ or $SU(n + 1)$. Thus for $n = 2$, $\dim M = 4$ and $G = SU(3)$. Let us consider the topology in more detail in these cases.

If we represent S^2 as $R^2 \cup \infty$ then removing ∞ we have a 2-parameter family of gauge-fields on M parametrized by R^2. Since this family was defined by a bundle over $M \times S^2$ the gauge-fields are all gauge-equivalent as we go to ∞ in R^2. In other words, taking a large circle in R^2 we get a closed path in the group $\mathcal{G}$ of gauge transformations (of the bundle on M). The degree $\mathcal{G}$ is essentially a homomorphism $\pi_1(\mathcal{G}) \to Z$ and if $\nu \neq 0$ the variation in phase of $\det D_\rho$ going round this path in $\mathcal{G}$ demonstrates the lack of gauge invariance. For $M = S^{2n}$ we have $\pi_1(\mathcal{G}) = \pi_{2n+1}(G)$ and one knows that

$$\pi_{2n+1}(SU(n + 1)) = Z$$

and the isomorphism is actually given by ν. This follows from the general index theorem for families [3].

The correspondence between $\pi_1(\mathcal{G})$ and bundles on $M \times S^2$ is quite general. Any loop in $\mathcal{G}$ applied to a given gauge field A generates a loop of gauge equivalent fields. This loop can be filled in to a 2-disc of gauge fields and this data enables us to construct a bundle on $M \times S^2$. Similarly higher homotopy groups $\pi_k(\mathcal{G})$ are related to bundles on $M \times S^{k+1}$.

The anomaly which we have been discussing, namely the obstruction to defining $\det D_A$ is usually computed infinitesimally along the loop in $\mathcal{G}$ which generates $\pi_1(\mathcal{G})$. Our approach here has been more global and topological.

§5. More on the index of families

The index of a single operator discussed in §2 and the degree of a 2-parameter family described in §3 appear to be quite different types of topological invariant. However, both involved the zero eigenvalues and they can in fact be put into the same overall picture as I shall now explain. As a first step suppose we have a 2-parameter family D_x with $N^+ = 1$, $N^- = 0$ for all x, so that we are in the situation where index $D_x = 1$. Then, for each x, we have a non-zero solution of $D_x\phi = 0$ unique up to a constant. As x varies these give a complex line-bundle over X and such line-bundles (over 2-dimensional X) are classified by a single integer invariant ν (the first Chern class or degree). If $\nu \neq 0$ the bundle is non-zero and it is impossible to make a global choice of ϕ varying continuously with x. In general if we try to choose ϕ continuously varying with x then it will vanish at a number of points and this number counted with appropriate multiplicities is just the invariant ν. Now if we replace the family $D_x : S_x^+ \to S_x^-$ by $\tilde{D}_x : S_x^+ \to S_x^- \oplus C$ where we have added a copy of C to match the function ϕ we see that $\tilde{D}$ is a family with index 0 and degree ν in the sense of §3.

This preliminary discussion was meant to show that the degree of §3 can also be defined when index $D_x \neq 0$, and it also prepares us to consider the more general case.

In general for any compact parameter space X let $\mathcal{N}_x^+$, $\mathcal{N}_x^-$ be the null-spaces of D_x and D_x^*. Assume first that these have <u>constant</u> dimension for all x in X. Then as x varies they define two complex vector bundles $\mathcal{N}^+$ and $\mathcal{N}^-$ over X of dimensions N^+ and N^-. Just as, for a single operator we defined the index as the integer $N^+ - N^-$, so for our family we define a generalized index to be the element

$$\mathcal{N}^+ - \mathcal{N}^- \quad \text{in} \quad K(X)$$

where $K(X)$ is the abelian group generated by vector bundles over X. If the dimensions of the null-spaces are not constant one can always reduce to the constant case by adding suitable fixed vector spaces, and the resulting index in $K(X)$ does not depend on any choices.

The general index theorem for families [3] computes this generalized index in topological terms. In particular various integer invariants such as Chern classes can be given by integrating suitable expressions over $M \times X$. For example when $X = S^2$ the degree of §3, which is just the first Chern class of the index in $K(X)$, is given by an integral over $M \times X$ which turns out to be just the same as that for the index of the Dirac operator of $M \times X$ extended to the bundle V. For further details see [6] and also [8].

§6. The real case

If we have a family of real differential operators then one can define a refinement of the index of §5 which is an element of $KR(X)$ the group generated by real vector bundles over X. For a single operator $KR(X) \cong K(X) \cong Z$, given by the dimension so that nothing is gained by this refinement. However, when $X = S^1$ we have

$$KR(S^1) = Z \oplus Z_2, \quad K(S^1) = Z$$

the extra Z_2 factor corresponding to the Möbius bundle over S^1. Thus for a family of real operators parametrized by the circle we have the index and a mod 2 degree. If the index is zero and D_x is invertible except at points $x_1, \ldots, x_k$ then the mod 2 degree ν is just the sum of the multiplicities (mod 2) of that described in §3 except that the real numbers replace the complex numbers throughout and invariants are only mod 2 because there are no natural orientations in the real case.

Moreover this mod 2 degree is the obstruction to defining a gauge invariant real determinant. The ambiguity is now one of sign, not phase, which is why our invariant is mod 2. This situation occurs for the Dirac operator when $X = S^4$ and $G = SU(2)$ (with V a C^2-bundle) and the obstruction is non-zero for this case. This was noted by Witten [14] where further details are explained. The computation of the mod 2 degree is also a special case of the general index theorem for real families [4].

§7. Odd dimensions and spectral flow

When dim M is odd the Dirac operator does not decompose into two parts and since it is self-adjoint its index is trivially zero. However, for a 1-parameter periodic family of such Dirac operators (parametrized by S^1) one can define a topological invariant called the spectral flow of the family [5]. This represents the number of negative eigenvalues that have become positive as the parameter traverses the circle (if the flow is negative the situation is reversed). This spectral flow can also be computed topologically. In fact it coincides with the index of the Dirac operator on $M \times S^1$ extended to the bundle V. This is quite analogous to the way the degree of a 2-parameter family can be computed (as noted in §5) by an index on $M \times S^2$.

This spectral flow invariant has recently been used by Vafa and Witten [13] to produce uniform upper bounds on the eigenvalues of Dirac operators.

§8. Conclusion

This has been a rather rapid survey of index theory and its relevance to Dirac operators and determinants. In particular we have seen that certain anomalies have a topological interpretation. However, I should emphasize that anomalies as understood by physicists are more precise algebraic objects than their topological counterparts. The precise relation between the algebraic and the topology is still a subject of investigation and for some new results in this direction I refer to [7].

References

1. M.F. Atiyah. The signature of fibre bundles, Collected Mathematical Papers in honour of K. Kodaira, Tokyo Univ. Press (1969).
2. M.F. Atiyah and I.M. Singer. The index of elliptic operators I, III, Ann. of Math. 87 (1968) 485-530, 546-604.
3. M.F. Atiyah and I.M. Singer. The index of elliptic operators IV, Ann. of Math. 93 (1971) 119-139.

4. M.F. Atiyah and I.M. Singer. The index of elliptic operators V, Ann. of Math. 93 (1971), 139-149.
5. M.F. Atiyah, V.K. Patodi and I.M. Singer, Spectral asymmetry and Riemannian geometry III, Math. Proc. Camb. Phil. Soc. 79 (1976), 71-99.
6. M.F. Atiyah and I.M. Singer, Dirac operators coupled to vector potentials, Proc. Nat. Acad. Sci. USA 81 (1984).
7. O. Alvarez, I.M. Singer and B. Zumino. Gravitational anomalies and the family's index theorem, Berkeley Preprints (1984).
8. L. Alverez-Gaumé and E. Witten, Gravitational Anomalies, Harvard Preprint, 1983 (to appear in Nuclear Physics B).
9. L. Alvarez-Gaumé and P. Ginsparg, Harvard preprint, 1983.
10. L. Baulieu, Algebraic construction of gauge invariant theories, Paris preprint, 1984 (to appear in Nuclear Physics).
11. L. Baulieu, Anomalies and gauge symmetry, Paris preprint, 1984.
12. T. Sumitani, Chiral anomalies and the generalized index theorem, Tokyo preprint, 1984.
13. C. Vafa and E. Witten, Eigenvalue inequalities for fermions in gauge theories, Princeton preprint, 1984.
14. E. Witten, A mod 2 anomaly, Princeton preprint.

STOCHASTIC QUANTIZATION, ASSOCIATED SUPERSYMMETRY AND NICOLAI MAP

Jnanadeva Maharana
Institute of Physics
Bhubaneswar 751005
India

ABSTRACT

In these lectures we present an introduction to the method of stochastic quantization due to Parisi and Wu and demonstrate the equivalence between stochastic and canonical methods of quantization in perturbation theory using the techniques due to Floratos and Iliopoulos. The lower critical dimension of spin systems with and without Gaussian random fields is discussed to illustrate the underlying supersymmetry of the system associated with stochastic differential equations. The Nicolai map is constructed for a simple model and its interpretation as a stochastic evolution equation is envisaged in a supersymmetric quantum mechanical model. The constraints on supersymmetry breaking, the Witten index and conjugation operations are discussed with some simple examples. The appendix contains relevant definitions and formulas of stochastic processes.

INTRODUCTION

The canonical and path integral methods of quantization are the two familiar prescriptions for quantization. They are equivalent and are adopted according to the problem under consideration. However, whenever they are employed to quantize gauge theories, gauge fixing becomes necessary. In case of non-Abelian gauge theories Faddeev-Popov ghosts[1] appear in the Lagrangian as a consequence of gauge fixing (of course, it is possible to choose ghost-free gauges which have their own complications). Furthermore, if one uses a covariant gauge the gauge fixing condition does not uniquely fix the gauge for large gauge potentials (this is the well-known Gribov ambiguity[2]).

Recently, a new method of quantization, the stochastic quantization has been introduced by Parisi and Wu[3]. This technique enjoys certain advantages over the conventional methods of quantization such as canonical and path integral prescription. One of the salient features of stochastic quantization method is that gauge fixing is not necessary and one can work in the nonperturbative domain without encountering problems associated with the Gribov ambiguity. Another important aspect of this formulation is that the symmetries of the Lagrangian are maintained as far as possible at each stage of computation in the new perturbative frame work (introduced by Parisi and Wu[3]).

The new quantization procedure is based on the Langevin equation[4] of non-equilibrium statistical mechanics. According to this approach the Langevin equation is a stochastic equation in a fictitious time dimension (we shall elaborate on this aspect later) and the new perturbation theory is applied in order to investigate how the system approaches the "equilibrium" configuration.

The method has also been applied to study the computer simulation of lattice gauge theories[5] and to get a better understanding of quenched-reduced models[6] besides gauge theories.[3,7-9]

Another important aspect of stochastic evolution equations is the presence of hidden supersymmetry associated with such equations.[10] It is well known that the lower critical dimensions of a spin system in the presence of a random magnetic field in D dimensions is the same as that of a spin system in D-2 dimensions without the random magnetic field.[11] Parisi and Sourlas discovered that the reason for this mysterious phenomena is the underlying supersymmetry of the system.

One of the important results for supersymmetric theories is the theorem due to Nicolai.[12] It has been shown that in a supersymmetric theory there is a mapping which renders the theory equivalent to a simpler theory where one has to deal with bosonic fields only. In specific cases of supersymmetric models there is a stochastic interpretation of the Nicolai mapping. Nicolai's method has proved to be very useful to facilitate nonperturbative construction of supersymmetric models.[13] In general the theorem is proved within the frame work of perturbation theory. The Nicolai mapping exists if and only if the vacuum energy of the supersymmetric theory is zero i.e. there is no spontaneous or dynamical breaking of supersymmetry. Therefore, it is important to investigate constraints on supersymmetry breaking although it may be argued that supersymmetry is not respected by Nature.

The purpose these lectures is to provide a pedagogical review of stochastic quantization, the Nicolai mapping and the Witten index.[14] There is an intimate relation between the Witten index and the index of Nicolai mapping. First we introduce the method of stochastic quantization due to Parisi and Wu. Then we follow the techniques of Floratos and Iliopoulos[8] to show the equivalence of stochastic and canonical quantization within the frame work of perturbation theory. Next, the hidden supersymmetry associated with stochastic equations is discussed. The theorem due to Nicolai, examples of Nicolai mapping and the relation of this mapping with stochastic evolution equations in case of specific supersymmetric models is discussed. Then we introduce the Witten index and constraints on supersymmetry breaking and demonstrate the relation of the Witten index with the index of the Nicolai mapping. We summarise definitions and some of the properties of stochastic processes in appendix A. In these lectures we have tried to provide several examples from quantum mechanics and field theory in 1+1 dimensions in order to illustrate various ideas.

IIa. STOCHASTIC QUANTIZATION

The method of stochastic quantization is based on the Langevin equation of non-equilibrium statistical mechanics. In case of conventional methods of quantization we are interested in computing Greens function (correlation functions) exactly or in some approximation scheme. Let us consider an Euclidean scalar field theory in D dimensions. The system is assumed to be at temperature T ($\beta = 1/kT$). The Euclidean correlation functions are defined as

$$<\phi(x_1)..\phi(x_n)> = \frac{\int d[\phi]\phi(x_1)..\phi(x_n)e^{-\beta S[\phi]}}{\int d[\phi]\exp(-\beta S[\phi])} \tag{1}$$

where $S[\phi]$ is the Euclidean action of the theory. Parisi and Wu have generalized the problem as follows. It is assumed that the field, ϕ, is a function of a fictitious time τ ($0 < \tau < \infty$) and it is coupled to a heat reservoir at temperature T. For large values of τ, $\phi(x,\tau)$ will reach the equilibrium configuration. Thus if one knows the evolution equation of $\phi(x,\tau)$ in τ then it is possible to obtain the equilibrium configuration (the limit $\tau \to \infty$) and consequently compute the equilibrium correlation functions given by eq.(1). We can assign $\phi(x,\tau)$ any "time" evolution equation such that we arrive at the equilibrium configuration for large τ.

The simplest evolution equation for a statistical system is the Langevin equation (see Appendix A),

$$\frac{\partial\phi(x,\tau)}{\partial\tau} = - \frac{\delta S[\phi]}{\delta\phi(x,\tau)} + \eta(x,\tau) \tag{2}$$

where $\eta(x,\tau)$ is Gaussian random noise with following properties

$$<\eta(x,\tau)> = 0 \tag{3}$$

$$<\eta(x,\tau)\eta(y,\tau')> = \frac{2}{\beta}\delta^D(x-y)\delta(\tau-\tau') \tag{4}$$

The above averages are Gaussian averages with respect to η i.e.

$$\langle\eta(x,\tau)\eta(x',\tau')\rangle = \frac{\int d(\eta)\,\eta(x,\tau)\eta(x',\tau')\,e^{-\frac{\beta}{2}\int d^Dx d\tau \eta^2(x,\tau)}}{\int d(\eta)\exp(-\frac{\beta}{2}\int d^Dx d\tau\ \eta^2(x,\tau))} \tag{5}$$

and the correlations satisfy

$$\begin{aligned}\langle\eta(x_1,\tau_1)\ \ldots\ \eta(x_4,\tau_4)\rangle &= \langle\eta(x_1,\tau_1)\ \ldots\ \eta(x_4,\tau_4)\rangle_C \\ &- \langle\eta(x_1,\tau_1)\eta(x_2,\tau_2)\rangle\langle\eta(x_3,\tau_3)\eta(x_4,\tau_4)\rangle \\ &- \langle\eta(x_1,\tau_1)\eta(x_3,\tau_3)\rangle\langle\eta(x_2,\tau_2)\eta(x_4,\tau_4)\rangle \\ &- \langle\eta(x_1,\tau_1)\eta(x_4,\tau_4)\rangle\langle\eta(x_2,\tau_2)\eta(x_3,\tau_3)\rangle\end{aligned} \tag{6}$$

The time evolution of $\phi(x,\tau)$ is specified uniquely once we are given the random noise $\eta(x,\tau)$ and impose the boundary conditions at $\tau=0$. To be specific let us assume $\phi(x,o)=0$, $S(0)=0$ in what follows. The stochastic correlation functions are defined as

$$\langle\phi^\eta(x,\tau)\phi^\eta(x',\tau')\rangle = \frac{\int d(\eta)\,\phi^\eta(x,\tau)\phi^\eta(x',\tau')\,e^{-\frac{\beta}{2}\int \eta^2(x,\tau)d^Dx d\tau}}{\int d(\eta)\exp(-\frac{\beta}{2}\int\eta^2(x,\tau)d^Dx d\tau)} \tag{7}$$

Here $\phi^\eta(x,\tau)$ denotes solution of the Langevin equation (2) for a specific $\eta(x,\tau)$.

Parisi and Wu[3] use the crucial theorem[15] of statistical mechanics that the equal time non-equilibrium correlation functions tend to the equilibrium correlation functions for large time. In other words

$$\langle\phi^\eta(x_1,\tau)\ \ldots\ \phi^\eta(x_n,\tau)\rangle_\eta \xrightarrow[\tau\to\infty]{} \langle\phi(x_1)\ \ldots\ \phi(x_n)\rangle \tag{8}$$

The LHS of eq.(8) is the Gaussian average with respect to η whereas the RHS is the average defined through eq.(1). We shall prove this result later in perturbation theory. In the rest of our article[16] we shall set $\beta=1$.

Alternatively, we can study the dynamics of the system given by eq.(2) through the Fokker-Planck equation. The time evolution of the Fokker-Planck probability function $P[\phi,\tau]$ is

$$\frac{\partial P[\phi,\tau]}{\partial\tau} = \int d^Dx\left(\frac{\delta^2 P[\phi,\tau]}{\delta\phi(x)^2} + \frac{\delta}{\delta\phi(x)}\left(\frac{\delta S}{\delta\phi(x)}\right)P[\phi,\tau]\right) \tag{9}$$

where

$$P[\phi,\tau] = \int d[\eta] e^{-\frac{1}{4}\int d^Dx d\tau\ \eta^2(x,\tau)} \prod_y \delta(\phi(y) - \phi^\eta(y,\tau)) \qquad (10)$$

$P[\phi,0]$ satisfies the initial boundary condition

$$P[\phi,0] = \prod_y \delta(\phi(y)) \qquad (11)$$

The stochastic correlation functions are given by

$$<\phi^\eta(x_1,\tau_1)\ldots\phi^\eta(x_n,\tau_n)>_\eta = \int d[\phi]\phi(x_1)\ldots\phi(x_n)P[\phi,\tau] \qquad (12)$$

which agrees with the earlier expression eq.(7) for the stochastic correlation function. This becomes transparent if we substitute eq. (10) for $P[\phi,\tau]$ in eq.(12). Let us define $P[\phi,\tau] = \psi[\phi,\tau]\exp(-S[\phi]/2)$. Then the Fokker-Planck equation (9) can be rewritten as

$$\frac{\partial\psi[\phi,\tau]}{\partial\tau} = \int d^Dx \left(- \frac{\delta^2}{\delta\phi(x)^2} - \frac{1}{2}\frac{\delta^2 S}{\delta\phi(x)^2} + \frac{1}{4}\left(\frac{\delta S}{\delta\phi(x)}\right)^2\right)\psi[\phi,\tau] \qquad (13)$$

We can write (13) as

$$-\frac{\partial\psi[\phi,\tau]}{\partial\tau} = H_{F.P.}\ \psi[\phi,\tau] \qquad (14)$$

With

$$H_{F.P.} = \int \alpha^\dagger(x)\alpha(x)d^Dx \qquad (15)$$

and

$$\alpha = -i\left(\frac{\delta}{\delta\phi(x)} + \frac{1}{2}\frac{\delta S}{\delta\phi(x)}\right) \qquad (16)$$

The Fokker-Planck Hamiltonian, $H_{F.P.}$ is formally self-adjoint and consequently we can expand $\psi[\phi,\tau]$ in a complete set of its eigenstates

$$\psi[\phi,\tau] = \sum_{n=0}^{\infty} a_n\, e^{-\lambda_n\tau}\ \psi_n[\phi,\tau] \qquad (17)$$

Since $H_{F.P.}$ is positive the eigenvalues, $\lambda_n \geq 0$. We find that $\exp(-S[\phi]/2)$ is annihilated by α and corresponds to a zero energy eigenstate $\lambda_o = 0$ of $H_{F.P.}$. If this state is not orthogonal to $\psi[\phi,\tau]$ and all other eigenvalues are nonzero then the stochastic correlation functions can be written as

$$<\phi^{\eta}(x_1,\tau)..\phi^{\eta}(x_n,\tau)> = \int d[\phi]\phi(x_1)..\phi(x_n)e^{-S[\phi]}$$
$$+ \int d[\phi]\phi(x_1)..\phi(x_n)\sum_{n=1}^{\infty} a_n e^{-\lambda_n\tau}\psi_n[\phi,\tau], \ \lambda_n > 0 \quad (18)$$

Notice that equal-time stochastic correlation functions go over to the Euclidean correlation functions in the $\tau \to \infty$ limit. Thus the validity of the equivalence between stochastic and canonical quantization as stated above depends on the crucial facts that the Fokker-Planck Hamiltonian is positive, that there exists a mass gap and the initial condition on $\psi[\phi,\tau]$ is such that it is not orthogonal to the ground state wave functional of $H_{F.P.}$. Now we are in a position to prove the equivalence between stochastic and canonical quantization in perturbation theory.

IIb. EQUIVALENCE BETWEEN STOCHASTIC AND CANONICAL QUANTISATION

Let us write the Lagrangian density for Euclidean self-interacting scalar field theory as sum of two terms[8]

$$L = L_o + g\, L_I \quad (19)$$

where L_o and L_I are free and interaction Lagrangian densities and g is the coupling constant. $S[\phi]$ is the corresponding action functional. The Green functions of the theory are vacuum expectation values of fields i.e. $<\phi(x_1)...\phi(x_n)>$ and are defined through path integral method in the standard manner. It is always possible to introduce an appropriate regulator such that the probability distribution $\exp(-S[\phi]) / \int d[\phi]\exp(-S[\phi])$ is regularized and is well defined order by order in perturbation theory. The Green functions are computed as power series in the coupling constant g. We can also expand the Fokker-Planck probability function $P[\phi,\tau]$ as a power series in g as follows:

$$P[\phi,\tau] = \sum_{n=0}^{\infty} g^n P_n[\phi,\tau] \quad (20)$$

The ℓ-point equal time stochastic correlation function at n^{th} order of perturbation theory is given by

$$<\phi^{\eta}(x,\tau)...\phi^{\eta}(x_\rho,\tau)> = g^n\int d[\phi]\phi(x_1)...\phi(x_\ell)P_n[\phi,\tau] \quad (21)$$

We may remark here that $P_n[\phi,\tau]$ defined in eq.(20) is to be understood in the sense that it is the probability function which defines the correlation functions (21) in perturbation theory.

Furthermore, inserting eq.(20) in the Fokker-Planck equation (9) we get the following equation for P_n

$$\frac{\partial P_n[\phi,\tau]}{\partial\tau} = \int d^Dx \, \frac{\delta}{\delta\phi(x)} \left(\frac{\delta}{\delta\phi(x)} + \frac{\delta S_o}{\delta\phi(x)}\right) P_n[\phi,\tau] + \int d^Dx \, \frac{\delta}{\delta\phi(x)} \frac{\delta S_I[\phi]}{\delta\phi(x)} P_{n-1}[\phi,\tau] \tag{22}$$

with $S_o = \int \mathcal{L}_o \, d^Dx$ and $S_I = \int \mathcal{L}_I \, d^Dx$. The initial conditions are

$$P_o[\phi,0] = \prod_y \delta(\phi(y)) \tag{23}$$

$$P_n[\phi,0] = 0 \qquad n = 1,2,\ldots \tag{24}$$

We shall prove that for $1 < k < n-1$, $P_k[\phi,\tau]$ is uniformly bounded and

$$\underset{\tau\to\infty}{\mathrm{W.Lim}} \, P_k[\phi,\tau] = P_k^{eq}[\phi] \tag{25}$$

in other words, in the large time limit the probability distribution $P_n[\phi,\tau]$ goes over to the equilibrium distribution $P_n^{eq}[\phi]$. This equation (25) has the interpretation that it is applied always to a string of fields and the actual meaning of the equation is that ℓ-point equal time stochastic correlation functions, $\langle\phi^\eta(x_1,\tau)\ldots\phi^\eta(x_\ell,\tau)\rangle$, in n^{th} order in the perturbation theory go over to corresponding Euclidean Green function in the regularized perturbation theory.

The proof of (25) will be inductive. We exclude k=0 due to the fact that $P_o[\phi,\tau]$ is not bounded as $\tau \to 0$ since it satisfies the initial boundary condition (23). However, we shall show $P_o[\phi,\tau] = P_o^{eq}[\phi]$ in the limit $\tau \to \infty$. Now we shall show that

$$\underset{\tau\to\infty}{\mathrm{W.Lim}} \, \frac{\partial}{\partial\tau} P_k[\phi,\tau] = 0 \tag{27}$$

Proof: Let us write solution to eq.(22) as

$$P_k[\phi,\tau] = \int d[\phi'] \int_0^\tau \mathbb{D}_o[\phi,\phi',\tau-\tau'] \int d^Dx \, \frac{\delta}{\delta\phi'(x)} \frac{\delta S_I}{\delta\phi'(x)} P_{k-1}[\phi',\tau'] \tag{28}$$

where $\mathbb{D}_o[\phi,\phi',\tau-\tau']$ is a Green functional which satisfies the equation

$$\frac{\partial}{\partial\tau}\mathbb{D}_o[\phi,\phi',\tau-\tau'] - \int d^Dx \frac{\delta}{\delta\phi(x)}\left(\frac{\delta}{\delta\phi(x)} + \frac{\delta S_o}{\delta\phi(x)}\right)\mathbb{D}_o[\phi,\phi',\tau-\tau']$$

$$= \delta(\tau-\tau')\prod_x \delta(\phi(x) - \phi'(x)) \qquad (29)$$

with the initial boundary condition

$$\mathbb{D}_o[\phi,\phi',\tau-\tau']\Big|_{\tau=\tau'} = \prod_x \delta(\phi(x) - \phi(x')) \qquad (30)$$

We can explicitly solve eq.(29). Let us first obtain an expression for $P_o[\phi,\tau]$ by solving eq.(22) for $g = 0$ and it follows from (10) that

$$P_o[\phi,\tau] = \int d[\eta] e^{-\frac{1}{4}\int\eta^2(x,\tau)d^Dxd\tau} \prod_y \delta(\phi(y) - \phi_o^\eta(y,\tau)) \qquad (31)$$

where the "free" field ϕ_o^η satisfies the Langevin equation

$$\frac{\partial\phi_o^\eta(x,\tau)}{\partial x} = (\Box - m^2)\phi_o^\eta(x,\tau) + \eta(x,\tau) \qquad (32)$$

Thus

$$\phi_o^\eta(x,\tau) = \int d^Dx' \int_o^\tau d\tau' \, G(x-x', \tau-\tau')\eta(x',\tau') \qquad (33)$$

with $\phi_o^\eta(x,0) = 0$. The retarded Green function $G(x-x', \tau-\tau')$ has the form[3]

$$G(x-x', \tau-\tau') = \frac{1}{(2\pi)^D}\int d^Dk \, e^{ik.(x-x')-(\tau-\tau')(k^2+m^2)} \qquad (34)$$

The correlation function D is defined as

$$D(x,x',\tau,\tau') \equiv \langle\phi^\eta(x,\tau)\phi^\eta(x',\tau')\rangle_\eta$$

$$= 2\int d^Du \int_o^{\min(\tau,\tau')} d\tau'' G(x-u,\tau-\tau'')G(x'-u,\tau'-\tau'') \qquad (35)$$

Eq.(21) can be re-expressed as

$$P_o[\phi,\tau] = \int d\,\eta\, e^{-\frac{1}{4}\int d^Dx\, d\tau\, \eta^2(x,\tau)}$$

$$e^{i\int d^Dx\xi(x)[\phi(x) - \phi_o^\eta(x,\tau)]}\, d[\xi] \qquad (36)$$

Here we have used the Fourier representation for the delta function $\prod_x\delta(\phi(x) - \phi_\eta^o(x,\tau))$. If we use the solution (33) for $\phi_o^\eta(x,\tau)$ in eq.(36)

then the Gaussian integrations over $d[\eta]$ and $d[\xi]$ can be performed exactly. We obtain following expression for $P_o[\phi,\tau]$.

$$P_o[\phi,\tau] = N_o^{-1}\exp(-\frac{1}{2}\int d^Dx d^Dx'\phi(x)D^{-1}(x,x',\tau,\tau)\phi(x')) \tag{37}$$

Here D^{-1} is the inverse of the correlation function D defined through eq.(35), and

$$\int d^Dx'' D(x,x'',\tau,\tau)D^{-1}(x'',x',\tau,\tau') = \delta^D(x-x') \tag{38}$$

If we use eq.(34) in (35) we obtain an expression for $D(x,x',\tau,\tau)$ and consequently $P_o[\phi,\tau]$ has the following form

$$P_o[\phi,\tau] = N_o^{-1}\, e^{-\frac{1}{2}\int d^Dk\phi(k)\frac{k^2+m^2}{1-e^{-2\tau(k^2+m^2)}}\phi(-k)} \tag{39}$$

N_o is a normalization factor. In order to obtain the Green functional $\mathbb{D}_o[\phi,\phi',\tau,\tau']$ we use the definition (see appendix A)

$$\mathbb{D}_o[\phi,\phi',\tau-\tau'] \equiv P_o[\phi,\phi',\tau,\tau']/P_o[\phi',\tau'] \tag{40}$$

as the conditional probability density, where

$$P_o[\phi,\phi',\tau,\tau'] = \int d[\eta] e^{-\frac{1}{4}\int\eta^2(x,\tau)d^Dx d\tau} \prod_y \delta(\phi(y) - \phi_o^\eta(y,\tau))\delta(\phi'(y) - \phi_o^\eta(y'\tau')) \tag{41}$$

It is now straight forward to obtain form for $\mathbb{D}_o[\phi,\phi',\tau]$

$$\mathbb{D}_o[\phi,\phi',\tau] = \Delta_o^{-1}\theta(\tau)e^{-\frac{1}{2}\int d^Dk\tilde{\phi}(k)D(\tau)\tilde{\phi}(-k)} \tag{42}$$

where $\tilde{\phi}(k)$ denotes $(\phi(k),\ \phi'(k))$ and $D(\tau)$ is the following 2x2 matrix

$$D_{11}(\tau) = (k^2+m^2)/[1-e^{-2\tau(k^2+m^2)}] \tag{43}$$

$$D_{12}(\tau) = D_{21}(\tau) = -\, D_{11}(\tau)e^{-\tau(k^2+m^2)} \tag{44}$$

$$D_{22}(\tau) = D_{11}(\tau)e^{-2\tau(k^2+m^2)} \tag{45}$$

The normalization constants N_o and Δ_o are determined from the condition that

$$\int d[\phi] P_o[\phi,\tau] = 1 \tag{46}$$

and

$$\int d[\phi]\, \mathbb{D}_o[\phi,\phi',\tau] = \theta(\tau) \tag{47}$$

We find that in the limit $t \to \infty$, P_o and D_o satisfy following relations.

$$P_o[\phi,\tau] \xrightarrow[\tau\to\infty]{} P_o^{eq}[\phi] = N_o^{eq^{-1}}\, e^{-\frac{1}{2}\int d^D k \phi(k)(k^2+m^2)\phi(-k)} \tag{48}$$

Note that $P_o^{eq}[\phi]$ can be used to obtain all free Green functions of the theory. Further,

$$\frac{\partial P_o}{\partial \tau} \xrightarrow[\tau\to\infty]{} O(e^{-m^2\tau}) \tag{49}$$

$$\mathbb{D}_o[\phi,\phi',\tau] \xrightarrow[\tau\to\infty]{} P_o^{eq}[\phi] \tag{50}$$

$$\frac{\partial \mathbb{D}_o}{\partial \tau} \xrightarrow[\tau\to\infty]{} O(e^{-m^2\tau}) \tag{51}$$

$$\frac{\delta \mathbb{D}_o}{\delta\phi'(x)} \xrightarrow[\tau\to\infty]{} O(e^{-m^2\tau}) \tag{52}$$

Having shown that $P_o[\phi,\tau] \to P_o^{eq}[\phi]$ as $\tau \to \infty$ now we focus our attention on $P_k[\phi,\tau]$. Partial functional integration of (28) gives following expression for $P_k[\phi,\tau]$

$$P_k[\phi,\tau] = -\int d[\phi']\int d^D x\, \frac{\delta S_I}{\delta\phi'(x)} \int_0^\tau d\tau'\, \frac{\delta \mathbb{D}_o[\phi,\phi',\tau-\tau']}{\delta'\phi'(x)}\, P_{k-1}[\phi',\tau'] \tag{53}$$

Similarly

$$\frac{\partial P_k[\phi,\tau]}{\partial\tau} = -\int d[\phi']\int d^D x\, \frac{\delta S_I}{\delta\phi'(x)} \frac{\partial}{\partial\tau} \int_0^\tau d\tau'\, \frac{\delta \mathbb{D}_o[\phi,\phi',\tau-\tau']}{\delta\phi'(x)}\, P_k[\phi',\tau'] \tag{54}$$

We can interchange the order of differentiation and integration over ϕ' and τ since above equations, (53) and (54), are to be understood only in presence of strings of fields. We have assumed that the theory is suitably regularized and such operations are permissible. Let us first consider the case k=1 in (53).

$$P_1[\phi,\tau] = -\int d[\phi']\int d^D x\, \frac{\delta S_I}{\delta\phi'(x)} \int_0^\tau d\tau'\, \frac{\delta \mathbb{D}_o[\phi,\phi',\tau-\tau']}{\delta\phi'(x)}\, P_o[\phi',\tau'] \tag{55}$$

Notice that $P_o[\phi,0] = \prod_\alpha \delta(\phi(x))$ due to boundary conditions; however, $\delta S_I/\delta\phi'(x)$ is monomial in ϕ' and consequently $P_1[\phi,0] = 0$. Further, if we use explicit form for $\mathbb{D}_o[\phi,\phi',\tau-\tau']$ and P_o in (55) it is easy to check that $P_1[\phi,\tau]$ is uniformly bounded in ϕ for all τ and in the limit $\tau \to \infty$ it is well defined. We can extend the proof for any $P_k[\phi,\tau]$ as follows: $P_{k-1}[\phi',\tau']$ is bounded and $\delta\mathbb{D}_o/\delta\phi'$ is integrable for all ϕ. Since $P_{k-1}[\phi',\tau]$ contains a Gaussian measure of the free theory, integration over ϕ' is always possible. Thus $P_k[\phi,\tau]$ is bounded in ϕ for all τ, $P_k[\phi,0] = 0$ and $P_k[\phi,\tau]$ exists in the limit $\tau \to \infty$. We can apply the same string of arguments to eq.(54) in order to prove that $\partial P_k/\partial\tau$ exists for all τ and ϕ and $\partial P_k/\partial\tau \to 0$ as $\tau \to \infty$.

Now we are in a position to prove our main result. $P_k[\phi,\infty] = P_k^{eq}[\phi]$. Let us consider the Fokker-Planck equation (22) and take the limit $\tau \to \infty$. The LHS vanishes as we have seen above. The second term in the RHS contains $P_{k-1}[\phi,\tau]$ and it is equal to $P_{k-1}^{eq}[\phi]$ in the limit $\tau \to \infty$ according to the inductive hypothesis. The result holds good for $P_1[\phi,\infty]$ independent of the inductive hypothesis since $P_o[\phi,\infty] = P_o^{eq}[\phi]$ from eq.(48). Thus $P_k[\phi,\infty]$ satisfies the following equation

$$\int d^Dx \frac{\delta}{\delta\phi(x)}\left(\frac{\delta}{\delta\phi(x)} + \frac{\delta S_o}{\delta\phi(x)}\right)P_k[\phi,\infty] = -\int d^Dx \frac{\delta}{\delta\phi(x)}\frac{\delta S_I}{\delta\phi(x)} P_{k-1}^{eq}[\phi] \tag{56}$$

The above equation is equivalent to the Schwinger-Dyson equation. Let us define $G_k[\phi]$

$$P_k[\phi,\infty] - P_k^{eq}[\phi] = e^{-S_o[\phi]} G_k[\phi] \tag{57}$$

Notice that $P_k[\phi,\infty]$ and $P_k^{eq}[\phi]$ satisfy eq.(56) and we can use this information to show that

$$\int d^Dx \frac{\delta}{\delta\phi(x)}\left(e^{-S_o[\phi]}\frac{\delta G_k}{\delta\phi(x)}\right) = 0 \tag{58}$$

If we multiply (58) by $G_k[\phi]$ and integrate functionally over ϕ we get (after partial integration)

$$\int d[\phi]\int d^Dx\, e^{-S_o[\phi]}\left(\frac{\delta G_k}{\delta\phi(x)}\right)^2 = 0 \tag{59}$$

Since the integrand is positive, $\delta G_k/\delta\phi(x) = 0$, and it follows that G_k is independent of ϕ. Thus $P_k[\phi,\infty]$ and $P_k^{eq}[\phi]$ differ by a constant multiplied by $\exp(-S_o)$. Further, $P_k[\phi,\infty]$ satisfies the Schwinger-Dyson equation. This arbitrariness can be removed if we use the initial condition $P_k^{eq}[\phi=0] = 0$ for $k = 0$ or the normalization condition $\int d[\phi]P^{eq}[\phi] = 0$ for $k = 0$. The same constraint is satisfied by $P_k[\phi,\tau]$ and consequently it follows that $G_k = 0$ and

$$P_k[\phi,\infty] = P_k^{eq}[\phi] \tag{60}$$

This completes the proof of equivalence between stochastic and canonical quantization, Eq.(57) has the interpretation that it defines the distribution $P_k[\phi,\infty]$ in terms of $P_{k-1}^{eq}[\phi]$ which generates the Feynman diagrams of the ordinary perturbation series. The associated Schwinger-Dyson equation has a unique solution in regularized unrenormalized perturbation theory. The same result holds good for the renormalized series up to a renormalization group transformation.

III. STOCHASTIC QUANTIZATION AND SUPERSYMMETRY

In this section we shall illustrate the existence of hidden supersymmetry of a spin system in presence of a random magnetic field. It is a well known result that the Ising spin system has lower critical dimensionality one. In other words, there is no order for one or lower dimensions for such a system whereas the higher dimensional spin systems have an ordered phase. However, if the spin system is subjected to a random magnetic field then the lower critical dimension becomes three as was shown by Parisi and Sourlas.[10] They used an elegant argument based on supersymmetry and showed that order-by-order the values of the most-infrared-divergent diagrams in D-dimensional spin systems with random fields are equivalent to the system without random fields in D-2 dimensions. They showed that this mysterious result has its origin in a simple geometrical interpretation. The spin system in D-dimensions can be described by a scalar field Lagrangian density near the critical temperature.

$$= -\frac{1}{2}\,\phi\Box\phi + V(\phi) \tag{61}$$

with

$$V(\phi) = \frac{1}{2}\,m^2\phi^2 + \frac{1}{4}\,g\phi^4$$

$$F[h] = \ln\int d[\phi]\exp[-\int d^Dx(\,\mathcal{L} + h(x)\phi(x))] \tag{62}$$

Here $F[h]$ is the free energy and $h(x)$ is the random magnetic field.

Define free energy F_R averaged over the Gaussian magnetic field is as follows.

$$F_R = \int d[h]F[h]e^{-\frac{1}{2}\int h^2(x)d^Dx} \tag{63}$$

We can construct perturbative expansion for F_R and it is well known from dimensional analysis that the most-infrared-divergent diagrams contain the maximum number of h^2 insertions. The tree approximation for F[h] is obtained when contribution of other diagrams are neglected. If we use the correspondence between tree approximation and the classical nonlinear differential equation[17] then the two point function in this limit is

$$<\phi(x)\phi(0)>_R = \int d[h]\phi_h(x)\phi_h(0)e^{-\frac{1}{2}\int h^2(x)d^Dx} \tag{64}$$

$\phi_h(x)$ satisfies the stochastic differential equation (h(x) being the random Gaussian field)

$$-\Box\phi + \frac{\delta V(\phi)}{\delta\phi} + h = 0 \tag{65}$$

We can rewrite (64) as

$$<\phi(x)\phi(0)> = \int d(h)d[\phi]\delta(\phi(x) - \phi_h(x))e^{-\frac{1}{2}\int h^2(x)d^Dx}\phi(x)\phi(0) \tag{66}$$

$$= \int d(h)d[\phi]\delta(-\Box\phi + \frac{\delta V}{\delta\phi} + h)\det(-\Box + \frac{\delta^2 V}{\delta\phi^2})$$
$$\times \exp[-\frac{1}{2}\int h^2(x)d^Dx]\phi(x)\phi(0) \tag{67}$$

Now the functional integration over h can be done by using the delta function. We introduce an anticommuting scalar field ψ in order to exponentiate the determinant in (67) and introduce another auxiliary scalar field A to re-express (67)

$$<\phi(x)\phi(0)> = \int d[\phi]d[\psi]d[A]\phi(x)\phi(0)e^{-\int \mathcal{L}_s d^Dx} \tag{68}$$

where

$$\mathcal{L}_s = -\frac{1}{2}A^2 + A(-\Box\phi + \frac{\delta V}{\delta\phi}) + \bar{\psi}(-\Box + \frac{\delta^2 V}{\delta\phi^2}) \tag{69}$$

The above Lagrangian is invariant under following supersymmetry transformations[18]

$$\delta\phi = - \bar{a}\varepsilon_\mu x_\mu \psi \tag{70}$$

$$\delta A = 2\bar{a}\varepsilon_\mu \partial_\mu \psi \tag{71}$$

$$\delta\psi = 0 \tag{72}$$

$$\delta\bar{\psi} = \bar{a}(\varepsilon_\mu x_\mu A + 2\varepsilon_\mu \partial_\mu \phi) \tag{73}$$

where $\bar{a}$ is an infinitesimal anticommuting number and ε_μ is an arbitrary vector. This is the hidden supersymmetry of the associated stochastic equation.

If we consider the Langevin equation (2) connected with the stochastic quantization then it is possible to show that there is an underlying supersymmetry of the theory. Let us construct a generating functional,[19] W, that gives all the stochastic correlation functions such that

$$\frac{\delta^n W(J)}{\delta J(x_1,\tau)\ldots\delta J(x_n,\tau)} = \langle \phi^\eta(x_1,\tau) \ \ldots \ \phi^\eta(x_n,\tau)\rangle_\eta \tag{74}$$

It is easy to check that W(J) with source J has the following form

$$W(J) = \int d[\phi]d[\eta]\Pi\delta(\phi-\phi^\eta)e^{-\frac{1}{4}\int \eta^2 d^Dx d\tau + \int J\phi d^Dx d\tau} \tag{75}$$

Using the same trick as in eq.(69) we can write (75) as

$$W(J) = \int d[\phi]e^{\{-\frac{1}{4}\int(\frac{\partial\phi}{\partial\tau} + \frac{\delta S[\phi]}{\delta\phi})^2 d^Dx d\tau + \int J\phi d^Dx d\tau\}} \det(\frac{\delta\eta}{\delta\phi}) \tag{76}$$

where we have used the fact that ϕ^η is a solution of the Langevin equation (2). Now one can introduce appropriate anticommuting variable with suitable boundary conditions to exponentiate the Jacobian determinant $\delta\eta/\delta\phi$. We can obtain a supersymmetric Lagrangian such that $W(J) \sim \int d[\phi]\exp[-\int \mathcal{L}^{F.P.} d^Dx d\tau + \int J\phi d^Dx d\tau$ and the corresponding Fokker-Planck Hamiltonian density has the structure

$$H_{F.P.} = - \frac{\delta^2}{\delta\phi(x)^2} + \frac{1}{4}(\frac{\delta S}{\delta\phi})^2 - \frac{1}{2}\sigma_3 \frac{\delta^2 S}{\delta\phi^2} \tag{77}$$

The supersymmetry of this Hamiltonian is not the conventional supersymmetry involving Lorentz generators. The supersymmetry is analogous to that of supersymmetric quantum mechanics[20] such that ϕ plays the role of configuration variable x and the stochastic time has role of ordinary time. The corresponding supercharges are

$$Q_1 = (\Pi_\phi + \frac{1}{2}\frac{\delta S}{\delta \phi})\psi \tag{78}$$

$$Q_2 = \bar{\psi}(\Pi_\phi - \frac{1}{2}\frac{\delta S}{\delta \phi}) \tag{79}$$

Π_ϕ is the canonical momentum of field ϕ. The supersymmetry transformations are

$$\delta\phi = \varepsilon_\psi[Q_1,\phi] + \varepsilon_{\bar{\psi}}[Q_2,\phi] \tag{80}$$

$$\delta\psi = \varepsilon_\psi\{Q_1,\psi\} + \varepsilon_{\bar{\psi}}\{Q_2,\psi\} \tag{81}$$

$$\delta\bar{\psi} = \varepsilon_\psi\{Q_,\bar{\psi}\} + \varepsilon_{\bar{\psi}}\{Q_2,\bar{\psi}\} \tag{82}$$

where ψ and $\bar{\psi}$ are the anticommuting fields similar to the fields introduced in (68) and ε_ψ, $\varepsilon_{\bar{\psi}}$ are two infinitesimal anticommuting parameters. It is possible to formulate this problem in superspace in terms of variables in superspace.[19]

To summarise, we have shown the existence of supersymmetry of stochastic equations. In the second example it is clear that the zero eigenvalue of the Fokker-Planck Hamiltonian implies that the underlying supersymmetry is unbroken. Indeed, there are intimate relations between supersymmetry and stochastic equations which arise in the context of the Nicolai mapping. This is the topic of discussion of the next section.

IV. NICOLAI MAP

Supersymmetric theories employ anticommuting numbers, the Grassmann variables, graded Lie algebras in an essential manner. Nicolai[12] has discovered a new characteristic of such theories that can be used to define supersymmetric theories without incorporating the concept of Grassmann variable and graded Lie algebras. His results hold good for scalar as well as gauge theories with supersymmetry. In the case of supersymmetric theories there exists a transformation which renders the theory equivalent to a theory without fermions.

To be specific let us consider a theory with scalars (or pseudoscalars) ϕ_i and some Majorana spinors ψ_i. It is understood that all auxiliary fields have been eliminated. The expectation values are defined through functional integrals with a suitable regularization prescription. We do not specify the space-time dimensions at this stage. The central result can be presented as a THEOREM (Nicolai[12]). Scalar supersymmetric models are characterised by existence of a generally non-linear and non-local transformation T_g of the boson fields ϕ_i , (g being coupling constant)

$$T_g: \phi_i(x) \rightarrow \phi_i'(x,g,\phi_i) \tag{83}$$

and T_g satisfies following properties.

(a) T_g is invertible in the sense of formal power series,

(b) $S[g,\phi] = S_o[\phi'(g,\phi)]$, where $S[g,\phi]$ is the full bosonic action of the original theory and S_o is its free bosonic part.

(c) The Jacobian of the transformation T_g is nothing but the Matthews-Salam determinant[21,22] that is obtained by integrating the fermions. Notice that if the fermions are self conjugate Majorana spinors the Matthews-Salam determinant is in fact the square root of a determinant. If there is a single scalar field ϕ and $D(g,\phi)$ is MS determinant then vacuum expectation values of a string of scalar fields are nothing but the moments of the functional measure

$$d\mu_g = e^{-S[g,\phi]} D(g,\phi) d[\phi] / \int d[\phi] e^{-S[g,\phi]} \tag{84}$$

Nicolai has shown that for supersymmetric theories there exists a transformation T_g such that

$$d\mu_g(T_g^{-1}\phi) = d\mu_o(\phi) = e^{-S_o[\phi]} d[\phi] / \int d[\phi] e^{-S_o[\phi]} \tag{85}$$

It is interesting to note that the moments of the non-Gaussian measure $d\mu_g$ have been expressed in terms of the transforms of moments of a Gaussian one. Does it mean that all supersymmetric theories are trivial and where are all interactions gone? In fact all the interactions lie in the transformation T_g. It was also shown that the vacuum energy of the theory must vanish to all orders in perturbation theory in order that T_g exists. Furthermore, if supersymmetry is not exact then it is not possible to find T_g satisfying properties (a), (b) and (c).

We present an example due to Nicolai in order to illustrate the theorem by constructing the most general supersymmetric Lagrangian with one scalar field ϕ and one Majorana spinor ψ. It is constructed by imposing a constraint that Matthews-Salam determinant be a Jacobian in two limiting cases. The model is considered in 1+1 dimensions.

The free Lagrangian for ϕ is

$$= \frac{1Z}{2}(\partial_\mu \phi)^2 + \frac{m^2}{2Z}\phi^2, \ Z > 0 \tag{86}$$

It follows from (86) that

$$\int \prod_x \frac{d\phi(x)}{\sqrt{2\pi Z}} e^{-\frac{1}{2}Z\int dx\phi(x)[-\Box + \frac{m^2}{Z^2}]\phi(x)} = \det(-Z^2\Box + m^2)^{-1/2} \tag{87}$$

We have introduced appropriate normalization factors in (87). If we introduce another term in (86) involving a two-component Majorana spinor ψ then it is possible to cancel the determinant on the LHS of eq.(87) since

$$\begin{aligned} \int \prod_x \prod_{\alpha=1,2} d\psi_\alpha \exp[-\frac{1}{2}\int dx\bar{\psi}(x)(Z\partial\!\!\!/+m)\psi(x)] &= \det(Z\partial\!\!\!/+m)^{1/2} \\ &= \det(-Z^2\Box + m^2)^{1/2} \end{aligned} \tag{88}$$

Thus if $\mathcal{L}$ in (86) is replaced by

$$\mathcal{L} \rightarrow \mathcal{L} + \frac{1}{2}\bar{\psi}(Z\partial\!\!\!/+m)\psi \tag{89}$$

then the two determinants appearing on LHS of (87) and (88) cancel. How does one construct an interacting theory? It can be constructed through following ansatz: the bosonic part of the Lagrangian, $_B$, is given by

$$_B = \frac{Z}{2}(\partial_\mu\phi)^2 + \frac{1}{2Z}V(\phi), \ V(\phi) = m^2\phi^2 + \ldots \ V(\phi) > 0 \tag{90}$$

whereas the fermionic Lagrangian is given by

$$_F = \frac{1}{2}Z\bar{\psi}\partial\!\!\!/\psi + \frac{1}{2}W(\phi)\bar{\psi}\psi, \ W(\phi) = m + \ldots \tag{91}$$

As $Z \rightarrow 0$ the kinetic energy terms can be neglected and we get

$$\begin{aligned} \lim_{Z\rightarrow 0} \frac{1}{\sqrt{2\pi Z}} e^{-V(\phi)/2Z} &= \delta(V(\phi))^{1/2} \\ &= \Sigma\delta(\phi-\phi_i)/|(V'(\phi_i))^{1/2}| \end{aligned} \tag{92}$$

where ϕ_i are zeros of $V(\phi)$, i.e. $V(\phi_i) = 0$. The fermionic functional integral is

$$\int \prod_{x} \prod_{\alpha=1,2} d\psi_{\alpha}(x) \exp[-\frac{1}{2}\int dx W(\phi)\bar{\psi}\psi = \prod_{x} W(\phi(x)) \tag{93}$$

Define $P(\phi) = V(\phi)^{1/2}$, and it follows that

$$|P'(\phi)| = W(\phi) > 0 \quad P(\phi) = \text{const} + m\phi + g\bar{\phi}^{2} + \ldots \tag{(4)}$$

in order to satisfy our constraints. It is obvious that either $P(\phi) > 0$ or $P(\phi) < 0$ for all ϕ. The local but nonlinear map $\phi \to P(\phi)$ can only be inverted if the potential has one and only one absolute minimum at $V = 0$. The total Lagrangian is

$$= \frac{1}{2} Z(\partial_{\mu}\phi)^{2} + \frac{1}{2Z} P(\phi)^{2} + \frac{1}{2} Z\bar{\psi}\partial\!\!\!/\psi + \frac{1}{2} P'(\phi)\bar{\psi}\psi \tag{95}$$

We can introduce an auxiliary field, A, to cast it in the familiar form

$$= \frac{1}{2} Z[(\partial_{\mu}\phi)^{2} + \bar{\psi}\partial\!\!\!/\psi + A^{2}] + iAP(\phi) + \frac{1}{2} P'(\phi)\bar{\psi}\psi \tag{96}$$

It is clear that when we integrate over the fields A

$$\int_{-\infty}^{+\infty} \frac{d[A]}{\sqrt{2\pi}} \exp[-ZA^{2} - iAP(\phi)] = \frac{1}{\sqrt{Z}} \exp[-P^{2}(\phi)/2Z)] \tag{97}$$

If we now choose $P(\phi) = m\phi + g\phi^{3}$ with $m > 0$ and $g > 0$ so that $P'(\phi) > 0$. The transformation T_{g} to order g^{3} is given by

$$\begin{aligned} \phi'(x,g,\phi) = {} & \phi(x) + mg\int G(x-x')\phi^{3}(x')dx' \\ & - \frac{3}{2} g^{2}Z^{2}\int \partial_{\mu}G(x-x')\phi^{2}(x')\partial_{\mu}G(x'-x'') \\ & \phi^{3}(x'')dx'dx'' + \frac{9}{2} g^{3}mZ^{2}\int dx'dx''dx''' \\ & \partial_{\mu}G(x-x')\phi^{2}(x')G(x'-x'')\phi^{2}(x'') \\ & \partial_{\mu}G(x''-x''')\phi^{3}(x''') \end{aligned} \tag{98}$$

Where

$$G(x-x') = \int \frac{d^{2}k}{(2\pi)^{2}} e^{ik.(x-x')}/(Z^{2}k^{2}+m^{2}) \tag{99}$$

Notice that

$$\frac{1}{2Z}\int dx\phi'(x,g,\phi)(-Z^2\Box+m^2)\phi'(x,g,\phi)$$

$$= \int dx[\frac{Z}{2}(\partial_\mu\phi(x))^2 + \frac{1}{2Z}(m\phi(x) + g\phi^3(x))^2 + O(g^4)] \quad (100)$$

and

$$(\det[\delta(x-y) + 3g(Z\not\partial+m)^{-1}(x-y)\phi^2(y)])^{1/2} = \det\frac{\delta\phi'}{\delta\phi(y)}(x,y,\phi) + O(g^4) \quad (101)$$

Equation (27) is a perturbative proof of the existence of the transformation. It should be noted that there must be a suitable regularization prescription consistent with supersymmetry in order to carry out above constructions. Notice that the determinants defined in equations (87), (88), (92) and (93) are to be understood in a regularized sense. Similarly eq.(100) and (101) are to be interpreted as properly regularized equations. Indeed, Nicolai[23] has given a suitable regularization prescription where these expressions are defined properly.

We have learnt from the above example that if there is a supersymmetric theory with vanishing vacuum energy then it is possible to find a transformation (in general nonlinear and non-local) such that the theory is equivalent to a simpler theory without fermions. In specific case of supersymmetric models there is a stochastic interpretation of this transformation. This special characteristic of the supersymmetry theory arises due to the fundamental properties of the functional measure. The measure, $d\Phi(x_\mu,\theta_\alpha)$, is independent of the particular super fields, $\Phi(x_\mu,\theta_\alpha)$ used as long as Φ respect the Grassmann structure of the superspace; consequently the measure is unique and universal. As we have described earlier we can transform the bosonic fields through the Nicolai map to arrive at a simpler theory. The stochastic nature of the Nicolai map is best demonstrated through the following example of supersymmetric quantum mechanics[24,25]. The Lagrangian is

$$L = \frac{1}{2}\dot{x}^2 + \frac{1}{2}(\frac{\partial V}{\partial x})^2 + \bar{\eta}i\frac{d}{dt}\eta + i\bar{\eta}\frac{\partial^2 V}{\partial x^2}\eta \quad (102)$$

where η, $\bar{\eta}$ are complex Grassmann variables. The Nicolai mapping is

$$\xi(t) = \dot{x}(t) + \frac{\partial V}{\partial x} \quad (103)$$

and

$$\frac{1}{2}\,\xi(t)^2 = \frac{1}{2}\,(\dot{x}(t) + \frac{\partial V}{\partial x})^2$$

$$= \frac{1}{2}\,\dot{x}^2 + (\frac{\partial V}{\partial x})^2 + \text{total time derivative} \tag{104}$$

The Jacobian is

$$\det(\frac{\delta\xi}{\delta x}) = \det(\frac{d}{dt} + \frac{\partial^2 V}{\partial x^2})$$

$$= \int d[\eta]d[\bar{\eta}]e^{-\int(\bar{\eta}i\frac{d}{dt}\eta + i\bar{\eta}\frac{\partial^2 V}{\partial x^2}\eta)dt} \tag{105}$$

Note that the determinant is to be defined with appropriate boundary condition for the differential equation involved. Furthermore, it is found that the Lagrangian of (102) reduces to a simple Gaussian in $\xi(t)$ after the Nicolai transformation.

It is, therefore, pertinent to ask following questions: What is the physical meaning of Nicolai map? Is there a physical interpretation of the Gaussian field $\xi(t)$?

We can get some insight into the problem if we recall Parisi-Sourlas supersymmetric Lagrangian for the spin systems given by eq.(69). If we eliminate the auxiliary field and use the following correspondence between the fields ϕ, ψ, $\bar{\psi}$ and $x(t)$, η, $\bar{\eta}$;

$$\Box \leftrightarrow \frac{d}{dt}\,,\ \phi \leftrightarrow x \text{ and } \psi,\ \bar{\psi} \leftrightarrow \eta,\ \bar{\eta} \tag{106}$$

We find the Lagrangian (69) after eliminating the field A) goes over to the Lagrangian of (102). Thus the Nicolai mapping is

$$\xi = -\Box\phi + \frac{\partial V}{\partial \phi} \tag{107}$$

and the measure is

$$d\mu(\xi) = d[\xi]\exp[-\frac{1}{2}\int \xi^2 d^D x \tag{108}$$

It is clear that ξ is just the random magnetic field (recall Parisi-Sourlas argument) with vanishing Gaussian average which interacts with a spin system whose Lagrangian is

$$= -\frac{1}{2}\,\phi\Box\phi + V(\phi) \tag{109}$$

The functional transformation considered above is a stochastic differential equation which connects ϕ with the random field ξ. Thus we can conclude from the above analogy that the Nicolai mapping transforms the original bosonic fields into random would-be magnetic fields. The transformation, indeed, is a stochastic equation of motion.

Notice that the transformation (103) is the well known Langevin equation encountered in stochastic processes (102) where t is the physical time.

In order to get further insight into the Nicolai transformation let us consider the analog of Newton's law for classical stochastic processes (particle mass is set equal to one).

$$\dot{x} = - \frac{\partial F}{\partial x} + \xi(t) \tag{110}$$

F(x) is a particular solution of the Hamilton-Jucobi equation and ξ is the random Gaussian field. Let us now consider the classical Euclidean Hamilton-Jucobi equation in the subspace of the phase space such that $H(x,p) = E = 0$ ($H(x,p)$ is the bosonic Hamiltonian). Thus we have

$$\frac{1}{2}\left(\frac{\partial F}{\partial x}\right)^2 - \frac{1}{2}\left(\frac{\partial V}{\partial x}\right)^2 = E = 0 \tag{111}$$

We can solve for F from above equation. Substituting the solution $F = \pm V + \text{const.}$ into (110) we find that it reduces to the Nicolai map if we choose $F = + V + \text{const.}$ Therefore the Nicolai mapping is the classical stochastic process associated with the classical Euclidean vacuum.

The system approaches equilibrium, i.e. the equilibrium correlation functions are recovered in the $t \to \infty$ limit only if $V(x) \to \infty$ as $|x| \to \infty$. If $V(x) \to -\infty$ as $|x| \to \infty$ then we have to use the solution $F = - V + \text{const.}$ in (110) and then we shall get a backward stochastic process. If the Lagrangian is to remain invariant under $V \to - V$ we have to interchange the Grassmann variables $\eta \leftrightarrow \bar{\eta}$ and the vacuum is annihilated by the fermionic creation operator instead of the fermionic annihilation operator. It has been claimed[24] that the theory can be solved exactly if the Nicolai mapping is (i) a stochastic differential equation i.e. the equation relating $\xi(t)$ and $x(t)$ is time local and (ii) the stochastic equation describes a non-equilibrium system which asymptotically reaches equilibrium configuration. This is possible only if supersymmetry remains an exact symmetry. An exact Nicolai map has been constructed for a 1+1 dimensional field theoretic model by Cecotti and Girardello which satisfy above constraints.

Therefore, it is important to investigate constraints on supersymmetry breaking. That is the topic of our next section.

V. THE WITTEN INDEX

It is well known that if supersymmetry remains unbroken then the ground state energy, E_o, must vanish. However, $E_o = 0$ is not a good criterion to decide whether supersymmetry is broken or not for a specific theory under consideration. Suppose $E_o = 0$ in some approximation for the theory at hand; we find that the correction to the approximation is infinitesimally small but nonzero and in this case it is obvious that supersymmetry is broken. Therefore vanishing of the ground state energy is not a suitable criterion to settle the issue since we cannot draw any definite conclusion, in cases where ground state energy can not be calculated exactly, even in the weak coupling limit; and the strong coupling domain poses a more formidable problem to compute E_o reliably.

In view of the above difficulties it is more convenient to introduce a parameter such that it is nonzero when supersymmetry is unbroken whereas if supersymmetry is broken the parameter vanishes. This leads us to define topological invariants of the field theory. In case of supersymmetric theories it is important to define topological quantum numbers that are properties of the entire theory rather than those of any particular field configurations. These are to be contrasted with topological objects such as monopoles or instantons in gauge theories which define the topological quantum number of individual classical field configurations.

Let us consider supersymmetric theories in a finite volume for the reasons stated below; (i) the spectrum of the Hamiltonian is discrete, (ii) states in the Hilbert space can be counted in a clear-cut, well-defined way and (iii) there are only a finite number of states whose energy is bounded from above.

The bosons and fermions must satisfy the same (periodic) boundary condition in the spatial directions since translation is a part of the supersymmetry algebra and this must be preserved in finite volume. Notice that, unlike internal symmetries, it is possible to break supersymmetry for a system with finite degrees of freedom in a finite volume i.e. it is possible to get $E_o(V) \neq 0$ in finite volume. It is also clear that if supersymmetry is unbroken in any finite arbitrary, volume, i.e. $E_o(V) = 0$, it will remain unbroken in the infinite volume limit; $E_o(V) = 0$ as $V \to \infty$. On the contrary if supersymmetry is broken in the finite volume theory, i.e. $E_o(V) \neq 0$, it is not necessarily broken in the infinite volume since if $E_o(V) \neq 0$, it is quite possible that $\lim_{V\to\infty} E_o(V) = 0$ (in fact energy density $E(V)/V$ is the relevant quantity).

Our purpose is to study a class of theories for which supersymmetry remains unbroken in any finite volume. We shall follow Witten's method[14] in order to investigate this problem.

Let us consider a theory defined in a volume, V, with associated Hilbert space, $\mathcal{H}$. We are interested in finding possible zero energy states in this space. Note that $E > |\vec{p}|$ for any state in a supersymmetric theory and $\vec{p} = 0$ for a zero energy state. Let us now focus our attention on $\vec{p} = 0$ sector of $\mathcal{H}$ since the supersymmetry algebra is simplified considerably in this sector. The algebra is

$$Q_1^2 = Q_2^2 = Q_k^2 = H \tag{112}$$

$$Q_i Q_j + Q_j \cdot Q_i = 0 \quad i \neq j \tag{113}$$

where $Q_1 \ldots\ldots Q_k$ are normalized hermitian supercharges. Let us consider a theory with a single supercharge Q for the sake of simplicity. One of the properties of Q is that it maps fermions $\rightleftarrows$ bosons. It is more appropriate, however, to talk of fermionic states $|f\rangle$ and bosonic $|b\rangle$ in the finite volume case rather than fermions and bosons since the particle concept is not well-defined in finite volume. The states $|b\rangle$ and $|f\rangle$ satisfy following properties under rotation around the z axis

$$\exp(2\pi i J_z)|b\rangle = |b\rangle \tag{114}$$

$$\exp(2\pi i J_z)|f\rangle = -|f\rangle \tag{115}$$

The generator of infinitesimal rotations is not a well-defined concept in finite volume; however, rotation by $\pi/2$ around the z axis can be defined. Thus the operations in eq.(114) and (115) are to be understood in this sense. The operator $(-1)^F$, defined as,

$$(-1)^F = \exp(2\pi i J_z) \tag{116}$$

plays a crucial role in distinguishing bosons from fermions.

Another important fact of the supersymmetric theories is that states of nonzero energy are paired under the action of Q. Let $|f\rangle$ be a normalized fermionic state of energy E; define a bosonic state by $|b\rangle = Q|f\rangle/\sqrt{E}$. Notice that

$$Q|f\rangle = \sqrt{E}\ |b\rangle \quad \text{and} \quad Q|b\rangle = \sqrt{E}\ |f\rangle \tag{117}$$

which follows from the fact $Q^2 = H$. Thus the states $|f\rangle$ and $|b\rangle$ appear in two dimensional multiplets.

It is easy to check that the zero energy states do not form such multiplets; since $Q^2 = H$ and if a state is annihilated by H (a zero energy state, either $|b\rangle$ or $|f\rangle$) then it is also annihilated by Q. Thus $Q|b\rangle = 0$ or $Q|f\rangle = 0$ if $|b\rangle$ or $|f\rangle$ is a zero energy state. As a consequence there may be an arbitrary number of zero energy bosonic states, n_o^B, and an arbitrary number of zero energy fermionic states n_o^F.

As we vary the parameters of the theory such as masses, m_i, coupling constants, g_i, and the volume, V n_o^B and n_o^F may change. It may so happen that some of the non zero energy states may come down to $E = 0$ as we vary the parameters. But they have to come down in fermion-boson pairs. On the other hand if zero energy states go up to non zero energy level, as parameters vary they also have to be in fermion-boson pairs. Therefore, although n_o^B and n_o^F may change under the variation of the parameters the difference $n_o^B-n_o^F$ remains unchanged. Notice that $n_o^B-n_o^F$ can be calculated reliably since it is independent of the parameters of the theory. Suppose we compute $n_o^B-n_o^F$ in some approximation. If we compute corrections to the approximation then n_o^B and n_o^F may change but the non zero energy states will be always paired. Thus any valid approximation will determine $n_o^B-n_o^F$ correctly.

If $n_o^B-n_o^F \neq 0$, then either $n_o^B \neq 0$ or $n_o^F \neq 0$ or both are non zero, consequently supersymmetry is not broken.

If $n_o^B-n_o^F = 0$ then there are the following two possibilities (i) $n_o^B = n_o^F \neq 0$ and (ii) $n_o^B = n_o^F = 0$. In case of the former supersymmetry is unbroken, whereas it is spontaneously broken in case of the latter. If $n_o^B-n_o^F = 0$ then it is possible to draw a qualitative conclusion that there are massless fermions in both the cases. In the first case there is a zero energy fermionic state which can be interpreted as a massless fermion in the infinite volume limit whereas in the second case supersymmetry is broken spontaneously and the massless Goldstino appears in the $V \to \infty$ limit.

It is important to note that $n_o^B-n_o^F$ can be identified as the trace of the operator $(-1)^F$. The nonzero energy states being paired, the contribution of the bosonic state of a given energy to $\mathrm{Tr}(-1)^F$ is cancelled by that of the fermionic partner. Therefore, only the zero energy states contribute to the above operator and $\mathrm{Tr}(-1)^F = n_o^B-n_o^F$. However, the trace is to be taken over an infinite set of states in the Hilbert space and we have to invoke a regularization prescription in order that the trace is well defined. This can be achieved if we consider $\mathrm{Tr}(-1)^F \exp(-\beta H)$, $\beta > 0$ instead of $\mathrm{Tr}(-1)^F$. The contribution of nonzero energy states to the trace of the regularized operator vanish pairwise and then we can take the limit $\beta \to 0$ to recover the result stated above.

In fact $\mathrm{Tr}(-1)^F$ has a deeper mathematical interpretation: it is index of an operator. We can split the Hilbert space H of the theory under consideration to fermionic and bosonic subspaces H_F and H_B respectively. The supercharge Q maps bosons $\rightleftarrows$ fermions and has the form

$$Q = \begin{pmatrix} 0 & M^* \\ M & 0 \end{pmatrix} \tag{118}$$

Here M^* is the adjoint of M and Q is a hermitian charge. The states can be arranged in the form $\begin{pmatrix} B \\ \vdots \\ F \end{pmatrix}$; where B is a bosonic state and F is the fermionic one. If ψ is a zero energy state it is annihilated by $H = Q^2$ as well as by Q. Therefore $M^*\psi_F = 0$, ψ_F is zero energy fermionic state. $n_o^B - n_o^F$ is nothing but difference of the number of solutions of $M\psi_B = 0$ and those of $M^*\psi_F = 0$. The latter quantity is the index of the operator M by definition. It follows from the general theorem that the index of an operator is invariant under small deformations and as a consequence $\mathrm{Tr}(-1)^F$ is independent of the parameters of the theory.

The following remarks are in order at this stage: (a) How do ultraviolet divergences affect the above arguments? Since $\mathrm{Tr}(-1)^F$ only involves the low-lying states, the ultraviolet divergences do not really affect above results. (b) The behaviour of the potential energy for large fields does affect $\mathrm{Tr}(-1)^F$. The underlying assumption is that when the parameters are varied the energy eigenvalues do not appear or disappear suddenly. Thus the variation of the parameters should be such that it acts as a perturbation. $\mathrm{Tr}(-1)^F$ remains invariant under any change in parameters such that for large value of fields, the Hamiltonian changes by terms no bigger than the terms already present. The index is independent of the numerical value of a parameter as long as it is not set equal to zero. We can set a parameter equal to zero if the behaviour of the Hamiltonian for large fields is not charged. Thus in a finite volume $n_o^B - n_o^F$ is invariant under arbitrary (but reasonable) changes in the parameters of the supersymmetric theory.

What can we say about invariance of n_o^B and n_o^F under change of parameters? In fact supersymmetry alone does not impose enough constraints to keep n_o^B and n_o^F invariant under the variation of parameters. However, it is possible to find a restrictive class of transformations which keeps n_o^B and n_o^F separately invariant.

Let us consider a simple theory which has only two supercharges Q_1 and Q_2 and define

$$Q_\pm = Q_1 \pm iQ_2 \tag{119}$$

The supersymmetry algebra is simplified considerably in the zero momentum sector which is of interest to us in what follows:

$$Q_+^2 = Q_-^2 = 0 \tag{120}$$

$$Q_+Q_- + Q_-Q_+ = H \tag{121}$$

Equation (120) implies that $Q_+\psi = 0$ has many solutions and Q_+ annihilates at least half of the states in the Hilbert space since if ψ is a state, either $Q_+\psi = 0$, or the supersymmetric partner $\chi = Q_+\psi$ is annihilated

$$Q_+\chi = 0, \; Q_+^2 = 0 \tag{122}$$

Any operator Q_+ satisfying eq.(122) can be put in Jordan canonical form

$$Q_+ = \begin{pmatrix} 0 & 1 & & & & & \\ 0 & 0 & & & & & \\ & & 0 & 1 & & & \\ & & 0 & 0 & & & \\ & & & & 0 & & \\ & & & & & 0 & \\ & & & & & & 0 \end{pmatrix} \tag{123}$$

There are arbitrary number of 2x2 block matrices with structure $\begin{pmatrix} 0 & 1 \\ 0 & 0 \end{pmatrix}$ and Q_+ acts on two dimensional multiplets as a raising operator with the property. $\psi = Q_+\chi$, $Q_+\psi = 0$. Further, there is an arbitrary number of unpaired zeros which correspond to zero energy states.

Let χ be any given state; if $\chi = Q_+\psi$ for some ψ then $Q_+\chi = 0$. Conversely given a state $Q_+\chi = 0$, can we write $\chi = Q_+\psi$ for some ψ? It is easy to see that it is possible to find a ψ if and only if χ corresponds to a state of nonzero energy. Therefore we characterize the zero energy states as those states χ such that $Q_+\chi = 0$ but it is not possible to write $\chi = Q_+\psi$ for any ψ.

Let $N = n_o^B + n_o^F$, then N is the dimensionality of the quotient space Ker Q_+/Im Q_+. Here Ker Q_+ is the kernel of Q_+, the space of solutions of $Q_+\chi = 0$ and Im Q_+ is the image of Q_+, the space of all states that can be written as $Q_+\psi$ for some ψ. The quotient space consists of all χ in Ker Q_+ such that χ and $\chi + Q_+\psi$ are equivalent for any ψ. As we have remarked earlier, the constraints due to supersymmetry are not enough to keep N invariant when we vary the parameters. Consider the example; where

$$Q_+ = \begin{pmatrix} 0 & \lambda \\ 0 & 0 \end{pmatrix} /\sqrt{2} \ , \text{ then } H = \begin{pmatrix} \lambda^2 & 0 \\ 0 & \lambda^2 \end{pmatrix}$$

where λ is arbitrary parameter.

If $\lambda \neq 0$ then there is no zero energy state and $N = 0$ while for $\lambda = 0$, Ker Q_+ is of dimension 2 and Im Q_+ is zero dimensional. Therefore $N = 2$ and it jumps from 0 to 2.

The number of zero energy states varies as we change the parameters of the supersymmetric theory. However, it is possible to consider transformations in the parameter space that keep the total number of zero energy states invariant. Let $\{Q_+, Q_-, H\}$ and $\{\tilde{Q}_+, \tilde{Q}_-, \tilde{H}\}$ be the initial and final set of operator, which are related to each other through following transformations:

$$\tilde{Q}_+ = M^{-1}Q_+M \tag{124}$$

$$\tilde{Q}_- = M^*Q_-M^{*-1} \tag{125}$$

$$\tilde{H} = Q_+Q_- + Q_-Q_+ \tag{126}$$

where M is an arbitrary invertible operator and M* is its adjoint. If M is unitary the two sets of operators differ from each other by a change of basis in the Hilbert space.

On the other hand if M not unitary and M^{-1} exists, then the spectra of non zero energy states of H and $\tilde{H}$ differ from each other in general. However, the number of zero energy states of H and $\tilde{H}$ remain unchanged.

The total number of zero energy states is the number of linearly independent solutions of $Q_+\chi = 0$ such that $\chi \neq Q_+\psi$ for any ψ. Now, if χ is such a state $\tilde{\chi} = M^{-1}\chi$ satisfies $\tilde{Q}_+\tilde{\chi} = 0$ but cannot be written as $\tilde{\chi} = \tilde{Q}_+\tilde{\psi}$ for any $\tilde{\psi}$. If it were allowed then we can conclude $\chi = Q_+(M\tilde{\psi})$ contrary to our hypothesis.

If $\tilde{Q}_+\tilde{\chi} = 0$ such that $\tilde{\chi} \neq \tilde{Q}_+\tilde{\psi}$ for any $\tilde{\psi}$ then $\chi = M\tilde{\chi}$ satisfies $Q_+\chi = 0$ such that $\chi \neq Q_+\psi$ for any ψ.

Therefore the mapping $\chi \leftrightarrow M^{-1}\chi$ is a one to one mapping. The transformations that satisfy this constraint are called conjugations by linear operators. When we bring about changes in the parameters of the theory by such operations we call them changes due to conjugation. Under a change in parameters Q_α changes by conjugation. Thus if supersymmetry is unbroken for some (m,g) it remains unbroken for all (m,g). If the supercharges $Q_\alpha(m_1,g_1) \to Q_\alpha(m_2,g_2)$ due to a change in the parameters under conjugation then the two charges are related as follows

$$Q_\alpha(m_2,g_2) = M^{-1}(m_1,g_1,m_2,g_2)Q_\alpha(m_1,g_1)M(m_1,g_1,m_2,g_2) \tag{127}$$

$$M = \exp[2\mathrm{Re}\int d^3x(W_2(\phi_i(x)) - W_1(\phi_i(x)))] \tag{128}$$

where W_1 and W_2 correspond to superpotentials for the set of parameters (m_1,g_1) and (m_2,g_2) respectively.

Let us consider a simple example due to Witten in supersymmetric quantum mechanics.

$$Q_1 = \frac{1}{2}[\sigma_1 p + \sigma_2 W(x)] \tag{129}$$

$$Q_2 = \frac{1}{2}[\sigma_2 p - \sigma_1 W(x)] \tag{130}$$

$$H = \frac{1}{2}[p^2 + W(x)^2 + \sigma_3 \frac{dW(x)}{dx}] \tag{131}$$

where σ_i are Pauli matrices and p is the momentum operator. Define

$$Q_\pm = (Q_1 \pm iQ_2)/\sqrt{2} \tag{132}$$

W(x) can be changed by conjugation

$$Q_+(\tilde{W}) = \exp[-F(x)]Q_+(W)\exp[+W(x)] \tag{133}$$

where F(x) is such that $\frac{dF}{dx} = \tilde{W}(x) - W(x)$. Let us consider $W(x)=x^2+a^2$ for sake of definiteness.

(i) $a^2 > 0$. Supersymmetry is broken at tree level since the classical potential $V(x) = W(x)^2$ is a positive function.

(ii) $a^2 < 0$. Supersymmetry is unbroken at the tree level. However, it is broken dynamically as is well known. The dynamical breaking is related to the fact that sign of a^2 can be changed by conjugation.

Note that in the first case $n_o^B + n_o^F = N = 0$. Further $Q_+(a^2)$ and $Q_+(-a^2)$ are related by conjugation,

$$Q_+(-a^2) = \exp(2a^2x)Q_+(a^2)\exp(-a^2x) \tag{134}$$

Thus N = 0 for the second case and consequently supersymmetry is broken dynamically, although N = 2 at the tree level and in perturbation theory calculations.

The operator, M, defined in (127) is required to satisfy following constraints.

In field theory M has to be suitably defined with supersymmetry preserving regularization prescription. Witten has proposed that all superfields $\phi_i(x)$ be smeared with an appropriate kernel $K(x,y)$ such that $\tilde{\phi}_i(x,t) = \int d^3yK(x,y)\phi_i(y,t)$. Thus the superspace interaction term is $\int d^4xW(\tilde{\phi}_i(x,t))$. This prescription takes care of ultraviolet divergences and W changes in the same way as before under conjugation. Finally, we can take the limit $K(x,y) \to \delta(x-y)$. In order to retain the canonical Hamiltonian frame work the smearing is done in the spatial directions alone.

There are further restrictions on the form of M when we take into account the behavior of the Hamiltonian for large values of the field. We require that if M acts on an energy eigenstate, it must give a normalizable, finite energy state. This imposes a constraint on allowed changes in the superpotential, W, under conjugation.

Let us first consider a quantum mechanical case where the bosonic potential, $V(x) \simeq x^2$. Now the ground state wave function $\psi_o \sim \exp(-x^2)$. We cannot introduce a term in the superpotential by conjugation such that $M \sim \exp(x^3)$ (this can be done if the change in superpotential is x^3), since M will not be normalizable. However, if $V(x) \sim x^2 + bx^4$ then a conjugation with $M \sim \exp(x^3)$ is allowed, and we can make the parameter b arbitrarily large or small by conjugation so long as we do not set b = 0. Thus the allowed conjugation operations are those whose behavior for large fields does not exceed that of the Hamiltonian of the theory.

VI. WITTEN INDEX AND NICOLAI MAPPING

It was observed in the last section that the Witten index is a topological object. The topological nature of $\mathrm{Tr}(-1)^F$ is best seen in a functional integral approach. The topological structure is not that of the classical configurations but of the full theory as emphasised earlier. Therefore, it involves the quantization procedure of the whole theory.

If we consider a supersymmetric theory in a box of finite volume, V, where both bosons and fermions satisfy periodic boundary condition the Witten index has the following path integral representation[26]

$$\Delta \equiv \mathrm{Tr}(-1)^F e^{-\beta H} = \int d[\Psi]\exp(-S[\Psi]) \tag{135}$$

where the path integral is over all fields Ψ, fermions and bosons, which satisfy same periodic boundary condition. Let us assume that there are only Majorana fermions in the theory. Then integrating over the fermions we get

$$\Delta = \int d[\phi]D[\phi,g]\exp(-S[\phi]) \tag{136}$$

where $D[\phi,g]$ is the appropriate Mathews-Salam determinant, g is the coupling constant and $S[\phi]$ is the bosonic action.

On the other hand under the Nicolai mapping the bosonic part of the action is Gaussian in the transformed field and the Jacobian is the Mathews-Salam determinant if supersymmetry is exact. Thus we can write

$$\Delta_N = \int d[\phi]D[\phi,g]\exp(-S[\phi]) \tag{137}$$

Δ_N is equal to number of times that the mapping $\phi \to \xi[\phi]$ covers the space of continuous functions. It is evident that Δ_N is the winding number of the Nicolai mapping, in other words, the number of distinct configurations of the original bosonic fields which are mapped in a given configuration of the Gaussian fields. Comparing eq.(136) and (137) we conclude that Δ is nothing but the winding number of the Nicolai mapping. Recently, the path integral representation of the Witten index and its topological structure for various models has been investigated by several authors.[27]

To summarise the content of the lectures: first we introduce the techniques of stochastic quantization and then presented a perturbative proof of the equivalence between stochastic and canonical quantization. Next, we discussed the underlying supersymmetry of the spin systems in the presence of random magnetic field and the supersymmetry associated with the Fokker-Planck Hamiltonian. The content of the theorem of Nicolai and an example of the Nicolai map was discussed in Sec.III. We constructed an exact Nicolai map for a supersymmetric quantum mechanical system and the stochastic nature of this map was discussed. This provides a physical interpretation of the map. Since the mapping exists if and only if supersymmetry is exact; we studied Witten index and its topological interpretation in some detail. Then we discussed a more restrictive class of transformations, the so called conjugation operation. It is observed that the Witten index can be identified with the winding number of the Nicolai mapping. The exact Nicolai map has been constructed for a field theoretic model in 1+1 dimensions. It will be interesting to see if such a construction exists for supersymmetric theories in higher dimensions with or without gauge fields, since the topological structure of the gauge group itself will play a crucial role in such constructions.

ACKNOWLEDGEMENTS

It is a great pleasure to acknowledge valuable discussions with S.C. Mishra who has patiently explained the results on lower critical dimensions for spin systems and stochastic processes. I would like to thank A. Khare and L.P. Singh for their suggestions and for carefully reading the manuscript and Probir Roy and P. Majumdar for valuable discussions. The criticism and discussions with members of the High Energy Physics Group at Bhubaneswar and with S.P. Misra are appreciated. I am grateful to T. Pradhan for encouragements and constructive remarks.

The gracious hospitality of the organizing committee of the Winter School at Mahabaleshwar is acknowledged.

APPENDIX A

In this appendix we shall recapitulate some of the results of stochastic processes[4] which are relevant for our purpose.

A random process $y(t)$ is a process in which the variable y does not depend on the independent variable, t, in a definite way. On the other hand one gets different functions $y(t)$ in different observations. Therefore, only certain probability distributions are observables. The random process is completely specified by the following set of probability distributions.

$W_1(y,t)dy$ = Probability of finding y in the interval y and $y + dy$ at time t.

$W_2(y,t_1,y_2t_2)dy_1dy_2$ = Joint probability of finding y, in the range $(y_1, y+dy_1)$ at time t_1 and in the range (y_2, y_2+dy_2) at t_2.

$W_3(y_1t_1, y_2t_2, y_3t_3)dy_1dy_2dy_3$ = Joint probability of finding three values of y in the ranges dy_1, dy_2 and dy_3 at times t_1, t_2 and t_3.

We can define other probability functions similarly. The probability distribution functions satisfy the following conditions

(i) $W_n \quad 0$

(ii) $W_n(y,t_1...y_nt_n)$ is a symmetric function in the set of variables $y_1t_1...y_nt_n$ since W_n is a joint probability.

(iii) $W_k(y_1,t_1...y_kt_k) = \int...\int y_{k+1}... dy_nW_n(y_1t_1...y_nt_n)$ since each W_n must imply all previous W_k with $k < n$. It is normally assumed that the probabilities are stationary in time. In other words the underlying mechanism which causes fluctuations does not change with time. Thus if we change the t-axis it does not influence W_n and consequently:

$W_1(y)dy$ = Probability of finding y in the interval (y, y+dy).

$W_2(y_1y_2t)dy_1dy_2$ = Joint probability of finding a pair of values of y in the range dy_1 and dy_2 which a time interval apart t. $(t = t_2 - t_1)$. We can define other probability functions similarly. In what follows we shall consider only stationary processes.

Conditional Probabilities: $P_2(y_1/y_2,t)dy_2$ is the probability that given y_1 one finds y in the range (y_2, y_2+dy_2) a time t later. P_2 is related to W_2 as follows

$$W_2(y_1y_2,t) = W_1(y_1)P_2(y_1/y_2,t) \tag{A1}$$

P_2 satisfies following constraints

$$P_2(y_1/y_2,t) > 0 \tag{A2}$$

$$\int dy_2P_2(y_1/y_2t) = 1 \tag{A3}$$

$$W_1(y_2) = \int W_1(y_i)P_2(y_1/y_2t)dy \tag{A4}$$

Markoff Process: A Markoff process is defined by the equation

$$P_n(y_1t_1...y_{n-1}, t_{n-1}/y_nt_n) = P_2(y_{n-n}, t_{n-1}/y_nt_n) \tag{A5}$$

where $P_n(y_1t_1\ldots y_{n-1}t_{n-1}/y_nt_n)$ is the conditional probability of finding y in the interval (y_n, y_n+dy_n) at time t_n given $y_1,\ldots y_{n-1}$ at $t_1\ldots y_{n-1}$ (with ordering $t_n > t_{n-1}\ldots > t_1$). It is clear from (A5) that all the information about the process is contained in W_2 and thus W_2 or P_2 completely describe the process. However, we cannot take P_2 as an arbitrary function of its variable since it satisfies

$$P_2(y_1/y_2t) = \int dy P_2(y_1/y, t_o)P_2(y/y_2, t-t_o) \tag{A6}$$

for $0 < t_o < t$ (A6) follows from the definition of a Markoff process and is known as the Smoluchowski equation.

Let us consider a stationary random process y(t) whose average is zero. If we consider a very long time interval T such that $y(t)=0$, $t > T$ then we can write y(t) in terms of a Fourier integral

$$y(t) = \int_{-\infty}^{+\infty} d\omega A(\omega)e^{2\pi i\omega t} \tag{A7}$$

where $A(\omega) = A^*(-\omega)$ due to reality of $y(t)\langle y^2\rangle_{av}$ is defined as follows

$$\langle y^2\rangle_{av} = \lim_{T\to\infty} \frac{1}{T} \int_{-\infty}^{+\infty} y^2(t)dt = \int_0^{\infty} d\omega G(\omega) \tag{A8}$$

with $G(\omega) \equiv \lim_{T\to\infty} \frac{2}{T} |A(\omega)|^2$ is called the spectral density. The correlation function, $\rho(\tau)$, defined as

$$\rho(\tau) = \langle y(t)y(t+\tau)\rangle_{av}/\langle y^2(t)\rangle_{av} \tag{A9}$$

plays an important role in the study of random processes.

The Gaussian Random Process: The Gaussian random process is such that all the basic distribution functions are Gaussian distributions. However, we can start with the Fourier development of Gaussian random function y(t) which is repeated periodically with period T. The Fourier series for y(t) is

$$y(t) = \sum_{n=1}^{\infty} (a_n \cos 2\pi\omega_n t + b_n \sin 2\pi\omega_n t) \tag{A10}$$

where $\omega_n = n/T$ and if $\langle y\rangle_{av} = 0$ the constant term does not appear. The Fourier coefficients a_n, b_n are random variables and are assumed to be independent of each other. They are Gaussian distributions with zero average. Let $W(a_1\ldots a_n, b_1\ldots b\ )$ be the probability of finding a_k in the range da_k and b_1 in the range db_1 then

$$W(a_1a_2\ldots,b_1b_2\ldots) = \prod_k \frac{1}{\sigma_k\sqrt{2\pi}} \exp[-(a_k^2+b_k^2)/2\sigma_k^2] \tag{A11}$$

where $\sigma_k^2 = \langle a_k^2\rangle_{av} = \langle b_k^2\rangle_{av} = G(\omega_k)/T$; $G(\omega_k)$ being the spectral density.

Let the variables $e_1\ldots x_n$ be distributed according to

$$W(x_1\ldots x_n) = \prod_{i=1}^{n} \frac{1}{\sigma_i(2\pi)^{1/2}} \exp(-\frac{x_i^2}{2\sigma_i}) \tag{A12}$$

Let $y_1\ldots y_s$ be linear combinations of x_i such that

$$y_k = \sum_{\ell=1}^{n} a_{k\ell}x_\ell \text{ , } k = 1,2,\ldots\ s < n \tag{A13}$$

where $a_{k\ell}$ are constants. The y_k will be distributed according to s-dimensional Gaussian distributions:

$$P(y_1\ldots y_s) = \frac{1}{(2\pi)^{s/2}B^{1/2}} \exp[-\frac{1}{2B} \sum_{k,\ell=1}^{s} B_{k\ell}y_ky_\ell] \tag{A14}$$

where $B_{k\ell}$ is the cofactor of the element in the matrix $b_{k\ell}$ defined by the equation

$$b_{k\ell} = \sum_{i=1}^{n} a_{ki}a_\ell\cdot\sigma_i^2 = \langle y_ky_\ell\rangle_{av} \tag{A15}$$

and B is the determinant of the matrix b. For S=2 we have

$$P(y_1y_2) = \frac{1}{2\pi\sigma\tau(1-\rho^2)^{1/2}} \exp\left\{- \frac{1}{2(1-\rho^2)} (\frac{y_1^2}{\sigma^2} + \frac{y_2^2}{\tau^2} - \frac{2\rho}{\sigma\tau} y_1y_2)\right\} \tag{A16}$$

where

$$\sigma^2 = \langle y_1^2\rangle_{av},\ \tau^2 = \langle y_2^2\rangle_{av}\ \langle y_1y_2\rangle_{av} = \sigma\tau\rho,\ \rho$$

is the correlation coefficient.

Now we are in a position to discuss the Langevin and Fokker-Planck equations. The general form of the Langevin equation of our interest is

$$\frac{dy(t)}{dt} + \beta y(t) = F(t) \tag{A17}$$

F(t) purely Gaussian random variable such that

$$\langle F(t)\rangle_{av} = 0 \tag{A18}$$

$$\langle F(t_1)F(t_2)\rangle_{av} = 2D\delta(t_1-t_2) \tag{A19}$$

Since F(t) is Gaussian it follows from (A17) that y(t) is also a Gaussian random process. The spectrum can be computed easily and it turns out to be

$$G_y(\omega) = 4D/(\beta^2 + (2\pi\omega)^2) \tag{A20}$$

The corresponding correlation function $\rho(t) = \exp(-\beta(t))$ and the second probability distribution is a two dimensional Gaussian distribution

$$W_2(y_1y_2t) = \frac{\beta}{2\pi D(1-\rho^2)^{1/2}} \exp\left(- \frac{\beta}{2D(1-\rho^2)} (y_1^2+y_2^2-2\rho y_1y_2)\right) \tag{A21}$$

Since $\langle y^2\rangle_{av} = \int_0^\infty d\omega G_y(\omega) = D/\beta$. It follows from Doob's theorem[28] that y(t) is a Markoff process and W_2 describes the process completely.

We give a derivation of the Fokker-Planck equation in what follows.

The Smoluchowski relation (A6) can be written as

$$P(x/y,\ t+\Delta t) = \int dz\ P(x/z,t)P(z/y,\Delta t) \tag{A22}$$

Consider the integral

$$\int dy\ R(y)\ \frac{\partial P}{\partial t}\ (x/y,\ t) \tag{A23}$$

where R(y) is an arbitrary function y such that $R(y) \to 0$ sufficiently fast for $y \to +\infty$. We can rewrite (A23) as

$$\int dyR(y)\ \frac{\partial P}{\partial t} = \lim_{\Delta t\to 0} \frac{1}{\Delta t} \int dyR(y)[P(x/y,\ t+\Delta t) - P(x/y,t)] \tag{A24}$$

Using (A22) in first term of the RHS of (A24) we get

$$\int dyR(y)\ \frac{\partial P}{\partial t} = \lim_{\Delta t\to 0} \frac{1}{\Delta t}\ [\int dyR(y)\int dzP(x/z,t)P(z/y,\Delta t) - \int dzR(z)P(x/z,t)] \tag{A25}$$

Let us interchange the order of integration in the first term of (A25) and expand R(y) in a power series in (z-y). We keep terms up to $(z-y)^2$ and neglect higher order terms in the limit $\Delta t \to 0$. Then it follows that

$$\int dyR(y)\ \frac{\partial P}{\partial t} = \int dzP(x/z,t)\,[R'(z)A(z) + \frac{1}{2} R''(z)B(z)] \qquad (A26)$$

where

$$A(z) = \lim_{\Delta t\to 0} \frac{1}{\Delta t} \int dy(y-z)P(z/y,\ \Delta t) \qquad (A27)$$

$$B(z) = \lim_{\Delta t\to 0} \frac{1}{\Delta t} \int dy(y-z)^2 P(z/y,\ \Delta t) \qquad (A28)$$

If we integrate partially RHS with respect to z then we arrive at

$$\int dyR(y)\left\{\frac{\partial P}{\partial t} + \frac{\partial}{\partial y}\,(AP) - \frac{1}{2}\,\frac{\partial^2}{\partial y^2}\,(BP)\right\} = 0 \qquad (A29)$$

The equation (A29) holds for arbitrary function R(y) and the quantity inside square bracket must vanish

$$\frac{\partial P}{\partial t} = - \frac{\partial}{\partial y}\,(AP) - \frac{1}{2}\,\frac{\partial^2}{\partial y^2}\,(BP) \qquad (A30)$$

which is the Fokker-Planck equation.

We can obtain the Fokker-Planck equation from the Langevin equation (A17) as follows. Integrating (A17) over a short time Δt we get

$$\Delta y = - \beta y\Delta t + \int_t^{t+\Delta t} d\xi F(\xi) \qquad (A31)$$

Then we can obtain

$$(y) = \lim_{\Delta t\to 0} \frac{\langle \Delta y\rangle_{av}}{\Delta t} = - \beta y$$

since $\langle F\rangle_{av} = 0$. Moreover

$$\langle \Delta y^2\rangle_{av} = \beta^2 y^2 \Delta t^2 + \iint_t^{t+\Delta t} d\xi d\eta \langle F(\xi)F(\eta)\rangle \qquad (A32)$$

The second term of (A34) is 2D t as follows from (A19). Therefore $B(y) = \lim_{\Delta t\to 0} \langle y^2\rangle_{av}/\Delta t = 2D$. The Fokker-Planck equation is

$$\frac{\partial P}{\partial t} = \beta\ \frac{\partial}{\partial y}\,(yP) + D\ \frac{\partial^2 P}{\partial y^2} \qquad (A33)$$

The fundamental solution is

$$P(y_o/y,t) = \frac{1}{(2\pi\sigma^2)^{1/2}}\ \exp[-(y-\bar{y})^2/2\sigma^2] \qquad (A34)$$

where $\langle y \rangle_{av} = y_o \exp(-\beta t)$, $\sigma^2 = \langle (y-\bar{y})^2 \rangle_{av}$. As $t \to \infty$ we get

$$W_1(y) = \lim_{t\to\infty} P(y_o/y,t) = \left(\frac{\beta}{2\pi D}\right)^{1/2} \exp\left(-\frac{\beta y^2}{2D}\right) \quad (A35)$$

The second probability distribution is

$$W_2(y_1y_2t) = W_1(y_1)P(y_1/y_2,t) \quad (A36)$$

We have made use of the generalizations of Langevin equation, Fokker-Planck equation and probability distributions discussed here in Section II and III in this article.

References

1. L.D. Faddeev and V.N. Popov, Phys. Lett. 25B, 29 (1967); R.P. Feynmann, Acta Phys. Polonica 26, 697 (1963); For a review see E. Abers and B.W. Lee, Phys. Rep. 9C, 1 (1973).
2. V.N. Gribov, Nucl. Phys. B139, 1 (1978).
3. G. Parisi and Y.-S. Wu, Sc. Sin. 24, 483 (1981).
4. P. Langevin, Comptes rendus 146, 530 (1908); Excellent exposition to stochastic processes relevant for physicist can be found in S. Chandrasekhar, Rev. Mod. Phys. 15, 1 (1943); M.C. Wang and G.E. Uhlenbeck, Rev. Mod. Phys. 17, 323 (1945).
5. G. Parisi, Nucl. Phys. B180, 378 (1981); B205, 337 (1982); F. Fucito and E. Marinari, Nucl. Phys. B190, (1981).
6. J. Alfaro and B. Sakita, Phys. Lett. 121B, 339 (1983).
7. D. Zwanziger, Nucl. Phys. B192, 259 (1981); L. Baulieu and D. Zwanziger, Nucl. Phys. B193, 161 (1981).
8. E. Floratos and J. Iliopoulos, Nucl. Phys. B214, 392 (1983).
9. J. Breit, S. Gupta and A. Zaks, IAS Princeton preprint. Also see, Chaturdevi, A.K. Kapoor and G. Srinivasan, Hyderabad Univ.Preprints.
10. G. Parisi and N. Sourlas, Phys. Rev. Lett. 43, 744 (1979).
11. Y. Imry and S.-K. Ma, Phys. Rev. Lett. 35, 1399 (1975).
12. H. Nicolai, Phys. Lett. 89B, 341 (1980); 101B, 396 (1981); Nucl. Phys. B176, 419 (1980).
13. S. Cecotti and L. Girardello, Nucl. Phys. B226, 417 (1983); N. Sakai and I. Sakamoto, Nucl. Phys. B229, 173 (1983).
14. E. Witten, Nucl. Phys. B202, 253 (1982).
15. For reviews see R. Graham, Springer Tracts in Modern Physics 66, 1 (1973); P.C. Hohenberg and B. Halperin, Rev. Mod. Phys. 49, 435 (1977); R.F. Fox, Phys. Rep. 48C, 179 (1978).
16. Discussion on the analogy between statistical mechanical system and quantum field theory can be found in A. Strominger, Ann. Phys. 146, 419 (1983), $\beta \leftrightarrow h$ Action $\leftrightarrow$ Energy. Also see J. DeAlfaro, S. Fubini and G. Furlan, CERN preprint 3426 (1982).
17. K. Symanzik in Lectures in theoretical physics III, Ed. E. Brittin et al. Int. Science New York 1961.
18. For an introduction to supersymmetry see S. Joglekar. These proceedings. Note that the supersymmetry of eqs.(70)-(73) is generated by an anticommuting four-vector. Thus it is different from the usual supersymmetry in field theory which is generated by an anticommuting spinor.
19. E. Gozzi preprint CCNY-HEP-83/4.

20. E. Witten, Trieste Lectures (1981); P. Salmanson and J.W. Van Holten, Nucl. Phys. B196, 509 (1982); F. Cooper and B. Freedman, Ann. Phys. 146, 262 (1983).
21. P.T. Matthews and A. Salam, Nuovo Cimento 2, 120 (1955).
22. F.A. Berezin, The Method of Second Quantization. Academic Press, New York, 1966.
23. H. Nicolai in Group theoretic methods in physics (Springer, Heidelberg 1978).
24. S. Cecotti and L. Girardello, Ann. Phys. 145, 81 (1983).
25. G. Parisi and N. Sourlas, Nucl. Phys. B206, 321 (1983). Also see J.L. Cardy, Phys. Letters 125B, 470 (1983); A. Klein and J.F. Perez, Phys. Letters 125B, 473 (1983); A. Niemi, Phys. Rev. Lett. 49, 1808 (1982).
26. S. Cecotti and L. Girardello, Phys. Lett. 110B, 39 (1982).
27. L. Alvarez-Gaumes, Commun. Math. Phys. 90, 161 (1983) and Harvard preprint HUTP-83/A035, L. Girardello, C. Imbimbo and S. Mukhi, Phys. Lett. 132B, 69 (1983).
28. J.L. Doob, Ann. Math. 43, 351 (1942).

CHIRAL SYMMETRY, 1/N EXPANSION AND LOW ENERGY QUANTUM CHROMODYNAMICS

Spenta R. Wadia
Tata Institute of Fundamental Research
Homi Bhabha Road, Bombay 400005, India

Recently there has been a revival of interest in old subjects like current algebra and strong coupling theory. The significance of this revival lies in the fact that many hitherto unrelated ideas are unified by virtue of certain general features and concepts of quantum chromodynamics, namely chiral symmetry, the renormalization group and the $\frac{1}{N}$ expansion. QCD dynamics can roughly be divided into 3 energy regimes. Short distances where asymptotic freedom enables a good description in terms of quark and gluon degrees of freedom. Intermediate distances of the order of the confinement scale present the most difficult dynamical problem. Here there is a mixing between quark-gluon and hadronic degrees of freedom. At long distances there are only hadronic degrees of freedom, so one expects once again a relatively simple description in terms of an effective lagrangian. Such a description is especially suited for static properties and low energy scattering processes.

In these notes we explore the long distance effective lagrangian for QCD. The general scenario is to look for the lagrangian that describes weakly coupled degrees of freedom. The non-linear sigma model has been widely discussed in the literature as a viable candidate. We shall discuss this but eventually argue that the linear sigma model can provide a better alternative. We discuss the following topics:

(i) QCD chiral symmetries and Landau-Ginzburg type effective Lagrangian
(ii) Baryons as solitons and the Skyrme model
(iii) Wess-Zumino term
(iv) Consequences of the Wess-Zumino term
(v) Quantization of the Skyrme soliton
(vi) Nambu-Jona-Lasinio Model.

(i) QCD Chiral Symmetries

QCD is described by the lagrangian

$$L = + \frac{N}{4g^2} \mathrm{tr} F_{\mu\nu} F_{\mu\nu} + \bar{\psi} \not{D} \psi + \sum_i m_i \bar{\psi}_i \psi_i \tag{1}$$

A_μ is an SU(N) gauge field. $F_{\mu\nu}$ is the field strength and $D_\mu = i\partial_\mu - A_\mu$. The index 'i' stands for the 3 quark flavours. In the first approximation we set $m_i = 0$, and the classical Lagrangian (1) has a $U(3)_L \times U(3)_R$ chiral symmetry. This is seen by writing the fermion part of (1) in terms of the left and right-chiral fields $\psi_L = \frac{1+\gamma_5}{2}\psi$ and $\psi_R = \frac{1-\gamma_5}{2}\psi$. Then

$$\bar{\psi}\not{D}\psi = \bar{\psi}_L\not{D}\psi_L + \bar{\psi}_R\not{D}\psi_R$$

$U(3)_L$ acts on ψ_L and $U(3)_R$ acts on ψ_R.

Not all classical symmetries are realized in the quantum theory. The left and right U(1) currents are not conserved due to the Adler-Bell-Jackiw anomaly [1]

$$\partial_\mu J^L_\mu = \frac{g^2}{N}\frac{1}{64\pi^2}\,\mathrm{tr}_c F_{\mu\nu}{}^*F_{\mu\nu}$$
$$\partial_\mu J^R_\mu = -\frac{g^2}{N}\frac{1}{64\pi^2}\,\mathrm{tr}_c F_{\mu\nu}{}^*F_{\mu\nu} \tag{2}$$

The currents corresponding to $SU(3)_L \times SU(3)_R$ are exactly conserved. However, at zero temperature, these symmetries are realized in the Nambu-Goldstone mode leading to 8 massless pions. This fact forms the basis of current algebra and the subject of low energy effective Lagrangians [2]. A convincing argument for the existence of the Nambu-Goldstone mode rests on the 't Hooft anomaly condition [3] and spectral properties of Green's functions at large N [4]. On the other hand for sufficiently high temperatures we would expect (from our experience in statistical mechanics) the realization of $SU(3)_L \times SU(3)_R$ in the Wigner-Weyl mode, meaning there is a phase transition between the broken and unbroken phases of chiral symmetry.

Armed with the above facts one may attempt to discuss QCD from the view point of the Landau theory of phase transitions, in analogy with the Landau-Ginzburg theory of superconductivity and its relation to the microscopic BCS theory. The Landau theory has 2 ingredients: a) the notion of an order parameter and b) a mechanism or expansion parameter which can control the fluctuations of the order parameter. In QCD a natural choice of the order parameter is the 3x3 complex matrix $M_{ij}(x) = \sum_{\alpha=1}^{N} \bar{\psi}^\alpha_i(x)(1+\gamma_5)\psi^\alpha_j(x)$. (Subsequently we will see that the situation is a bit more subtle) It is local, gauge and Lorentz invariant and transforms as $M(x) \to U_L M(x) U_R^\dagger$. If we assume that $M(x)$ is a slowly varying field, then the simplest Lagrangian involving $M(x)$ and incorporating the symmetries we have discussed is

$$L = a\,\mathrm{tr}\partial_\mu M \partial_\mu M^\dagger + b\,\mathrm{tr}M^\dagger M + c\,\mathrm{tr}MM^\dagger MM^\dagger +$$

$$+ d(\mathrm{tr}MM^{\dagger})^2 + e|\ell n\, \det M|^2 \quad (3)$$

a,b,c,d,e are coefficients which are determined by the underlying microscopic dynamics. The last term, which is the square of the U(1) phase of M, is the mass term of the η'. Its inclusion explicitly violates the axial U(1) chiral symmetry as required by the A-B-J anomaly [5]. Further as is clear from (2), the A-B-J anomaly vanishes for large N as $\frac{1}{N}$. Hence the coefficient 'e' is of order $\frac{1}{N}$, giving η' a mass of order $\frac{1}{N}$ compared to the massless pion octet. In the broken symmetry phase the coefficients b, c and d are such that the potential energy has a unique minimum at $M = H_o$. This ensures that the chiral symmetry $SU(3)_L \times SU(3)_R$ is spontaneously broken to diagonal SU(3). This and the fact that η' is massive leads to the SU(3) chiral model for massless Nambu-Goldstone bosons.

$$L' = \frac{F_\pi^2}{16}\, \mathrm{tr}\partial_\mu U \partial_\mu U^\dagger \quad (4)$$

where the SU(3) matrix U(x), is defined by $M(x) = H_o e^{i\phi} U$, $\phi = \ell n \det M$ and $F_\pi^2/16 = aH_o^2$. The pion field is defined by $U(x) = \exp(i2\pi^a t^a/F_\pi)$. F_π is the pion coupling constant.

The dependence of F_π on the number of colours can be easily obtained by appealing to QCD planar diagrams [6]. Consider the correlation function of the axial current $j_\mu(x) = \bar{\psi}\gamma_\mu\gamma_5\psi$, $G(x) = \langle j_\mu(x) j_\mu(0)\rangle$. For large x,

$$G(x) = |\langle\pi|j_\mu(0)|0\rangle|^2 e^{-m_\pi x}$$

m_π is the lowest mass pseudo-scalar meson. On the other hand for large N, $G(x) \propto N$ because internal quark loops are suppressed. Hence if the large N limit is smooth $F_\pi^2 = |\langle\pi|j_\mu(0)|0\rangle|^2 \propto N$ and m_π is N independent. This simple but important fact, introduces the QCD parameter N in the chiral model. One immediate consequence of this is that for large N a perturbation expansion of (4) is possible:

$$L' = \mathrm{tr}\partial_\mu \pi \partial_\mu \pi + \frac{1}{F_\pi^2}\, \mathrm{tr}\partial_\mu \pi^2 \partial_\mu \pi^2 + \ldots \quad (5)$$

Pions are weakly interacting and infact free particles at $N = \infty$.

There are 2 basic problems with the chiral Lagrangian (4). One concerns dynamics and the other kinematics. In the next section we discuss the dynamics problem.

(ii) Baryons as Solitons and the Skyrme Model

Some time ago Witten [7] argued that baryons in QCD appear as solitons in a weakly coupled theory of mesons. The qualitative reason is very simple. A baryon is a bound state of N quarks plus the interaction energy of gluon exchange by quarks. The gluon-quark vertex (in QCD) is $g/\sqrt{N}$. Hence the interaction energy due to 1 gluon exchange is g^2/N. However inside a baryon every quark interacts with (N-1) other quarks, giving rise to a total interaction energy $\frac{N(N-1)}{2}\frac{g^2}{N} \approx \frac{g^2}{2}N$ (for large N). Therefore the baryon mass is seen inversely proportional to $\frac{1}{N}$, the coupling constant of the field theory of mesons. This relation suggests that the baryon is a soliton of the field theory of mesons viz. the chiral model (4).

The above conclusion leads to the first of 2 problems we mentioned at the end of the last section. The chiral model (4) being a second derivative scalar field theory in 4 dimensions cannot have a soliton solution. In order to get a soliton solution in the chiral model one needs to introduce higher derivative terms in the Lagrangian. The precise form of these higher derivative terms is an issue we will discuss in later sections. For the moment let us adopt the simplest model which was proposed by Skyrme [8] more than 20 years ago. Recent interest in this model was revived by Balachandran et al. [9]

$$L_S = \frac{F_\pi^2}{16\pi^2} \, tr\partial_\mu U \partial_\mu U^\dagger - \frac{1}{32e^2} \, tr[\partial_\mu U U^\dagger, \partial_\nu U U^\dagger]^2 \qquad (6)$$

(e, like F_π is a phenomenological coupling proportional to $\frac{1}{\sqrt{N}}$).

Skyrme also introduced the topological classification of the configuration space of the chiral field. The reasoning is as follows: In the broken symmetry phase U(x) must tend to a constant for large x , because the symmetry of the vacuum is the diagonal part of $SU(3)_L \times SU(3)_R$. This means that the space of fields U(x) at a fixed time considered as a mapping from R^3 into SU(3), by virtue of this constraint for large x , is a mapping from the compact space S_3 (3-sphere) into SU(3). Such mappings fall into topologically distinct classes characterized by the homotopy group $\pi_3(SU(3))$ = integers. Skyrme proposed to identify the elements of this group with the Baryon number [8,9] given by the formula

$$B(U) = \frac{i}{24\pi^2} \int d\vec{x} \; tr\varepsilon_{ijk}(\partial_i U U^\dagger \partial_j U U^\dagger \partial_k U U^\dagger) \; ... \qquad (7)$$

Skyrme also demonstrated a classical solution with B = 1. His original discussion was in the SU(2) chiral model but this can be trivially extended to SU(3). The static classical solution has the form

$$U_o(\vec{x}) = \left(\begin{array}{cc|c} e^{i\hat{x}\cdot\vec{\tau}f(r)} & & 0 \\ & & 0 \\ \hline 0 & 0 & 1 \end{array}\right) \tag{8}$$

the function f(r), satisfies the boundary condition $f(0) = \pi$ and $f(\infty) = 0$ and looks like

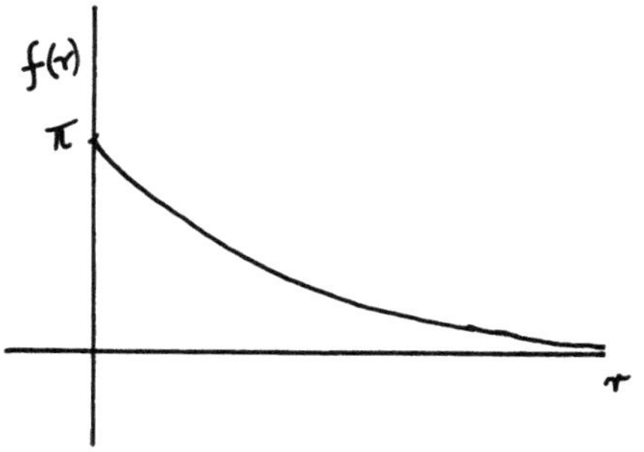

If we substitute (8) in (7) we get $B = \frac{1}{\pi}(f(0) - f(\infty)) = 1$. Before we discuss further implications of the soliton solution we turn to the 'kinematical' problem with the Skyrme model.

(iii) Wess-Zumino term

In the above we introduced pions in terms of the phase of the complex order parameter $M_{ij}(x) = \sum_\alpha \bar{\psi}_i^\alpha(x)(1+\gamma_5)\psi_j^\alpha(x)$. The parity transformation of $M(\vec{x})$ can be deduced from that of the quark fields. Since $\psi_i^\alpha(\vec{x}) \xrightarrow{P} \gamma_4\psi_i^\alpha(-\vec{x})$, we get $M(\vec{x},t) \xrightarrow{P} M^\dagger(-\vec{x},t)$, which says that pions are pseudo scalars: $\pi(\vec{x}) \xrightarrow{P} -\pi(-\vec{x})$. However the Skyrme model (6), is separately invariant under 2 discrete symmetries $\pi(x) \xrightarrow{P_1} -\pi(\vec{x})$ and $\pi(x) \xrightarrow{P_2} \pi(-\vec{x})$. The P_1 symmetry would forbid a process like $K^+K^- \to \pi^+\pi^o\pi^-$ and would also contradict the observed decay of ϕ into both K^+K^- and $\pi^+\pi^o\pi^-$. Only the combination $P = P_1P_2$ is a symmetry of QCD. In the spirit of phenomenological Lagrangians it is natural to inquire whether one can add a $SU(3)_L \times SU(3)_R$ term to (6) which is invariant only under the combination $P = P_1P_2$. This question was raised and answered by Witten [10]. The answer is non-trivial because the obvious candidate $\mathrm{tr}L_\mu L_\nu L_\rho L_\sigma \varepsilon_{\mu\nu\rho\sigma}$ ($L_\mu = i\partial_\mu UU^\dagger$) vanishes in 4-dimensions. The additional term turns out to be the Wess-Zumino [11] term of current algebra.

To describe Witten's construction it is best to work in Euclidean space. In the vacuum sector we have the boundary condition $U(x \to \infty) = I$, which identifies the boundary of 4-dimensional space-time with a single point. Hence $\{U(x)\}$ can be considered as a mapping from S_4 (the 4-sphere) into SU(3). Now introduce a fifth co-ordinate x_5 $(0 \leq x_5 \leq 1)$ and consider the function $U(x, x_5)$ such that $U(x,0) = I$ and $U(x,1) = U(x)$. Such a continuous function exists because $\pi_4(SU(3)) = 0$. The set of points (x_1,x_2,x_3,x_4,x_5) takes values in a 5-dimensional disc D whose boundary is S_4 (compactified 4-dim. space). The image of D in SU(3) is the upper hemisphere Q of a 5 dim. sphere S_5 in SU(3). The boundary of the disc mapps onto the equator of S_5.

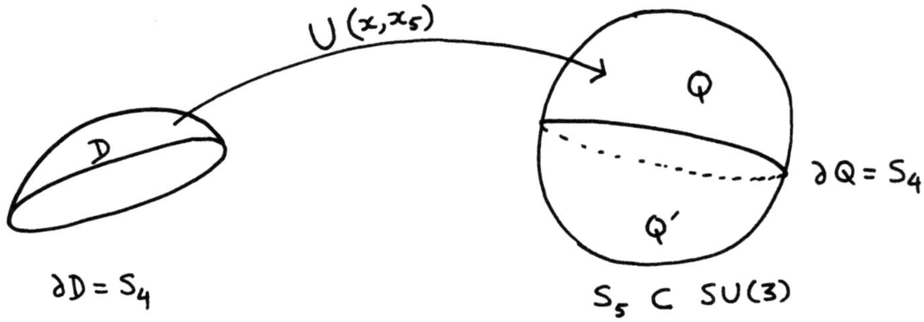

Now using the exponential parametrization $U = e^{it_a\theta_a}$, we can construct the following 5-form on Q,

$$\omega = \omega_{abcde} d\theta_a \wedge d\theta_b \wedge d\theta_c \wedge d\theta_d \wedge d\theta_e$$

where

$$\omega_{abcde} = \frac{-i}{240\pi^2} \operatorname{tr}\left\{\frac{\partial}{\partial\theta_a} UU^{-1} \frac{\partial}{\partial\theta_b} UU^{-1} \frac{\partial}{\partial\theta_c} UU^{-1} \frac{\partial}{\partial\theta_d} UU^{-1} \frac{\partial}{\partial\theta_e} UU^{-1}\right\} \tag{9}$$

(a,b,c,d,e antisymmetrized).

The additional term to be added to the Skyrme action, which is not invariant under P_1 and P_2 separately but only under $P = P_1P_2$, is the integral of the 5-form ω over Q

$$\Gamma = \lambda \int_Q \omega \tag{10}$$

λ for the moment is an arbitrary coupling constant. (10) can equivalently be written as a integral over the 5 dim. disc D

$$\Gamma = - \frac{\lambda}{240\pi^2} \int_D d\Sigma_{ijk\ell m} \, \mathrm{tr} L_i L_j L_k L_\ell L_m \tag{11}$$

where $L_i = i\partial_i U U^{-1}$ and $d\Sigma_{ijk\ell m}$ is the volume element of the disc D.

In our discussion we specified the choice of the image Q, of the disc D, as the upper hemisphere of the sphere S_5 in SU(3). However small deformations of Q will not change the value of Γ, because ω_{abcde} (being an anti-symmetric 5-tensor in 5-dim.) is curl free. It is physically reasonable to demand that in the path integral (where one averages over the random field U(x)) we could have as well used the lower hemisphere Q' of S_5, with

$$\Gamma' = - \lambda \int_{Q'} \omega \tag{12}$$

The minus sign indicates that S_4 bounds Q' and Q with opposite orientations. This physical requirement amounts to the condition

$$\exp(i\lambda \int_{Q+Q'} \omega) = 1 \tag{13}$$

Since $Q+Q' = S_5$, $\frac{1}{2\pi} \int_{Q+Q'} \omega$ counts the number of wrappings of S_5 onto SU(3), hence

$$e^{i\lambda 2\pi (\text{integer})} = 1$$

implying λ is an integer. Subsequently we will identify λ with the number of colours N of the underlying quark model. The Witten construction of the Wess-Zumino term and the quantization of λ has much in common with the dynamics of a charged particle in the field of a monopole. The quantization of λ is analogous to the Dirac quantization condition for the magnetic charge in units of $\frac{1}{e}$.

(iv) Consequences of the Wess-Zumino term

(i) New vertices for Nambu-Goldstone bosons

Let us expand the chiral field $U = e^{i\theta_a t_a}$, $\theta_a = \frac{2\pi_a}{F_\pi}$ in powers of $\frac{1}{F_\pi}$. To leading order $U \simeq 1 + \frac{2i}{F_\pi} \pi_a t_a$ and

$$\Gamma = N \frac{2}{15\pi^2 F_\pi^5} \int_D d\Sigma_{ijk\ell m} \, \mathrm{tr}(\partial_i \pi \partial_j \pi \partial_k \pi \partial_\ell \pi \partial_m \pi) + O(\frac{1}{F_\pi^6})$$

$$= N \frac{2}{15\pi^2 F_\pi^2} \int_D d\Sigma_{ijk\ell m} \partial_i \; tr(\pi\partial_j\pi\partial_k\pi\partial_\ell\pi\partial_m\pi)$$

$$= N \frac{2}{15\pi^2 F_\pi^2} \int_{S_4} d^4x \; \varepsilon_{\mu\nu\rho\sigma} tr\partial_\mu\pi\partial_\nu\pi\partial_\rho\pi\partial_\sigma\pi \qquad (14)$$

The above vertex describes the decay $K^+K^- \to \pi^+\pi^o\pi^-$ in the low energy limit when kaons and pions are both considered as goldstone bosons. This description is unrealistic though, due to the mass difference between kaons and pions.

(ii) Electromagnetic and weak interactions of Goldstone bosons

Consider the electric charge matrix of quarks

$$Q = diag(\tfrac{2}{3}, -\tfrac{1}{3}, -\tfrac{1}{3})$$

Since this is one of the generators of the unbroken flavour group, the chiral Lagrangian is invariant under the global transformations $\delta U = i\varepsilon[Q,U]$. If we introduce the photon field A_μ, a locally gauge invariant chiral model can be easily constructed replacing ∂_μ by the covariant derivative $D_\mu = \partial_\mu - iA_\mu[Q,]$ in (6). Since the Wess-Zumino term is not the integral of a 4 dim. $SU(3)_L \times SU(3)_R$ invariant density, it cannot be made gauge invariant by such a substitution. A trial and error procedure leads to the following gauge invariant Wess-Zumino term [10]

$$\tilde{\Gamma} = \Gamma(U) - Ne\int d^4x A_\mu J_\mu$$

$$+ \frac{ie^2N}{24\pi^2} \int d^4x \varepsilon_{\mu\nu\alpha\beta} \partial_\mu A_\nu A_\alpha Tr[Q^2\partial_\beta UU^{-1} + Q^2U^{-1}\partial_\beta U + QUQU^{-1}\partial_\beta UU^{-1}] \qquad (15)$$

and $j_\mu = \varepsilon_{\mu\nu\alpha\beta} Tr[Q\partial_\nu UU^{-1}\partial_\alpha UU^{-1}\partial_\beta UU^{-1} + QU^{-1}\partial_\nu UU^{-1}\partial_\alpha UU^{-1}\partial_\beta U]$ is the U(1) current that follows from the Noether prescription applied to $\Gamma(U)$.

$\tilde{\Gamma}$ is expected to contain information about QCD triangle anomalies. To see this let us expand the last term of (15) in powers of $1/F_\pi$. Since $Q^2 = diag(\frac{4}{9}, \frac{1}{9}, \frac{1}{9})$, there is a piece proportional to $\lambda_3 =$ diag(1,-1,0) giving rise to the $\pi_o \to 2\gamma$ vertex.

$$A = \frac{Ne^2}{48\pi^2} \pi_o \varepsilon_{\mu\nu\alpha\beta} F_{\mu\nu} F_{\alpha\beta}$$

We see that N must be equal to the number of colours because this agrees with calculations done at the quark level. The third term in (15) contains the $\gamma\pi^+\pi^o\pi^-$ vertex

$$B = -\frac{2}{3} ie \frac{N}{\pi^2 F_\pi^3} \varepsilon_{\mu\nu\alpha\beta} A_\mu \partial_\nu \pi^+ \partial_\alpha \pi^- \partial_\beta \pi^o \qquad (17)$$

which also agrees with the QCD VAAA anomaly. It turns out that the Wess-Zumino term $\tilde{\Gamma}$ embodies all effects of QCD anomalies in low energy processes involving photons and Nambu-Goldstone bosons, with precise numerical coefficients.

This agreement of anomalies at the level of the chiral model and the quark level is a deep fact. It tells us that anomalies embody universal information which is independent of the description of the physical system at any particular length scale. This information always seems to be of a topological nature and perhaps these two facts are always intimately related.

Witten has also discussed the Wess-Zumino term when other flavour subgroups H of $SU(3)_L \times SU(3)_R$ are gauged. It is beyond the scope of these notes to discuss this effective action in any detail. We shall however note that a generalization of Γ, which is gauge invariant under H, exists if and only if H is an anomaly free subgroup and if $\pi_4(H)$ is trivial.

(iii) The baryon current

Another interesting implication of Γ is the formula for the baryon current. If we associate a baryon number of $\frac{1}{N}$ with each quark turn the baryon current is by definition

$$B_\mu = \frac{1}{N} \operatorname{tr}(J_\mu^L + J_\mu^R) \qquad (18)$$

where $J_\mu^{L,R}$ are Noether currents in the chiral model corresponding to $SU(3)_{L,R}$ rotations. In the absence of electromagnetism, B_μ receives contribution only from the Wess-Zumino term Γ (11) and

$$B_\mu = \frac{i}{24\pi^2} \varepsilon_{\mu\nu\rho\sigma} \operatorname{tr} \partial_\nu U U^{-1} \partial_\rho U U^{-1} \partial_\sigma U U^{-1} \qquad (19)$$

This current is identically conserved and leads to the topological charge (7). In the presence of the electromagnetic field the baryon number that results from $\tilde{\Gamma}$ (15) has additional pieces which destroys its topological nature in the presence of singular fields making it possible for the soliton to 'unwind' into pions. In particular it is of interest to examine the consequences of this for the Rubakov process in the presence of U(1) monopoles [12].

(iv) Quantum states of the Skyrme soliton

Let us now discuss the implications of the Wess-Zumino term for soliton states [13,14,15]. This application is of particular importance since it clearly shows that the Wess-Zumino term contains information about underlying quark representations.

We collect the 2 pieces of the chiral model (6) and (11) in

$$S_x = \frac{F_\pi^2}{16} \int \mathrm{tr}\,\partial_\mu U \partial_\mu U^\dagger + \frac{1}{32e^2} \int \mathrm{tr}[\partial_\mu U U^\dagger,\ \partial_\nu U U^\dagger]^2 + \Gamma \tag{20}$$

Since F_π, $\frac{1}{e}$ and Γ are all proportional to N, there is a systematic WKB expansion in $\frac{1}{N}$ around the Skyrme soliton (8), which is a classical solution of (20). Note that since the soliton sits in a SU(2) subgroup of SU(3), the Wess-Zumino term does not contribute at the classical level.

To quantize the soliton we introduce collective co-ordinates in the broken symmetry phase, by parametrizing a field configuration (to leading order) by

$$U(\vec{x},t) = \Omega(t) U_o(\vec{x}) \Omega^{-1}(t),\quad \Omega(t) \in SU(3) \tag{21}$$

$\Omega(t)$ are the collective co-ordinates. The action for the collective co-ordinates is obtained by substituting (21) into (20). A straight forward calculation yields for the first two terms the action of a SU(3) rigid rotator,

$$S_o = M_o - \frac{1}{2} \int dt\ I_{ab} \omega^a \omega^b \tag{22}$$

Here M_o is the classical mass of the soliton, $\omega_a = \mathrm{tr}\,t_a \Omega^{-1} \frac{d}{dt} \Omega$ are the angular velocities and I_{ab} is the moment of inertia matrix, given as a function of the classical solution.

The form of the classical solution (8) has important consequences for the structure of I_{ab}. Since the solution sits in the SU(2) subgroup, $I_{a8} = 0$. Also since $U_o(\hat{R}\vec{x}) = V U_o(\vec{x}) V^\dagger$, i.e. a space rotation is equivalent to a flavour rotation, I_{ab} is rotationally invariant in the sense that $I = RIR^T$, where $R_{ab} = \mathrm{tr}\,t_a V t_b V^\dagger$, $V = \exp(i \sum_{i=1}^{3} t_i \theta_i)$ is a reducible representation of the rotation group in the SU(3) octet. The matrices R are block diagonal, the blocks constituting three, four and one dimensional irreducible representations of SU(2). Therefore I is diagonal and has the form $I = \mathrm{diag}(c_1,c_1,c_1,c_2,c_2,c_2,c_2,0)$. $c_1 = \frac{1}{3}\mathrm{tr}_{SU(2)} I$ and $c_2 = \frac{1}{4}(\mathrm{Tr}_{SU(3)} I - \mathrm{Tr}_{SU(2)} I)$, where $\mathrm{Tr}_{SU(3)}$ means trace over all SU(3) generators and $Tr_{SU(2)}$ means trace over the first three SU(2) generators of SU(3). Explicitly, in terms of (8),

$$c_1 = \frac{2N\pi}{3e'^2 f_\pi} \int_o^\infty \sin^2 f[r^2+4(r^2 f'^2 + \sin^2 f)]dr$$

and

$$c_2 = \frac{N\pi}{4e'^3 f_\pi} \int_o^\infty \sin^2 f[r^2+r^2 f'^2+2\sin^2 f]dr,$$

where $e' \equiv \frac{e}{\sqrt{N}}$ and $f_\pi \equiv \frac{F_\pi}{\sqrt{N}}$ are independent of N. Then (22) can be written as the Lagrangian of a symmetrical top,

$$L = -\frac{1}{2}\left\{c_1(\omega_1^2+\omega_2^2+\omega_3^2) + c_2(\omega_4^2+\omega_5^2+\omega_6^2+\omega_7^2)\right\} \tag{23}$$

We now calculate the Wess-Zumino term. In the vacuum sector we had the boundary condition $U(x \to \infty) = I$ and we chose space-time to be S_4. This boundary condition is incompatible with the presence of a soliton, since $U(\vec{x}, t \to \pm\infty) = U_o(\vec{x})$. A natural choice of the topology of space-time in the soliton sector is $S_3 \times S_1$. S_3 corresponds to the compactification of space and S_1 represents the circle of periodic time. The map $U(\vec{x},t)$ from $S_3 \times S_1$ into SU(3) is extended to a map $U(\vec{x},t,s)$ from $S_3 \times D_2$ into SU(3). D_2 is a 2-dim. disc of unit radius with S_1 as boundary. Such an extension is possible because $\pi_1(SU(3)) = 0$. More explicitly we extend the collective co-ordinates $\Omega(t)$ defined on S_1 to $\tilde{\Omega}(t,s)$ defined on the disc D_2 such that $\tilde{\Omega}(t,o) = 1$ and $\tilde{\Omega}(t,s=1) = \Omega(t)$.

With this geometry in mind, the Wess-Zumino term can be evaluated

$$\Gamma = \frac{N}{48\pi^2} \int_{D_2} dtds\ \mathrm{tr}K[\tilde{\omega},\tilde{\omega}'] \tag{24}$$

where

$$\tilde{\omega} = \tilde{\Omega}^{-1} \frac{\partial}{\partial t} \tilde{\Omega},\ \tilde{\omega}' = \tilde{\Omega}^{-1} \frac{\partial}{\partial s} \tilde{\Omega} \text{ and}$$

$$K = i\int d\vec{x}\ \varepsilon_{ijk}[\partial_i U_o U_o^{-1} \partial_j U_o U_o^{-1} \partial_k U_o U_o^{-1} + U_o^{-1}\partial_i U_o U_o^{-1} \partial_j U_o U_o^{-1} \partial_k U_o] \tag{25}$$

Since $\tilde{\omega}$ and $\tilde{\omega}'$ are pure gauge the curl free condition

$$\frac{\partial}{\partial s}\tilde{\omega} - \frac{\partial}{\partial t}\tilde{\omega}' - [\tilde{\omega}, \tilde{\omega}'] = 0$$

leads to

$$\Gamma = -\frac{N}{48\pi^2}\int_{D_2} dtds \left\{\frac{\partial}{\partial t}(\mathrm{tr}K\tilde{\omega}') - \frac{\partial}{\partial s}(\mathrm{tr}K\tilde{\omega})\right\} \quad (26)$$

Using Stokes theorem in 2 dimensions we get

$$\Gamma = -\frac{N}{48\pi^2}\int_{\partial D_2} dt\ \mathrm{tr}(K\omega) \quad (27)$$

a term linear in the angular velocity in the direction of the constant matrix K. To calculate K note that $[K,t_i] = 0$, $i = 1,2,3$. Hence $K = i24\pi^2 \mathrm{diag}(1,1,0)$. With this value of K (27) becomes

$$\Gamma = -\frac{iN}{2\sqrt{3}}\int dt\omega_8 \quad (28)$$

where we have used $t_8 = \frac{1}{\sqrt{3}}\mathrm{diag}(1,1,-2)$ which is proportional to the hypercharge generator $Y = 1/3\ \mathrm{diag}(1,1,-2)$.

Collecting all terms, the collective co-ordinate Lagrangian is

$$L = -\frac{1}{2}\left\{c_1\sum_{i=1}^{3}\omega_i^2 + c_2\sum_{i=4}^{7}\omega_i^2\right\} - \frac{iN}{2\sqrt{3}}\omega_8 \quad (29)$$

The Hamiltonian is defined using the parametrization $\Omega = e^{i\theta_a t_a}$ and treating θ_a as canonical co-ordinates. A simple calculation yields

$$H\ \frac{1}{8c_1}\sum_{i=1}^{3}E_i^2 + \frac{1}{8c_2}\sum_{i=4}^{7}E_i^2 \quad (30)$$

with the first class constraint

$$E_8 = \frac{N}{\sqrt{3}}\ . \quad (31)$$

E_i are the generators of right rotations defined by $[E_i, \Omega] = \Omega t_i$. Further (30) can be written as a sum of 2 casimirs

$$H = \frac{1}{2}\left(\frac{1}{c_1} - \frac{1}{c_2}\right)\sum_{i=1}^{3}\frac{E_i^2}{4} + \frac{1}{2c_2}\left(\sum_{a=1}^{8}\frac{Q_a^2}{4} - \frac{N^2}{12}\right) \quad (32)$$

where Q_a are generators of left rotations defined by $[Q_a,\Omega] = t_a\Omega$. We have used

$$\sum_{i=1}^{8} E_i^2 = \sum_{a=1}^{8} Q_a^2.$$

The problem is now easy to solve. The eigenfunctions of H (32) are SU(3) representation matrices

$$\psi^{(p,q)}_{(I,I_3,Y,I',I'_3,Y')}(\Omega) = < I,I_3,Y|D(\Omega)|I',I'_3,Y' > \tag{33}$$

with the restriction

$$Y' = \frac{N}{3} \tag{34}$$

on the right hypercharge from the Wess-Zumino term. The eigenvalues are

$$E = \frac{1}{2}\left(\frac{1}{c_1} - \frac{1}{c_2}\right) I'(I'+1) + \frac{1}{2c_2}\left(C_{p,q} - \frac{N^2}{12}\right)$$

where I' is the spin of the above state and $c_{p,q}$ is the Casimir of the (p,q) representation of SU(3).

Since flavour rotations in the space of collective co-ordinates are given by $\Omega \to V_F\Omega$, V_F SU(3), I, I_3 and Y are flavour quantum numbers. Since space rotations are given by $\Omega \to \Omega V_R$, $V_R = \exp i \sum_{i=1}^{3} t_i\theta_i$, I' and I'_3 are spin quantum numbers. The constraint (34) coming from the Wess-Zumino term leads to the following restriction on the allowed states:

$$I'_3 = \frac{N}{2} + \text{(integer)}$$

$$p-q = N + 3 \times \text{(integer)} \tag{34}$$

This means that the solitons are fermions or bosons depending upon whether N is an odd or even integer. Further if N is a multiple of 3 the SU(3) representations have zero triality namely octet, decuplet etc. A quark like representation is possible if N is not a multiple of 3. In particular for the QCD case of N=3, the quantum states (33) describe particle states with the quantum numbers of the known baryons.

Before we close this section we wish to emphasise the qualitative features once more. The fact that chiral symmetry is spontaneously broken lead to the topological classification of the configuration space of the chiral field. Details of the potential which supports a soliton solution depend on the QCD dynamics. The baryon number of the soliton is topological and at large N we do not expect any fluctuation which can invalidate this fact. 'N' also provides systematics of the quantum meaning of the soliton solution. Soliton quantum numbers are fixed by the Wess-Zumino term. The collective co-ordinate method also tells us that baryon-baryon couplings are of order N and baryon-meson couplings are of order $\sqrt{N}$. The picture of the baryon state is reminiscent of Wentzel's strong coupling theory [16,17], if strong coupling means large N.

In closing we mention that the Skyrme model and variants may have other non-trivial classical solutions. Time dependent solutions would describe soliton-anti-soliton bound states and also soliton scattering.

(vi) The Nambu-Jona-Lasinio model [18]

In the previous sections we have outlined some of the main features of strong interaction dynamics using the concepts of chiral symmetry and the large N expansion. However the connection of all this with quark-gluon dynamics is far from clear. In these last 2 sections we take a deeper look at this question.

The form of the QCD low energy Lagrangian and its couplings are likely to be determined by the fact that chiral symmetry is spontaneously broken and by the way the low energy degrees of freedom couple to slowly varying renormalized quark fields below the confinement length scale. The difficult question is the identification of the low energy degrees of freedom. That pions are included is obvious because chiral symmetry is spontaneously broken. However there can be scalar and vector mesons also. We can argue that in the large N limit the only relevant degree of freedom in the scalar channel is the flavour singlet σ-meson field. Also at large N transverse vector degrees of freedom are suppressed. We then assume that there is no light particle state with mass between the pion mass and H_o ($2H_o$ is the σ-mass). Our low energy Lagrangian then involves fields which are slowly varying over distances of order $1/H_o$.

A phenomenological model which incorporates all these features is the large N limit of the Nambu-Jona-Lasinio model which has two important features common with QCD. (i) its degrees of freedom are fermions carrying flavour and colour and (ii) it exhibits chiral symmetry breaking. In the large N limit it exhibits only colour singlet excitations.

The model is described by the Lagrangian

$$L = \bar{\psi}_i^\alpha i\not{\partial}\psi_i^\alpha + \frac{g^2}{N}\left(\bar{\psi}_i^\alpha\psi_i^\alpha\bar{\psi}_j^\beta\psi_j^\beta - \bar{\psi}_i^\alpha\gamma_5\psi_i^\alpha\bar{\psi}_j^\beta\gamma_5\psi_j^\beta\right) \tag{35}$$

$\alpha = 1, \ldots N$ and $i,j = 1,2,3$

(35) can also be written in terms of scalar and pseudoscalar fields σ and π

$$L = \bar{\psi}_i^\alpha (i\not{\partial} - i(\sigma + i\gamma_5\pi))\psi_i^\alpha - \frac{N}{2g^2}(\sigma^2+\pi^2) \tag{36}$$

By virtue of the equations of motion

$$\sigma = -i\,\frac{g^2}{N}\,\bar{\psi}_i^\alpha \psi_i^\alpha$$

$$\pi = \frac{g^2}{N}\,\bar{\psi}_i^\alpha \gamma_5 \psi_i^\alpha$$

In this form it is easy to integrate over the fermions in (36) leading to the action

$$S[\sigma+i\gamma_5\pi] = 3N\mathrm{Tr}\,\ell n(i\not{\partial}-i(\sigma+i\gamma_5\pi)) - \frac{3N}{2g^2}\int(\sigma^2+\pi^2)d^4x \tag{37}$$

The partition function of the NJL model is

$$Z = \int d\sigma d\pi e^S \tag{38}$$

For large N, the above integral is dominated by a saddle point determined by the gap equation $\pi = 0$, $\sigma = H_o$ and

$$\int \frac{d^4k}{(2\pi)^4}\,\frac{1}{k^2+H_o^2} = \frac{1}{g^2} \tag{39}$$

The NJL model must be considered as phenomenological field theory with a cut off that does not go to infinity. Hence in (39) we prescribe the propagator to be defined by the proper time integral with a cut off $\varepsilon = \frac{1}{\Lambda}$. (39) becomes

$$\int_{\varepsilon^2}^{\infty} dSe^{-SH_o^2} \int \frac{d^4k}{(2\pi)^4}\, e^{-Sk^2} = \frac{1}{g^2} \tag{40}$$

The above integral can be evaluated for small ε to give

$$1 - H_o^2\varepsilon^2(\ell n\,\frac{1}{\varepsilon^2 H_o^2} - \gamma) \qquad \frac{E^2}{\pi^2 g^2} \tag{41}$$

γ in (41) is Eulers constant. A solution for $H_o^2\varepsilon^2$ exists provided the dimensionless coupling $\frac{g}{\varepsilon} > \frac{1}{\pi}$.

We have just demonstrated that chiral symmetry is spontaneously broken in the NJL model provided the coupling of the 4-fermion term exceeds a critical value.

The question is about the low energy spectrum of the NJL model [19]. By low energy we mean that the effective fields describe the low energy dynamics are very slowly varying over distances of order $\frac{1}{H_0}$. From hindsight we know that these slowly varying fields are the pion fields. The relevant operators which create pions from the vacuum are the left and right handed flavour currents

$$(J_\mu^{L,R})_{ij} = \sum_{\alpha=1}^{N} \bar{\psi}_i^\alpha \gamma_\mu P_{L,R} \psi_j^\alpha \ , \ P_{L,R} = \frac{1 \pm \gamma_5}{2} \tag{42}$$

An effective field theory incorporating these collective degrees of freedom can be constructed as follows. We begin with the partition function of the NJL with (37) enforced in the functional measure

$$Z[J_\mu^L, J_\mu^R] = \int \prod_x [d\psi d\bar{\psi} \delta(J_\mu^{L,R} - \bar{\psi}\gamma_\mu P_{L,R}\psi) d\sigma d\pi] e^{\int L} \tag{43}$$

where L is the Lagrangian (36). Exponentiating the δ-function in (43) by vector fields ℓ_μ and R_μ and integrating over the fermions we get

$$Z[J_\mu^{L,R}] = \int \prod_x [d\ell_\mu dR_\mu d\sigma d\pi] e^{Ns_{eff} + i\int (J_\mu^R R_\mu + J_\mu^L \ell_\mu)} \tag{44}$$

$$s_{eff} = \mathrm{Tr}\ell n(i\not\partial - i\not\ell P_L - i\not R P_R - i(\sigma + i\gamma_5 \pi)) - \frac{1}{2g^2}\int (\sigma^2 + \pi^2)$$

Note that Z is the generating functional for Greens functions involving the vector fields ℓ_μ and R_μ. If we introduce the complex number $m(x) = \sigma + i\pi$, then

$$S_{eff} = \mathrm{Tr}\ell n(i\not\partial - i[\not\ell + m]P_L - i[\not R + m^*]P_R) - \frac{1}{2g^2}\int m^* m \tag{45}$$

In the absence of the mass term, the effective action has a formal $SU(3)_L \times SU(3)_R$ gauge invariance. Let us extract the longitudinal degrees of freedom of the vector fields ℓ_μ and R_μ using the Faddeev-Popov method by fixing a gauge. Hence insert the usual identity

$$\Delta[\ell] \int \prod_x d\Omega \delta(F(\ell^\Omega)) = 1 \tag{46}$$

and a corresponding one for R_μ into the path integral. F is a gauge fixing function and Δ is the Faddeev-Popov determinant. In the generalized axial gauge $\Delta[\ell]$ = constant. From now on for simplicity of presentation we will set the external sources $J_\mu^{L,R} = 0$. Non-zero sources can be trivially incorporated at any step.

We change variables in the usual fashion of the Faddeev-Popov method to arrive at

$$Z = \int d\mu \; e^{NS_{eff} - \frac{N}{2g^2}\int m^* m d^4x}$$

$$d\mu = \prod_x d\ell_\mu dR_\mu d\Omega_1 d\Omega_2 d\sigma d\pi \Delta[\ell_\mu]\Delta[R_\mu]\delta(F(\ell))\delta(F(R))$$

$$S_{eff} = \mathrm{Tr}\ln(i\not\partial - i[\not\ell^{\Omega_1} + m]P_L - i[\not R^{\Omega_2} + m^*]P_R) \qquad (47)$$

where

$$\ell_\mu^{\Omega_1} = \Omega_1 \ell_\mu \Omega_1^\dagger + \partial_\mu \Omega_1 \Omega_1^\dagger$$

$$R_\mu^{\Omega_2} = \Omega_2 R_\mu \Omega_2^\dagger + \partial_\mu \Omega_2 \Omega_2^\dagger$$

In (47) if we fix the generalized axial gauge then the Faddeev-Popov determinants are constants and ℓ_μ and R_μ are gauge invariant co-ordinates. In this case we see that as $N \to \infty$, S_{eff} has a minimum at the gauge invariant points $\ell_\mu = R_\mu = 0$. We also realize that had we started with flavour non-singlet σ and π we would have picked up only the singlet part in this limit. Hence in the large N limit we have

$$Z = \int \prod_x d\Omega_1 d\Omega_2 dm dm^* \; e^{-\frac{N}{2g^2}\int m^* m d^4x} \; \det D$$

where

$$D = i\not\partial - i\not\partial\Omega_1\Omega_1^{-1}P_L - i\not\partial\Omega_2\Omega_2^{-1}P_R - imP_L - im^*P_R$$

D is an elliptic operator in euclidean space with vector and axial vector couplings and its eigenvalues are in general complex [20]. Hence

$$\det D = e^{i\Phi}\sqrt{\det D^\dagger D} \qquad (48)$$

where we have used $|\det D| = \det D^\dagger D$. $D^\dagger$ being the adjoint of D. It is not difficult to prove directly that

$$D^\dagger D = -\partial^2 + \not\partial M P_R + \not\partial M^\dagger P_L + M^\dagger M \qquad (49)$$

where $M = m\Omega_1^{-1}\Omega_2$. This result can also be derived by writing a fermion representation for $D^\dagger D$ and noting that fermion change of variables do not produce anomalous Jacobians for real positive operators $D^\dagger D$. The anomalous part of detD is contained entirely in its phase. This is seen by establishing the anomalous Ward identify for $\ln \det D$. We write

$$\det D = \int d\psi d\bar{\psi}\; e^{\int\bar{\psi}(i\not\partial - \not{L}P_L - \not{R}P_R - imP_L - im^*P_R)\psi} \tag{50}$$

and make the change of variables $\psi_R \to V\psi_R$ and $\bar{\psi}_L \to \bar{\psi}_L V^\dagger$ and absorb these into R_μ and m. Then for $V \simeq 1+i\xi$, we get the anomaly equation for the right handed current

$$\lim_{m\to 0}\,[\mathrm{tr}\,\xi\nabla_\mu(R)\,\frac{\delta}{\delta R_\mu}\,\ln\det D] = i\,\mathrm{tr}\,\gamma_5\xi \tag{51}$$

A similar equation results for the left handed current with opposite sign for the anomaly term.

The 'i' in (51) is crucial because it says that only the phase of detD is anomalous i.e.

$$\lim_{m\to 0}\,[\mathrm{tr}\,\xi\nabla_\mu\,\frac{\delta}{\delta R_\mu}\,\Phi] = \mathrm{tr}\,\gamma_5 \tag{52}$$

To calculate the right hand side we regularize it using the eigenvalues of the massless Dirac operator

$$\mathcal{D} = (i\not\partial - \not{L}P_L - \not{R}P_R)$$

$$[\mathrm{tr}\,\gamma_5\xi]_{\mathrm{reg.}} = \lim_{\varepsilon\to 0}\,\mathrm{Re}[\mathrm{tr}\,\gamma_5\xi e^{-\varepsilon\mathcal{D}^2}] \tag{53}$$

Re stands for the real part of the trace. ε is the inverse cut-off and in principle in a theory with a finite cut-off the usual anomaly (53) has corrections proportional to the inverse cut-off. However since we are eventually interested in the long wavelength limit we will neglect these terms. (53) can be calculated by standard methods [21,22] and we get

$$[\mathrm{tr}\,\gamma_5\xi]_{\mathrm{reg.}} = \frac{1}{48\pi^2}\,a_\mu a_\nu a_\rho a_\sigma \varepsilon_{\mu\nu\rho\sigma} \tag{54}$$

$$a_\mu = i\partial_\mu\Omega_1\Omega_1^{-1} - i\partial_\mu\Omega_2\Omega_2^{-1}$$

Collecting formulae we get

$$Z = \int\prod_x d\Omega_1 d\Omega_2 dm dm^*\; e^{-\frac{N}{2g^2}\int m^* m}\,.\,e^{iN\Phi(a_\mu) + \frac{N}{2}\mathrm{Tr}\ln D^\dagger D} \tag{55}$$

In (55) it is easy to integrate over one of the 2 gauge functions Ω_1 or Ω_2. In terms of the coset variable $\Omega_1^{-1}\Omega_2 = U$

$$Z = \int \prod_x dUdmdm^* \; e^{-\frac{N}{2g^2}\int m^* m} \cdot e^{iN\Phi + \frac{N}{2}\mathrm{Tr}\ln D^\dagger D} \tag{56}$$

$$D^\dagger D = -\partial^2 + \partial\!\!\!/ M P_R + \partial\!\!\!/ M P_L + M^\dagger M, \quad M(x) = m(x)U(x)$$

$U(x)$ is the 'chiral field' and the phase satisfies the differential equation

$$\nabla_\mu \frac{\delta}{\delta L_\mu}\Phi = \frac{1}{48\pi^2}\varepsilon_{\mu\nu\rho\sigma}L_\mu L_\nu L_\rho L_\sigma \tag{57}$$

where $L_\mu = i\partial_\mu U U^\dagger$. This equation can be integrated to give the Wess-Zumino term (11) [19]

$$\Phi = \frac{1}{240\pi^2}\int_D d\Sigma_{ijk\ell m}\,\mathrm{tr}L_i L_j L_k L_\ell L_m \tag{58}$$

We calculated the phase of the determinant without approximation. We now evaluate the absolute value of the determinant of D in the long wavelength approximation, using the proper time representation.

$$\frac{1}{2}\ln \det D^\dagger D = -\frac{1}{2}\int_{\varepsilon^2}\frac{dS}{S}\,\mathrm{Tr}e^{-SD^\dagger D} \tag{59}$$

In $D^\dagger D$ we add and subtract a constant mass parameter H_o. This mass parameter will subsequently be determined self-consistently. Doing this we can perform a systematic $\frac{1}{H_o}$ expansion of (59). The calculation is straightforward and the answer is

$$\frac{1}{2}\ln \det D^\dagger D = -\frac{1}{2}\left[4n_f V C_o + \int d^4x\, L_\chi\right] \tag{60}$$

$$\begin{aligned} L_\chi &= \frac{c_1}{H_o^2}\mathrm{tr}\partial_\mu^2 M \partial_\mu^2 M^\dagger + \frac{c_1}{H_o^4}\mathrm{tr}[2(\partial_\mu M\partial_\mu M^\dagger)^2 - (\partial_\mu M \partial_\nu M^\dagger)^2] \\ &+ \frac{c_3}{H_o^2}\mathrm{tr}\partial^2 M\partial^2 M^\dagger + \frac{c_1}{H_o^4}\mathrm{tr}(M^\dagger M - H_o^2)^2 + \frac{c_3}{H_o^6}\mathrm{tr}(M^\dagger M - H_o^2)^3 \\ &+ \frac{c_2}{H_o^8}\mathrm{tr}(M^\dagger M - H_o^2)^4 + \text{higher order.} \end{aligned} \tag{61}$$

The coefficients are given as functions of H_o and $\varepsilon = \frac{1}{\Lambda}$,

$$c_o = \frac{H_o^4}{96\pi^2}\int_x^\infty dt\,\frac{e^{-t}}{t^3}, \quad c_1 = \frac{H_o^2}{64\pi^2}\int_x^\infty \frac{dt}{t}e^{-t}, \quad c_2 = \frac{1}{384\pi^2}\int_x^\infty dt\, e^{-t}t$$

$$c_3 = -\frac{1}{96\pi^2}\int_x^\infty e^{-t}dt.$$

where $x = H_o^2\varepsilon^2$. $\Lambda = \varepsilon^{-1}$ is the cut-off.

Now the self-consistent evaluation of H_o is the same as determining H_o from the gap equation (39) of the NJL model, which follows from the effective potential in the vacuum sector. From (56) and (60) this effective potential turns out to be

$$V_{eff}(H_o,\varepsilon,g^2) = c_o(H_o,\varepsilon) + \frac{H_o^2}{g^2} \tag{62}$$

$\frac{\partial V_{eff}}{\partial H_o} = 0$ leads to (41).

The effective Lagrangian (61) contains both the density field $m(x)$ and the phase field $U(x)$. To leading order in large N we must determine both of these from the classical equations of motion, with the boundary condition $m(x >> H_o^{-1}) = H_o$.

For such distances L is determined only in terms of the chiral field $U(x)$

$$L = c_1 \mathrm{tr} L_\mu L_\mu + c_2 \mathrm{tr} L_\mu^2 L_\nu^2 + \frac{c_2}{2}\mathrm{tr}[L_\mu,L_\nu][L_\nu,L_\mu]$$
$$+ c_3(\mathrm{tr} L_\mu^2 L_\nu^2 + (\partial_\mu L_\mu)^2) + O(\frac{1}{H_o^2}) \tag{63}$$

The first and third terms are those of the Skyrme Lagrangian, but there are additional terms as well.

Summarizing our discussion we note that the form of the Lagrangian (61), coefficients c_i and the Wess-Zumino term (58) emerged in the large N limit entirely from the coupling of vector, axial-vector, scalar and pseudo-scalar fields to effective quark fields. The determination of H_o is connected to the fact that chiral symmetry is spontaneously broken.

It is very likely that the linear σ-model emerges as the low energy Lagrangian for QCD in the large N limit. Further vector mesons can be systematically incorporated as subleading effects in $\frac{1}{N}$.

In conclusion we mention that the ideas presented here may be useful in studying finite temperature QCD. Such consideration may also be relevant to predict new particle states in unified theories.

Acknowledgement

The work reported in sections V and VI is in collaboration with Avinash Dhar, Sanjay Jain and R. Shankar. I thank Virendra Singh for support and encouragement and Probir Roy for gently persuading me to write these notes. My thanks to S.K. Bhonslay for typing the manuscript.

REFERENCES

1. S. Adler, Phys. Rev. 177, 2426 (1969); J.S. Bell and R. Jackiw, Nuovo Cimento 60, 47 (1969).
2. S. Adler and R. Dashen, 'Current Algebras'.
3. G. 't Hooft, Proceedings of the 1979 Cargese School; T. Banks, Y. Frishman, A. Schwimmer and S. Yankielowicz, Nuc. Phys. B177, 157 (1981).
4. S. Coleman and E. Witten, Phys. Rev. Lett. 45, 100 (1980).
5. G. 't Hooft, Phys. Rev. Lett. 37, 8 (1976); C. Rosenzweig, J. Schechter and G. Trahern, Phys. Rev. D21, 3388 (1980); P. Nath and R. Arnowitt, Phys. Rev. D23, 473 (1981).
6. G. 't Hooft, Nuc. Phys. B72, 461 (1974).
7. E. Witten, Nucl. Phys. B160, 57 (1979).
8. T.H.R. Skyrme, Proc. Roy. Soc. A260, 127 (1961).
9. A.P. Balachandran, V.P. Nair, S.G. Rageev and A. Stern, Phys. Rev. Lett. 49
10. E. Witten, Nuc. Phys. B223, 422 (1983).
11. J. Wess and B. Zumino, Phys. Lett. 37B, 95 (1971).
12. C.G. Callan and E. Witten, Princeton Preprint (1984).
13. S. Jain and S. Wadia, "Large N Baryons; collective co-ordinates of the topological soliton in SU(3) chiral model, TIFR preprint 1984 (To appear in Nuc. Phys.).
14. A.P. Balachandran and Collaborators (To appear).
15. G. Guadagini, Princeton Preprint.
16. J.L. Gervais and B. Sakita, Phys. Rev. Lett. 52, 87 (1984).
17. K. Bardakci, Berkeley Preprint; V. Singh (unpublished).
18. Y. Nambu and G. Jona-Lasinio, Phys. Rev. 122, 345 (1961).
19. A. Dhar and S. Wadia, Phys. Rev. Lett. 52, 959 (1984). Our discussion here differs in some details from this reference.
20. E. Witten and L. Alvarez-Gaume, Nuc. Phys. B234, 269 (1984).
21. W. Bardeen, Phys. Rev. 184, 1848 (1969).
22. K. Fujikawa, Phys. Rev. D21, 2848 (1980); A.P. Balachandran, G. Marmo, V.P. Nair and C.G. Trahern, Phys. Rev. D25, 2713 (1982); A. Dhar and S. Wadia (unpublished).

NUMERICAL SIMULATIONS - CANONICAL AND MICROCANONICAL

Gyan Bhanot
The Institute for Advanced Study
Princeton, NJ 08540, USA

ABSTRACT

A brief review of numerical simulation methods is given. After describing convergence criteria for the usual (canonical ensemble) Monte-Carlo methods, a numerical study of the microcanonical ensemble for Ising-like systems is described.

Any field theory on the lattice [1] is defined by an action $S(\sigma)$ which is a function of the dynamical variables $\sigma(i)$. 'i' labels sites (or links) on a lattice as well as any internal degrees of freedom (e.g. color in the case of QCD). The lattice is usually hypercubic with periodic boundary conditions to minimize finite size effects.

If $O(\sigma)$ is some function of the σ's, its expectation value (Green's function) is defined by,

$$\langle 0 \rangle_\beta = \frac{1}{Z} \int O(\sigma) e^{-\beta S(\sigma)} d\sigma \tag{1}$$

with,

$$Z = \int e^{-\beta S(\sigma)} d\sigma \tag{2}$$

where $\beta = 1/T$ is the inverse temperature.

In numerical simulations [2-5], one works on a finite lattice (usually with limitations both on computer memory and time). One generates a sequence of configurations of the σ fields in which averages like the one in Eqn. (1) can be computed. In principle, these configurations could be generated completely randomly. However, most of these would have a very small Boltzman weight (i.e. $e^{-\beta S(\sigma)}$ would be very small for them). The main idea of numerical simulation techniques is importance sampling. That is, one tries to generate configurations distributed as $e^{-\beta S(\sigma)}$ so that in the averaging (Eqn. (1)) each of these is equally important.

The basic problem is to find $\{\sigma\}$ distributed as,

$$P_{eq}(\sigma) = \frac{1}{Z} e^{-\beta S(\sigma)} d\sigma \tag{3}$$

Let $W(\sigma\to\sigma')$ be a transition matrix that generates a configuration $P(\sigma')$ from $P(\sigma)$. Thus

$$P(\sigma') = \sum_{\sigma} W(\sigma\to\sigma')\ P(\sigma) \tag{4}$$

W has the properties

$$W(\sigma\to\sigma') \geq 0 \tag{5a}$$

and

$$\sum_{\sigma'} W(\sigma\to\sigma') = 1 \quad \text{(normalization)} \tag{5b}$$

In Eqn. (4), $P(\sigma)$ is the probability distribution of any particular value of each of the σ's in the configuration $\{\sigma\}$. $W(\sigma\to\sigma')$ defines a Markov chain. We want to have $P_{eq}(\sigma)$ be an eigenvector of this chain. This is guaranteed if detailed balance is satisfied; i.e. if,

$$P_{eq}(\sigma)\ W(\sigma\to\sigma') = P_{eq}(\sigma')\ W(\sigma'\to\sigma) \tag{6}$$

The proof that P_{eq} is an eigenvector of W is trivial. We have

$$\begin{aligned}\sum_{\sigma} W(\sigma\to\sigma')\ P(\sigma) &= \sum_{\sigma} W(\sigma'\to\sigma)\ P(\sigma') \\ &= P(\sigma')\end{aligned} \tag{7}$$

where we have used Eqns. (6) and (5b). Notice that P_{eq} has eigenvalue unity. A theorem due to Frobenius and Perron [6] guarantees that this eigenvalue is non-degenerate. This theorem states that:

Theorem: A stochastic matrix with positive entries has a non-degenerate maximum eigenvalue whose eigenvector has only positive entries.

This proves that P_{eq} is the unique distribution. To motivate this theorem, let us prove the following:

Theorem: The Markov chain defined by W never diverges away from P_{eq}.

Let us define the norm N between two ensembles E and E' where $P(\sigma)$ and $P'(\sigma)$ are the probability distributions of configurations $\{\sigma\}$ in these two ensembles. Let N be given by

$$N \overset{\text{def}}{=} \| E-E' \| = \sum_{\sigma} | P(\sigma) - P'(\sigma) | \tag{8}$$

If E' is the ensemble obtained from E by applying W

$$P'(\sigma) = \sum_{\sigma'} W(\sigma'\to\sigma)\, P(\sigma') \tag{9}$$

Then

$$\begin{aligned} \| E'-E_{eq} \| &= \sum_{\sigma} | P'(\sigma) - P_{eq}(\sigma) | \qquad \text{(def)} \\ &= \sum_{\sigma} \Big| \sum_{\sigma'} W(\sigma'\to\sigma)(P(\sigma') - P_{eq}(\sigma')) \Big| \\ &\le \sum_{\sigma\sigma'} W(\sigma'\to\sigma)\, |P(\sigma') - P_{eq}(\sigma')| \\ &= \| E-E_{eq} \| \end{aligned} \tag{10}$$

This proves the theorem.

What is done in numerical simulations to implement this Markov chain is to start with some distribution of the σ's on the lattice and generate new configurations using one of two popular choices for W.

a) The Metropolis method

$$W(\sigma\to\sigma') = \begin{cases} \exp[\beta(S(\sigma) - S(\sigma'))] & \text{if } S(\sigma') > S(\sigma) \\ 1 & \text{otherwise} \end{cases} \tag{11}$$

and

b) The Heat bath method

$$W(\sigma\to\sigma') = \exp[-\beta S(\sigma')] \tag{12}$$

Both of these satisfy Eqn. (6) and will result in configurations distributed as Eqn. (3). For more detail on the numerical simulation procedure, see Refs. [2-5,7].

We now turn to a more exotic way of doing numerical simulations in the microcanonical ensemble [8-10]. Here, instead of a transition probability matrix $W(\sigma\to\sigma')$, one uses a deterministic map ϕ which generates states invariant under its mapping. The density function of these states is the microcanonical ensemble. In most systems, there are conservation laws. Whenever they exist, one must restrict the map to fixed values of the conserved quantities. Let us consider a concrete example, the 2-d Ising model. The action is

$$S = \sum_i (\sigma_i \sigma_{i+1}) \tag{13}$$

One way to do numerical simulations is to use the canonical ensemble Monte-Carlo methods described before. A more novel way is to generate states for which S is fixed. Thus, define the "microcanonical" partition function,

$$Z_{MC} = \sum_{\{\sigma\}} \delta_{S(\sigma),S_0} \tag{14}$$

The "average" of any operator $O(\sigma)$ is

$$\bar{O} = \sum_{\{\sigma\}} O(\sigma)\, \delta_{S(\sigma),S_0}/Z_{MC} \tag{15}$$

The crucial question is: How does $\bar{O}$ depend on the initial state $\{\sigma\}_0$ from which we start the evolution at fixed $S = S_0$? A system is ergodic iff $\bar{O}$ is independent of $\{\sigma\}_0$. This happens if a single cycle traces through the entire space of states at constant S_0. It is clear that in a finite system, it is impossible to avoid Poincare recurrences. The hope is that when the volume tends to infinity, ergodicity holds. However, no proofs exist for this except for very special systems.

A possible way out would be to introduce a small amount of randomness in the way the space of states at constant S_0 is mapped out. This randomness would have the effect of ensuring that one does not get stuck in cycles from a bad choice of initial state. In a sense, the randomness would allow an average over all possible initial states and this would certainly increase the possibility of ergodic behavior.

A sample microcanonical simulation for the 2-d Ising model was done recently [10] using a method originally proposed by Creutz [9]. In this method [11], instead of generating states with S = constant, the sum of S plus a 'demon' action S_d is kept constant. This allows a local algorithm to be constructed to update the lattice spins. The demons are a set of variables that act as heat exchangers between different parts of the lattice. The demon attempts to flip the spin locally and the spin flip is accepted if the sum $S_{Tot} = S_d + S$ can be kept fixed with the flipped spin. The partition function of this system is

$$Z = \sum_{Cc} \delta_{S(C)+S_d(c),S_{Tot}} \tag{16}$$

where C and c label lattice and demon configurations respectively. Averages of a lattice quantity O in this partition function are defined by

$$\bar{\bar{O}} = \sum_{Cc} O(C)\ \delta_{S(C)+S_d(c),S_{Tot}}/Z \tag{17}$$

Given a demon energy and the configuration around a particular spin, the algorithm that decides whether the spin flips is totally deterministic. However, a certain amount of randomness can be introduced by making the choice of sites to update randomly over the lattice.

The result of the study [10] showed that if this was done, problems with ergodicity were not encountered. Microcanonical averages (Eqn. (17)) can then be related to canonical expectation values if 1/V effects are properly taken into account. The relation between these quantities (correct to $(1/V^2)$) is,

$$\langle O\rangle_{\beta^*} = \bar{\bar{O}}_{S_T} - \frac{1}{2V}\frac{\partial}{\partial\beta}\left[\frac{\partial\langle O\rangle/\partial\beta}{\partial\langle e\rangle/\partial\beta}\right] \tag{18}$$

where β^* is defined implicitly by

$$\langle e\rangle_{\beta^*} = \bar{\bar{e}}_{S_T} \tag{19}$$

and

$$e = S/V \tag{20}$$

In general, there might be several advantages to the microcanonical over the canonical formulation for numerical simulation.

1) The deterministic nature of the evolutions allows very fast computer codes to be written for discrete systems (such as the Ising model and possibly also SU(2) gauge theory with the Icosahedral subgroup as gauge group).
2) The renormalization group can probably be implemented more directly in the microcanonical ensemble.
3) In certain situations, the canonical ensemble does not exist while the microcanonical does [12]. For such systems, there is no choice but to use the microcanonical formulation.

Acknowledgements

I thank the National Science Foundation for a travel grant that made it possible for me to attend the conference in Mahabaleshwar and the organizers of the school for their hospitality.

References

[1] K. Wilson, Phys. Rev. D10 (1974) 2445

[2] M. Creutz, "Quarks, Gluons and Lattices" (Cambridge Univ, Press, 1983)

[3] M. Creutz, L. Jacobs and C. Rebbi, Phys. Rep. 95 (1983) 201

[4] C. Rebbi, "Lattice Gauge Theory and Monte Carlo Simulations" (World Scientific, Singapore, 1983)

[5] K. Binder in "Phase Transitions and Critical Phenomena", C. Domb and M. S. Green, eds. Vol. 5B (Academic Press, New York, 1976)

[6] F. R. Gantmacher, "Applications of the Theory of Matrices", (Interscience, N.Y. 1959). Translated by J. L. Brenner.

[7] G. Bhanot, "A Review of Numerical Simulation Methods", ed. A. Faessler, in Progress in Particle and Nuclear Physics, (Pergamon Press Ltd., 1984)

[8] D. Callaway and A. Rahman, Phys. Rev. Lett. 49 (1982) 613

[9] M. Creutz, Phys. Rev. Lett. 50 (1983) 1411

[10] G. Bhanot, M. Creutz and H. Neuberger, IAS preprint, "Microcanonical Simulation of Ising Systems", Nov. 1983, to appear in Nucl. Phys. B [FS]

[11] See [10] for details

[12] A. Strominger, Ann. Phys. 146 (1983) 419

LIST OF PARTICIPANTS

S.No.	Name	Address
1.	M. Atiyah	Oxford University, U.K.
2.	M. Azam	Tata Institute of Fundamental Research, Bombay
3.	G. Bhanot	Institute of Advanced Study, Princeton, U.S.A.
4.	S.K. Bose	Gorakhpur University, Gorakhpur
5.	T. Curtright	University of Florida, Gainseville, U.S.A.
6.	A. Datta	Jadavpur University, Calcutta
7.	P.P. Divakaran	Tata Institute of Fundamental Research, Bombay
8.	A. Dhar	Tata Institute of Fundamental Research, Bombay
9.	T. Eguchi	Tokyo University, Tokyo, Japan
10.	S. Ferarra	CERN, Geneva, Switzerland
11.	Rohini Godbole	Bombay University, Bombay
12.	T.R. Govindarajan	Loyola College, Madras
13.	V. Gupta	Tata Institute of Fundamental Research, Bombay
14.	L. Hall	Harvard University, Cambridge, U.S.A.
15.	S. Jain	Tata Institute of Fundamental Research, Bombay
16.	T. Jayaraman	Madras University, Madras
17.	S.D. Joglekar	Indian Institute of Technology, Kanpur
18.	A.S. Joshipura	Indian Institute of Technology, Kanpur
19.	D. Lohia	Delhi University, Delhi
20.	A.K. Kapoor	Central University of Hyderabad, Hyderabad
21.	A. Khare	Institute of Physics, Bhubaneshwar
22.	P. Majumdar	Saha Institute of Nuclear Physics, Calcutta
23.	J. Maharana	Institute of Physics, Bhubaneshwar
24.	S.P. Misra	Institute of Physics, Bhubaneshwar
25.	B.P. Mohapatra	Sambalpur University, Orissa
26.	R.N. Mohapatra	University of Maryland, College Park, U.S.A.
27.	A. Mukherjee	Indian Institute of Technology, Kanpur
28.	Pran Nath	Northeastern University, Boston, U.S.A.
29.	T. Padmanabhan	Tata Institute of Fundamental Research, Bombay
30.	L.K. Pandit	Tata Institute of Fundamental Research, Bombay
31.	S.K. Paul	Institute of Physics, Bhubaneshwar
32.	T. Pradhan	Institute of Physics, Bhubaneshwar
33.	S. Raby	Los Alamos National Laboratory, Los Alamos, U.S.A.
34.	A.K. Raina	Tata Institute of Fundamental Research, Bombay
35.	Probir Roy	Tata Institute of Fundamental Research, Bombay
36.	S.M. Roy	Tata Institute of Fundamental Research, Bombay

S.No.	Name	Address
37.	M. Sabir	Cochin University, Cochin
38.	D. Sankaranarayana	Tata Institute of Fundamental Research, Bombay
39.	U. Sarkar	Calcutta University, Calcutta
40.	L.P. Singh	Utkal University, Bhubaneshwar
41.	V. Singh	Tata Institute of Fundamental Research, Bombay
42.	B. Vijalakshmi	Madras University, Madras
43.	Spenta Wadia	Tata Institute of Fundamental Research, Bombay

Administrative Staff

44.	T. Balan	Tata Institute of Fundamental Research, Bombay
45.	S.K. Bhonslay	Tata Institute of Fundamental Research, Bombay

Vol. 173: Stochastic Processes in Quantum Theory and Statistical Physics. Proceedings, 1981. Edited by S. Albeverio, Ph. Combe, and M. Sirugue-Collin. VIII, 337 pages. 1982.

Vol. 174: A. Kadić, D.G.B. Edelen, A Gauge Theory of Dislocations and Disclinations. VII, 290 pages. 1983.

Vol. 175: Defect Complexes in Semiconductor Structures. Proceedings, 1982. Edited by J. Giber, F. Beleznay, J. C. Szép, and J. László. VI, 308 pages. 1983.

Vol. 176: Gauge Theory and Gravitation. Proceedings, 1982. Edited by K. Kikkawa, N. Nakanishi, and H. Nariai. X, 316 pages. 1983.

Vol. 177: Application of High Magnetic Fields in Semiconductor Physics. Proceedings, 1982. Edited by G. Landwehr. XII, 552 pages. 1983.

Vol. 178: Detectors in Heavy-Ion Reactions. Proceedings, 1982. Edited by W. von Oertzen. VIII, 258 pages. 1983.

Vol. 179: Dynamical Systems and Chaos. Proceedings, 1982. Edited by L. Garrido. XIV, 298 pages. 1983.

Vol. 180: Group Theoretical Methods in Physics. Proceedings, 1982. Edited by M. Serdaroğlu and E. İnönü. XI, 569 pages. 1983.

Vol. 181: Gauge Theories of the Eighties. Proceedings, 1982. Edited by R. Raitio and J. Lindfors. V, 644 pages. 1983.

Vol. 182: Laser Physics. Proceedings, 1983. Edited by J. D. Harvey and D. F. Walls. V, 263 pages. 1983.

Vol. 183: J. D. Gunton, M. Droz, Introduction to the Theory of Metastable and Unstable States. VI, 140 pages. 1983.

Vol. 184: Stochastic Processes – Formalism and Applications. Proceedings, 1982. Edited by G. S. Agarwal and S. Dattagupta. VI, 324 pages. 1983.

Vol. 185: H. N. Shirer, R. Wells, Mathematical Structure of the Singularities at the Transitions between Steady States in Hydrodynamic Systems. XI, 276 pages. 1983.

Vol. 186: Critical Phenomena. Proceedings, 1982. Edited by F.J.W. Hahne. VII, 353 pages. 1983.

Vol. 187: Density Functional Theory. Edited by J. Keller and J.L. Gázquez. V, 301 pages. 1983.

Vol. 188: A. P. Balachandran, G. Marmo, B.-S. Skagerstam, A. Stern, Gauge Symmetries and Fibre Bundles. IV, 140 pages. 1983.

Vol. 189: Nonlinear Phenomena. Proceedings, 1982. Edited by K. B. Wolf. XII, 453 pages. 1983.

Vol. 190: K. Kraus, States, Effects, and Operations. Edited by A. Böhm, J. W. Dollard and W. H. Wootters. IX, 151 pages. 1983.

Vol. 191: Photon Photon Collisions. Proceedings, 1983. Edited by Ch. Berger. V, 417 pages. 1983.

Vol. 192: Heidelberg Colloquium on Spin Glasses. Proceedings, 1983. Edited by J. L. van Hemmen and I. Morgenstern. VII, 356 pages. 1983.

Vol. 193: Cool Stars, Stellar Systems, and the Sun. Proceedings, 1983. Edited by S. L. Balliunas and L. Hartmann. VII, 364 pages. 1984.

Vol. 194: P. Pascual, R. Tarrach, QCD: Renormalization for the Practitioner. V, 277 pages. 1984.

Vol. 195: Trends and Applications of Pure Mathematics to Mechanics. Proceedings, 1983. Edited by P. G. Ciarlet and M. Roseau. V, 422 pages. 1984.

Vol. 196: WOPPLOT 83. Parallel Processing: Logic, Organization and Technology. Proceedings, 1983. Edited by J. Becker and I. Eisele. V, 189 pages. 1984.

Vol. 197: Quarks and Nuclear Structure. Proceedings, 1983. Edited by K. Bleuler. VIII, 414 pages. 1984.

Vol. 198: Recent Progress in Many-Body Theories. Proceedings, 1983. Edited by H. Kümmel and M. L. Ristig. IX, 422 pages. 1984.

Vol. 199: Recent Developments in Nonequilibrium Thermodynamics. Proceedings, 1983. Edited by J. Casas-Vázquez, D. Jou and G. Lebon. XIII, 485 pages. 1984.

Vol. 200: H. D. Zeh, Die Physik der Zeitrichtung. V, 86 Seiten. 1984.

Vol. 201: Group Theoretical Methods in Physics. Proceedings, 1983. Edited by G. Denardo, G. Ghirardi and T. Weber. XXXVII, 518 pages. 1984.

Vol. 202: Asymptotic Behavior of Mass and Spacetime Geometry. Proceedings, 1983. Edited by F. J. Flaherty. VI, 213 pages. 1984.

Vol. 203: C. Marchioro, M. Pulvirenti, Vortex Methods in Two-Dimensional Fluid Dynamics. III, 137 pages. 1984.

Vol. 204: Y. Waseda, Novel Application of Anomalous (Resonance) X-Ray Scattering for Structural Characterization of Disordered Materials. VI, 183 pages. 1984.

Vol. 205: Solutions of Einstein's Equations: Techniques and Results. Proceedings, 1983. Edited by C. Hoenselaers and W. Dietz. VI, 439 pages. 1984.

Vol. 206: Static Critical Phenomena in Inhomogeneous Systems. Edited by A. Pękalski and J. Sznajd. Proceedings, 1984. VIII, 358 pages. 1984.

Vol. 207: S. W. Koch, Dynamics of First-Order Phase Transitions in Equilibrium and Nonequilibrium Systems. III, 148 pages. 1984.

Vol. 208: Supersymmetry and Supergravity/Nonperturbative QCD. Proceedings, 1984. Edited by P. Roy and V. Singh. V, 389 pages. 1984.

Selected Issues from

Lecture Notes in Mathematics

Vol. 909: Numerical Analysis. Proceedings, 1981. Edited by J.P. Hennart. VII, 247 pages. 1982.

Vol. 912: Numerical Analysis. Proceedings, 1981. Edited by G. A. Watson. XIII, 245 pages. 1982.

Vol. 920: Séminaire de Probabilités XVI, 1980/81. Proceedings. Edité par J. Azéma et M. Yor. V, 622 pages. 1982.

Vol. 921: Séminaire de Probabilités XVI, 1980–81 Supplément: Géométrie Différentielle Stochastique. Proceedings. Edité par J. Azéma et M. Yor. III, 285 pages. 1982.

Vol. 922: B. Dacorogna, Weak Continuity and Weak Lower Semicontinuity of Non-Linear Functionals. V, 120 pages. 1982.

Vol. 923: Functional Analysis in Markov Processes. Proceedings, 1981. Edited by M. Fukushima. V, 307 pages. 1982.

Vol. 926: Geometric Techniques in Gauge Theories. Proceedings, 1981. Edited by R. Martini and E.M.de Jager. IX 219 pages. 1982.

Vol. 927: Y. Z. Flicker, The Trace Formula and Base Change for GL (3). XII, 204 pages. 1982.

Vol. 928: Probability Measures on Groups. Proceedings 1981. Edited by H. Heyer. X, 477 pages. 1982.

Vol. 929: Ecole d'Eté de Probabilités de Saint-Flour X – 1980. Proceedings, 1980. Edited by P.L. Hennequin. X, 313 pages. 1982.

Vol. 930: P. Berthelot, L. Breen, et W. Messing, Théorie de Dieudonné Cristalline II. XI, 261 pages. 1982.

Vol. 931: D.M. Arnold, Finite Rank Torsion Free Abelian Groups and Rings. VII, 191 pages. 1982.

Vol. 932: Analytic Theory of Continued Fractions. Proceedings, 1981. Edited by W.B. Jones, W.J. Thron, and H. Waadeland. VI, 240 pages. 1982.

Vol. 934: M. Sakai, Quadrature Domains. IV, 133 pages. 1982.

Vol. 935: R. Sot, Simple Morphisms in Algebraic Geometry. IV, 146 pages. 1982.

Vol. 936: S.M. Khaleelulla, Counterexamples in Topological Vector Spaces. XXI, 179 pages. 1982.

Vol. 937: E. Combet, Intégrales Exponentielles. VIII, 114 pages. 1982.

Vol. 938: Number Theory. Proceedings, 1981. Edited by K. Alladi. IX, 177 pages. 1982.

Vol. 942: Theory and Applications of Singular Perturbations. Proceedings, 1981. Edited by W. Eckhaus and E.M. de Jager. V, 363 pages. 1982.

Vol. 953: Iterative Solution of Nonlinear Systems of Equations. Proceedings, 1982. Edited by R. Ansorge, Th. Meis, and W. Törnig. VII, 202 pages. 1982.

Vol. 956: Group Actions and Vector Fields. Proceedings, 1981. Edited by J.B. Carrell. V, 144 pages. 1982.

Vol. 957: Differential Equations. Proceedings, 1981. Edited by D.G. de Figueiredo. VIII, 301 pages. 1982.

Vol. 963: R. Nottrot, Optimal Processes on Manifolds. VI, 124 pages. 1982.

Vol. 964: Ordinary and Partial Differential Equations. Proceedings, 1982. Edited by W.N. Everitt and B.D. Sleeman. XVIII, 726 pages. 1982.

Vol. 968: Numerical Integration of Differential Equations and Large Linear Systems. Proceedings, 1980. Edited by J. Hinze. VI, 412 pages. 1982.

Vol. 970: Twistor Geometry and Non-Linear Systems. Proceedings, 1980. Edited by H.-D. Doebner and T.D. Palev. V, 216 pages. 1982.

Vol. 972: Nonlinear Filtering and Stochastic Control. Proceedings, 1981. Edited by S.K. Mitter and A. Moro. VIII, 297 pages. 1983.

Vol. 978: J. Ławrynowicz, J. Krzyż, Quasiconformal Mappings in the Plane. VI, 177 pages. 1983.

Vol. 979: Mathematical Theories of Optimization. Proceedings, 1981. Edited by J.P. Cecconi and T. Zolezzi. V, 268 pages. 1983.

Vol. 982: Stability Problems for Stochastic Models. Proceedings, 1982. Edited by V. V. Kalashnikov and V. M. Zolotarev. XVII, 295 pages. 1983.

Vol. 989: A.B. Mingarelli, Volterra-Stieltjes Integral Equations and Generalized Ordinary Differential Expressions. XIV, 318 pages. 1983.

Vol. 994: J.-L. Journé, Calderón-Zygmund Operators, Pseudo-Differential Operators and the Cauchy Integral of Calderón. VI, 129 pages. 1983.

Vol. 999: C. Preston, Iterates of Maps on an Interval. VII, 205 pages. 1983.

Vol. 1000: H. Hopf, Differential Geometry in the Large, VII, 184 pages. 1983.

Vol. 1003: J. Schmets, Spaces of Vector-Valued Continuous Functions. VI, 117 pages. 1983.

Vol. 1005: Numerical Methods. Proceedings, 1982. Edited by V. Pereyra and A. Reinoza. V, 296 pages. 1983.

Vol. 1007: Geometric Dynamics. Proceedings, 1981. Edited by J. Palis Jr. IX, 827 pages. 1983.

Vol. 1015: Equations différentielles et systèmes de Pfaff dans le champ complexe – II. Seminar. Edited by R. Gérard et J.P. Ramis. V, 411 pages. 1983.

Vol. 1021: Probability Theory and Mathematical Statistics. Proceedings, 1982. Edited by K. Itô and J.V. Prokhorov. VIII, 747 pages. 1983.

Vol. 1031: Dynamics and Processes. Proceedings, 1981. Edited by Ph. Blanchard and L. Streit. IX, 213 pages. 1983.

Vol. 1032: Ordinary Differential Equations and Operators. Proceedings, 1982. Edited by W. N. Everitt and R. T. Lewis. XV, 521 pages. 1983.

Vol. 1035: The Mathematics and Physics of Disordered Media. Proceedings, 1983. Edited by B.D. Hughes and B.W. Ninham. VII, 432 pages. 1983.

Vol. 1037: Non-linear Partial Differential Operators and Quantization Procedures. Proceedings, 1981. Edited by S.I. Andersson and H.-D. Doebner. VII, 334 pages. 1983.

Vol. 1041: Lie Group Representations II. Proceedings 1982–1983. Edited by R. Herb, S. Kudla, R. Lipsman and J. Rosenberg. IX, 340 pages. 1984.

Vol. 1045: Differential Geometry. Proceedings, 1982. Edited by A.M. Naveira. VIII, 194 pages. 1984.

Vol. 1047: Fluid Dynamics. Seminar, 1982. Edited by H. Beirão da Veiga. VII, 193 pages. 1984.

Vol. 1048: Kinetic Theories and the Boltzmann Equation. Seminar, 1981. Edited by C. Cercignani. VII, 248 pages. 1984.

Vol. 1049: B. Iochum, Cônes autopolaires et algèbres de Jordan. VI, 247 pages. 1984.

Vol. 1054: V. Thomée, Galerkin Finite Element Methods for Parabolic Problems. VII, 237 pages. 1984.

Vol. 1055: Quantum Probability and Applications to the Quantum Theory of Irreversible Processes. Proceedings, 1982. Edited by L. Accardi, A. Frigerio and V. Gorini. VI, 411 pages. 1984.

GPSR Compliance

The European Union's (EU) General Product Safety Regulation (GPSR) is a set of rules that requires consumer products to be safe and our obligations to ensure this.

If you have any concerns about our products, you can contact us on ProductSafety@springernature.com

In case Publisher is established outside the EU, the EU authorized representative is:

Springer Nature Customer Service Center GmbH
Europaplatz 3
69115 Heidelberg, Germany

Batch number: 10838062

Printed by Printforce, the Netherlands